RELATIVITY

Dedication

This book is dedicated to my parents.

RELATIVITY

AN INTRODUCTION TO SPACE-TIME PHYSICS

Steve Adams

UK Taylor & Francis Ltd, 1 Gunpowder Square, London, EC4A 3DE
USA Taylor & Francis Inc., 1900 Frost Road, Suite 101, Bristol, PA 19007

British Library Cataloguing in Publication Data

A catalogue record for this book is available from the British Library.

ISBN 0-7484-0621-2 (paperback)

Library of Congress Cataloguing Publication data are available

Cover design by Hybert Design and Type, cover artwork by Nick Adams.

Printed in Great Britain by T.J. International Ltd, Padstow, UK.

CONTENTS

Preface

This book is written for those who would like an introduction to relativistic physics that emphasises the physical concepts and explains the mathematics. When I moved from high school to undergraduate physics I was disappointed by the dry mathematical presentation of one of the most revolutionary sets of ideas of all time. It seemed we were being trained in technique and were neglecting meaning, so that we became proficient at solving numerical problems but remained incapable of interpreting them. Most text books reinforce this, deriving results by mathematical manipulations that are neat but less than obvious. On the other hand, popularisations enthuse about ideas but do not dare approach the mathematics. This book chooses the middle way, providing an accessible route to the theory of relativity without neglecting the essential mathematics. It will appeal to university students studying a first course in relativity and to high school students who wish to pursue the ideas beyond the bounds of their syllabus.

Chapter one reviews *Classical Physics*: this sets the historical and conceptual context for what follows. Chapter 2 looks at consequences of Einstein's principle of relativity. It does this twice: initially deriving results (time dilation etc.), from specific examples; and then from the Lorentz transformation. The chapter ends with worked examples from particle physics. Chapter 3 adopts Minkowski's space-time interpretation and shows how the use of 4 vectors can give an accurate pictorial representation of relativistic processes. This prepares the way for Chapter 4, which begins with the equivalence principle and introduces Einstein's interpretation of gravity in general relativity as intrinsic space-time curvature. The importance of these ideas for modern cosmology is emphasised throughout. The relativistic interpretation of magnetism is discussed in an appendix.

The text is illustrated with a large number of line drawings and several colour plates. There are also many worked examples throughout the book and several sets of problems for students to work through. Some of these are straightforward derivations or calculations, but a significant number are more discursive: these should encourage students to think *about* the ideas rather than just use them. The structure of each chapter is summarised on its first page, this is intended to help those who, like me, rarely read a book from cover to cover.

The book itself grew from courses I gave to students at Manchester Grammar School and Westminster School between 1982 and 1997. I am grateful for their enthusiasm and all their penetrating questions, many of which I hope I have answered here. Thanks to James Gazet at Westminster, who made several helpful mathematical suggestions for the chapters on special relativity. Thanks to John Thomson at the Rutherford Appleton Laboratories, who located many of the particle physics pictures, and Sue Tritton at the Royal Observatory, Edinburgh, who did the same for the astronomical images. A big thank you to Nick, my brother, for his wonderful pen and ink portraits of the major physicists.

Steve Adams 1997

1

CLASSICAL PHYSICS

1. 1 WHAT IS CLASSICAL PHYSICS?

1.1.1 A Historical Perspective.

Pre-Classical science was an attempt to understand the world in terms of qualitative models based on the religious and philosophical ideas of the middle ages. Established opinion and tradition had greater authority than empirical evidence. The symbolism of science and superstition overlapped and there was no clear way for science to progress.

Four physicists seem to stand out from the rest: Galileo Galilei (1564-1642) and Isaac Newton (1642-1727) for classical mechanics, and Michael Faraday (1791-1867) and James Clerk Maxwell (1831-1879) for classical electromagnetism, but many others made essential contributions to the evolution of the classical world-view. As Newton said:

'If I have seen further it is by standing on the shoulders of giants.'

Among these 'giants' was Ptolemy who, in the second century AD, explained the observed motion of the heavens by a series of cyclical and epicyclical motions centred on the Earth, and Copernicus who suggested (heretically) that the system might be simplified if the Sun rather than the Earth were at its centre. Kepler developed Copernicus's ideas and suggested elliptical orbits for the planets, supporting his theory with mathematical arguments based on the time periods and average radii of their orbits. It was left to Newton to explain these ideas in terms of a central force (gravitation) originating in the mass of all bodies.

Medieval theories and descriptions of motion (based mainly on the ideas of Aristotle) were challenged by Galileo who clarified the underlying concepts of velocity, acceleration, force and inertia and so cleared the way for Newton's Laws.

These were published in the 'Principia' (in 1687) which included 'definitions' of space and time, the three laws of motion, and the theory of gravitation. Much of what followed for the next two hundred and fifty years, much of what we understand as 'physics', derives from the Principia.

fig 1.1 Aristotle (384-322 BC)
'If everything that is in motion is being moved by something, how comes it that certain things.... that are not self moving nevertheless continue with their motion.... when no longer in contact with the agent that gave them motion?'
Artwork by Nick Adams © 1996

One of the consequences of Newton's laws is the fact that certain quantities, for example total energy, are in some circumstances conserved; despite interactions and transformations within a system these quantities do not change. Once Joule and others had established that heat is a form of energy, the law of conservation of energy became one of the most important principles in physics. It is the first law of thermodynamics. (The second law, that the entropy of the universe tends to a maximum, explains why, although the total energy of the universe remains constant, the available energy for useful work is continually diminished. It is in this sense that the world faces an 'energy crisis'.)

Classical optics made several attempts to explain the nature and properties of light. Huygens (1629-1693) proposed a wave theory as did Hooke (1635-1703) in opposition to Newton's own particle model. In fact it was Newton's authority that delayed the general acceptance of a wave model until the accumulated experimental evidence seemed overwhelming. Grimaldi (1618-1663) reported the diffraction of light in work published posthumously in 1665 but it was a century and a half later that Young (1773-1829) made the crucial measurement of the wavelength of visible light from the interference fringes formed in a double slit experiment. This seemed to prove that light was by nature a wave-like phenomenon. The velocity of light was measured with increasing accuracy by Romer (1644-1710), Fresnel (1788-1827), Foucault (1819-1868) and Michelson (1852-1931) and was found to be about three hundred million metres per second in air. Physicists speculated on the possible nature of the all-pervading medium that might support the rapid oscillations of a light wave and called it the 'luminiferous ether'.

In the seventeenth century, technological advances (such as the improvement of optical instruments and the development of accurate clocks) together with a greater emphasis on experimental methods, created ideal conditions for the development of a new type of theory - a theory that was both empirical and quantitative. Classical

fig 1.2 By 1900 Classical physics had unified many diverse phenomena.

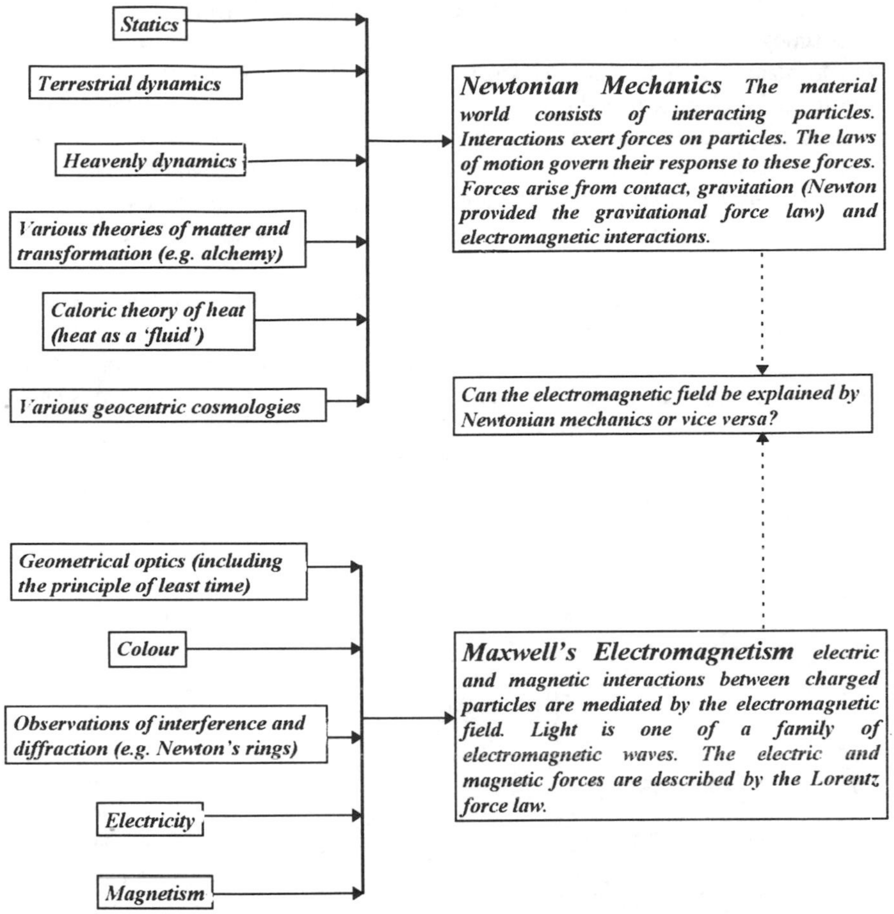

Statics

Terrestrial dynamics

Heavenly dynamics

Various theories of matter and transformation (e.g. alchemy)

Caloric theory of heat (heat as a 'fluid')

Various geocentric cosmologies

Newtonian Mechanics *The material world consists of interacting particles. Interactions exert forces on particles. The laws of motion govern their response to these forces. Forces arise from contact, gravitation (Newton provided the gravitational force law) and electromagnetic interactions.*

Can the electromagnetic field be explained by Newtonian mechanics or vice versa?

Geometrical optics (including the principle of least time)

Colour

Observations of interference and diffraction (e.g. Newton's rings)

Electricity

Magnetism

Maxwell's Electromagnetism *electric and magnetic interactions between charged particles are mediated by the electromagnetic field. Light is one of a family of electromagnetic waves. The electric and magnetic forces are described by the Lorentz force law.*

physics was developed between about 1650 and 1900 and is based on idealized mechanical models that can be subjected to mathematical analysis and tested against observation. It provides us with accurate and compelling explanations of matter, motion, gravity, electricity, magnetism and light on scales of magnitude extending from the invisibly small to the unimaginably large.

During the same period a vast amount of observational data concerning electricity and magnetism had accumulated. Coulomb (1736-1806) had formulated laws of force for charges and magnetic poles, Oersted (1777-1851) demonstrated the essential link between electricity and magnetism, and Biot (1774-1862), Savart (1791-1841), and Ampère (1775-1836) developed a mathematical theory of the magnetic effects of an electric current. Faraday (1791-1867) discovered electromagnetic induction and introduced the pictorial 'lines of force' description of electromagnetic interactions. He also showed that magnetic fields could rotate the plane of polarization of light and so linked optics and electromagnetism for the first time. Maxwell (1831-1879) developed a beautiful mathematical theory of the electromagnetic field (the Maxwell equations). His equations predicted the existence of electromagnetic waves which would travel at three hundred million metres per second in a vacuum - the speed of light - so the Maxwell equations unified electricity, magnetism and optics in an elegant mathematical model. The equations also predicted that radiation would exert a pressure and so can transfer energy and momentum between charged particles. Lorentz (1853-1928) showed how the fields exert forces on charges and Thomson (1856-1940) discovered the electron in 1897. Attempts to 'complete' physics by unifying electromagnetism and mechanics by finding either a mechanical basis for electromagnetism (pursued by Maxwell) or an electromagnetic basis for matter (e.g. the Abraham-Lorentz theory of the electron) failed, but eventually led to the crisis in classical physics from which special relativity emerged.

fig 1.3 Isaac Newton (1642-1727)
'We are to admit no more causes of natural things than such are both true and sufficient to explain their appearances.'
Artwork by Nick Adams © 1996

It is important to realize that, ever since Newton invented the differential and integral calculus and demonstrated its awesome power to formulate and solve problems in physics, new and increasingly sophisticated mathematical models were invented and used. Maxwell's electromagnetic equations are an example of this, as are the elegant generalized mechanics of Hamilton (1805-1865) and Lagrange (1736-1813). Often the symmetries of the mathematical models themselves revealed previously unsuspected relationships in physics. It is interesting that Maxwell's equations themselves already contain, in essence, much of special relativity, and Hamiltonian mechanics can be regarded as a 'template' for quantum theory.

By 1900 the classical world-view was established: matter consists of mechanical particles subject to Newton's laws of motion acted upon by gravitational and electromagnetic forces, and visible light is a narrow range of wavelengths within the much broader electromagnetic spectrum.

1.2 NEWTONIAN MECHANICS.

'The purpose of mechanics is to describe how bodies change their position in space with "time"...'
(Einstein, Relativity)

'The question what is place? presents many difficulties. An examination of all the relevant facts seems to lead to divergent conclusions.'
(Aristotle, Physics Book IV Chapter 1)

1.2.1 Space And Time.

'Absolute, true and mathematical time, of itself and from its own nature, flows equably without relation to anything external....'

'Absolute space, in its own nature, without relation to anything external, remains always similar and immovable....'
(Newton, Principia)

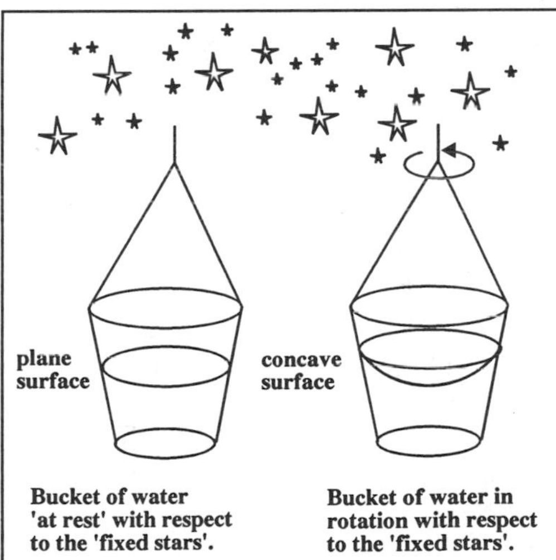

fig 1.4 Newton's bucket. The left hand bucket is at rest with respect to 'the distant stars' and the water surface is horizontal. The right hand bucket rotates with respect to the 'fixed stars' and the water surface is concave.

Newton's laws operate against a background of absolute space and absolute time. Following the motion of objects in Newton's universe is like plotting the paths of insects on uniform graph paper and timing them all with a single clock. Each particle occupies a particular position at each instant and instants follow one another in a unique sequence agreed by all observers.

In a famous experiment with a bucket of water Newton showed that the surface of the water would only lie flat when it was stationary with respect to the observed heavens. This suggested that absolute space might be identified with the frame of reference of the 'fixed stars', an observation apparently confirmed by Foucault's pendulum, which retains its plane of oscillation with respect to these 'fixed stars' despite the rotation of the Earth.

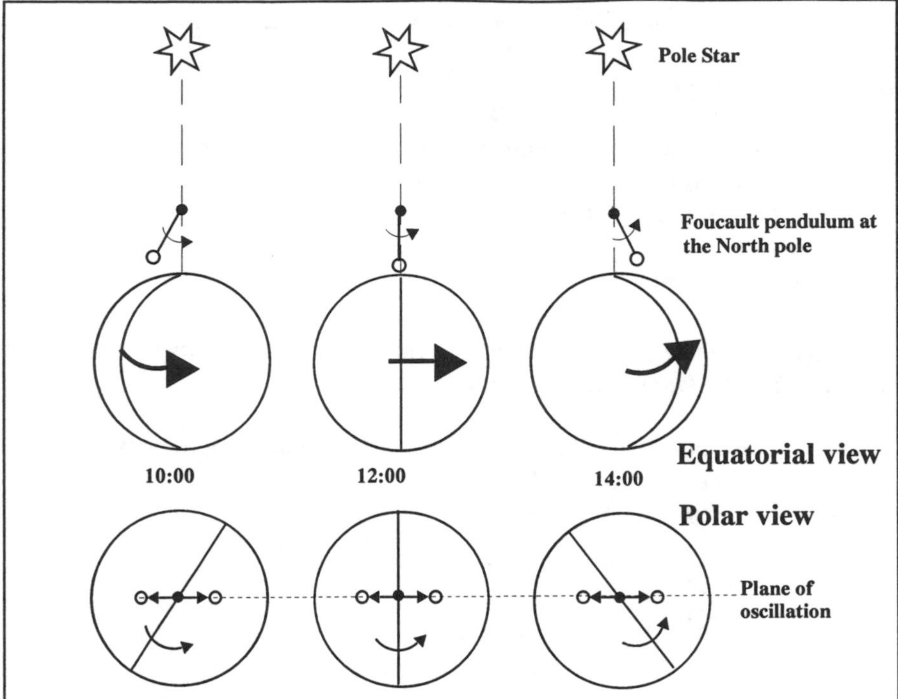

fig 1.5 Foucault's pendulum. A pendulum suspended from a perfect pivot above the North pole oscillates in a fixed plane with respect to the distant stars as the Earth rotates beneath it.

Newton was well aware that absolute measures of space and time were inaccessible and emphasized that we can only make relative measurements by comparison against chosen standards. For example, the standard unit of length might be that of a certain rod, whilst the unit of time might be related to an apparently regular periodic motion (perhaps the Earth's rotation on its axis). The choice of standards then and now is arbitrary, but it is guided by the need to relate physical change to phenomena which generate simple descriptions and reveal the regularities of nature. In other words the physicist chooses standards likely to lead to the most economical model of the physical world whilst at the same time yielding values which can be accurately reproduced in laboratory situations. In the Newtonian world standard rods have an absolute length and all similar clocks remain synchronized regardless of position or motion.

There are few pieces of evidence more beautiful and compelling than photographs of star trails above the Earth's poles. These photographs are taken with a stationary camera whose aperture is held open for several hours. The rotation of the Earth carries the camera with it and each star inscribes an arc of a circle on the film. It is

fig 1.6 Star trails around the South Pole.

very easy to agree that the Earth is rotating and the vast bulk of matter in the visible universe defines a stationary background, an absolute reference frame for any motion in space.

The position of a particle in three dimensional space can be given by three numbers. Cartesian co-ordinates give the position relative to an arbitrary origin by quoting the distances x, y, and z parallel to each of three mutually perpendicular axes. The distance between two points is then calculated using Pythagoras's theorem. Of course, the co-ordinate system chosen to measure position should not affect the measured or calculated values for the actual separation of objects in the universe. This means that different observers may well disagree on co-ordinate values but will still be able to measure or calculate values on which they agree. These invariant values are of central importance in relativity.

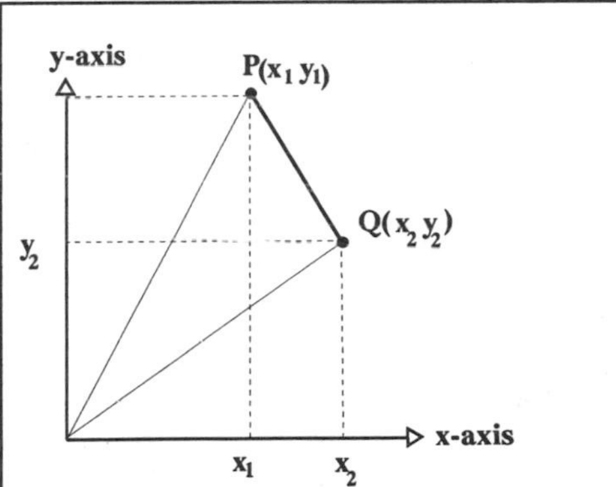

fig 1.7 2-D Cartesian co-ordinates of P and Q. Although different axes could be used the value of PQ is independent of this choice. PQ is an invariant quantity.

1.2.2 Matter.

The ultimate constituents of matter are particles which possess properties of extension, mass and motion and which can exert forces on one another through gravitation and contact. Some particles are charged and so exert electromagnetic forces on one another as well. Complex structures and materials reduce to basic particles bound together in various ordered and disordered arrangements.

1.2.3 Motion.

Newton's three laws of motion define the concept of force and provide a logical axiomatic structure from which all possible trajectories are in principle derivable.

For example, given the mathematical form of a particular force law such as gravitation, Newton's laws can then be used to derive the allowed orbits for planetary motion. Given the present positions and motions of all particles their future positions are determined by the equations of motion - Newton's world is mechanistic and deterministic even if, in practice, it is impossible to determine present locations and motions accurately enough to make precise predictions about the future. (In fact some systems produce deterministic chaos, an extreme sensitivity to initial conditions such that very similar starting configurations diverge rapidly from one another.)

- *'**LAW 1:** Every body continues in its state of rest or of uniform motion in a right line unless it is compelled to change that state by forces impressed upon it.'*
- *'**LAW 2:** The change of motion* is proportional to the motive force impressed and is made in the direction of the right line in which that force is impressed.'*
- *'**LAW 3:** To every action there is always opposed an equal reaction; or the mutual actions of two bodies upon each other are always equal and directed to contrary parts.'*

(Newton, Principia)

*Quantity of motion is what we understand as linear momentum (p), the product of a body's mass (m) and its velocity (v): $p = mv$.

The first law tells us the natural state of motion of a body upon which no resultant (or unbalanced) force acts - straight line motion at constant velocity. It also defines what we mean by a force, that is as the cause of a changing state of motion. The effect of a resultant force is to change the magnitude or direction of this motion. It was by applying this logic to falling bodies and to the Moon that Newton came to the conclusion that both were acted upon by a resultant force toward the centre of the Earth (since both have an acceleration towards the centre of the Earth) and so might have a common cause (the Earth's gravitational attraction). The second law gives us the measure of force. A unit force (1 N) changes motion at a rate of one unit of momentum per second (1 kgm s^{-2}) and is parallel to the change of momentum.

The third law is probably the most subtle: it implies that forces always arise in pairs, that is all individual forces are just one end of an interaction. The equality of action and reaction tells us that the change of momentum of one body is equal and opposite to that of the body it is interacting with, so the net change of momentum resulting from any interaction is always zero. Therefore, for a collection of many bodies interacting with one another, the overall momentum vector of the whole system will not change (momentum is a vector quantity like velocity and so has components in each of the three space dimensions). This argument is the extremely important law of conservation of linear momentum which applies to any closed system (i.e. one in which all action/reaction pairs are included).

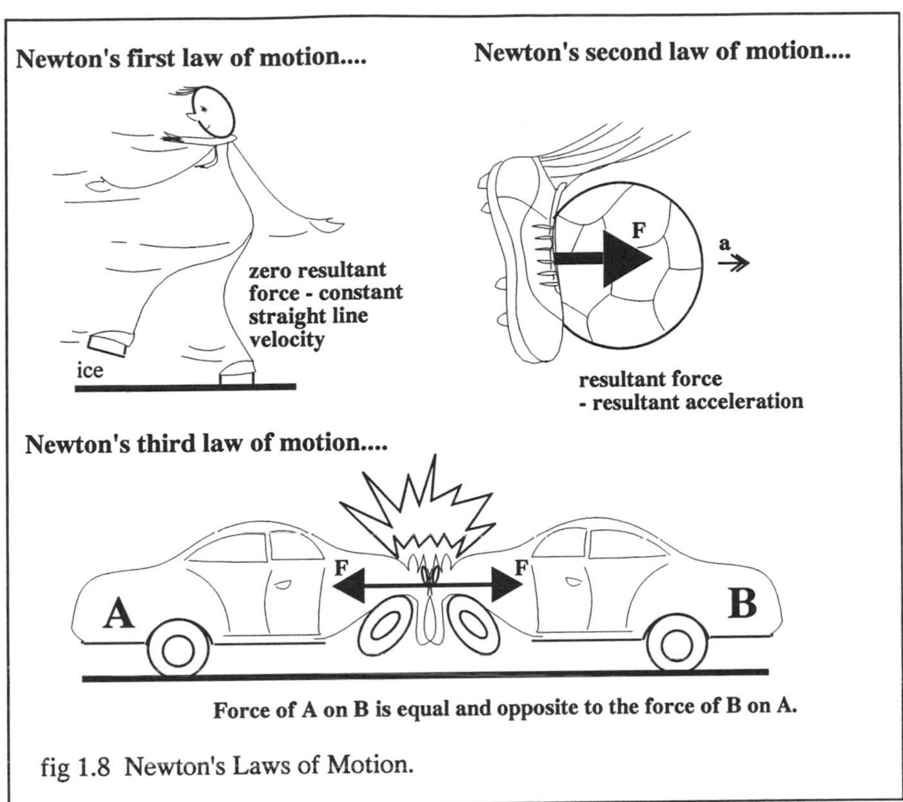

fig 1.8 Newton's Laws of Motion.

1.2.4 Mass And Inertia.

The mass of a body is related to the amount of matter it contains and in the Newtonian world this would be simply related to the sum of fundamental particles in the body. This amount of matter also determines the strength of gravitational forces it can exert on another massive object and how difficult it will be to change its state of motion. This latter property is called the body's inertia. The inertia of different bodies can be compared by applying the same force to each and comparing the ratio of their accelerations. One way to achieve this might be to compress a spring between a pair of bodies and let it push them apart. The body of greater mass (and hence greater inertia) acquires a smaller final velocity and we can define the ratio of masses to be the inverse ratio of final velocities. For example, if one body is a standard 1 kg mass then a body acquiring half the velocity of this mass in such an experiment is defined to have twice its mass, 2 kg.

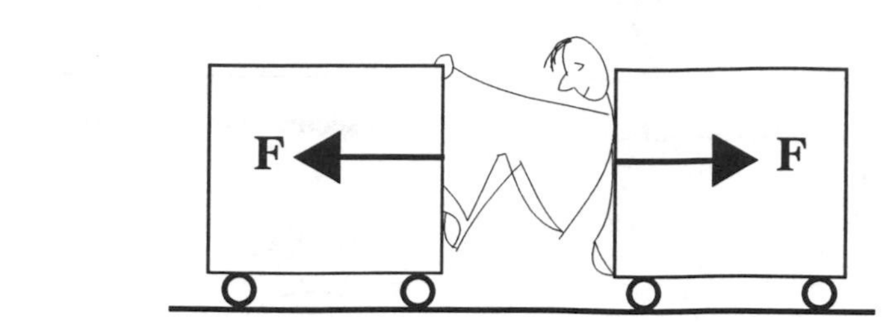

(1) **Masses A and B are given equal impulses.**

(2) **The ratio of acquired velocities is the inverse ratio of the masses:** $m_A/m_B = v_B/v_A$

fig 1.9 An operational procedure for comparing masses.

1.2.5 A Mathematical Summary Of Newtonian Mechanics.

Primitive concepts:

mass		m
spatial position co-ordinates		(x, y, z)
time		t

Derived concepts:

velocity = rate of change of position $v = ds/dt$

acceleration = rate of change of velocity $a = dv/dt = d^2s/dt^2$

linear momentum = mass × velocity $p = mv$

Force:

force = rate of change of linear momentum $F = dp/dt$

if mass is constant this leads to : $F = ma$

impulse = change of momentum $F\delta t = \delta p$

Interactions:

force of A on B = –(force of B on A) $F_{AB} = -F_{BA}$

Momentum conservation:

$$\delta p_A = F_{BA}\,\delta t = -F_{AB}\,\delta t = -\delta p_B$$
$$\delta p_{TOT} = \delta p_A + \delta p_B = 0$$

this can be generalized to a system of interacting bodies since all forces arise from interactions. The law of conservation of linear momentum applies to 'closed systems' only - that is systems which contain all action/reaction pairs.

Force laws:

Newton's laws tell us how bodies react when resultant forces are applied. Force laws describe the variation of force with position for each type of interaction:

e.g. Newton's law of gravitation: $F = -\dfrac{Gm_1m_2}{r^2}$

Coulomb's law for electrostatics: $F = \dfrac{Q_1Q_2}{r^2}$

Work and Energy:

work = constant force $\times$ parallel displacement $W = F \times s$

if force varies with position: $\delta W = F \times \delta s$

(this last relation should be compared with $\delta p = F \times \delta t$: these are crucial formulae which express the effects of forces in space and time. Relativity links space and time and the connection between energy and momentum follows naturally.)

Energy conservation:

Energy may be transformed but cannot be created or destroyed. The total energy of a closed system is constant.

Reference frames:

Newton's laws are referred to a reference frame in which a free body maintains constant linear momentum (i.e. obeys the first law). These frames are assumed to be at rest or moving uniformly through absolute space and are referred to as inertial reference frames. If we observe particle trajectories from a non-inertial frame we see free particles spontaneously change their momentum and need to introduce imaginary or inertial forces to explain their motions (e.g. the coriolis force on Earth that affects global weather systems).

Problem solving:

Many dynamics problems involve a common approach:

(i) Identify the forces acting (perhaps by drawing a free-body diagram).

(ii) Determine the resultant force (if any).
(iii) Solve the equation of motion (N2) to predict the particle trajectory.
OR (iv) Use conservation laws to determine future values of energy and momentum.

1.2.6 Problems.

The examples that follow could serve as a brief review of Newtonian mechanics but are chosen to emphasize mechanical principles that will be used or modified in Special Relativity.

1.1. A glass lift operates up and down the outside of a tall building. It is moving up at a constant 3 m s^{-1} when a passenger inside the lift drops her bag. The bag was exactly 1 m above the floor of the lift when it was released.
   a. How long will it take for the bag to fall to the lift floor?
   b. How far does the bag fall relative to the lift?
   c. How far does the bag fall relative to an outside observer?
   d. How would your answers change if the lift descends at 3 m s^{-1}?
   e. How would your answers change if the lift were free falling down the lift shaft?

1.2. Two cars collide and lock together. They are each of mass 800 kg and were travelling at a steady 20 m s^{-1} in opposite directions before the collision.
   a. What is the kinetic energy of the system before and after the collision?
   b. Analyse a similar collision in which one of the cars initially approaches the other (which is at rest) at velocity 40 m s^{-1}.
   c. Are these two situations equivalent?
   d. What do you conclude about kinetic energy?

1.3. A dense object is dropped vertically from a height of 100 m above a point on the Earth's equator. Where will it land? (Ignore atmospheric drag effects.)

1.4. A ping pong ball is held underwater in a glass.
   a. What happens when it is released?
   b. What would happen if the glass were in free fall at the instant of release?

1.5. The Earth moves in an almost circular orbit around the Sun. This means its linear momentum with respect to the Sun reverses every six months. Does this violate the law of conservation of linear momentum?

1.6. Two points have Cartesian co-ordinates (1,2) and (2,3) relative to perpendicular axes x and y. The axes are then rotated about the origin by 30° in a clockwise direction.

a. Use Pythagoras's theorem to calculate the separation of the two points.
b. Calculate the co-ordinates of the points relative to the new axes.
c. Use Pythagoras's theorem with the new co-ordinates to calculate the separation of the points.
d. Comment on the behaviour of co-ordinates and distances with respect to rotation of axes.
e. Prove that the same behaviour results for any arbitrary pair of points and any rotation angle.

1.7. A plane passenger with a fear of flying is thinking what he could do if the plane crashes. He convinces himself he can survive a crash if, the instant before the plane hits the ground, he leaps into the air and loses contact with the aircraft. He will then land on the stationary wreckage and walk free.
a. What is wrong with this argument?
b. Will leaping in the air make any difference whatsoever?

1.8. Are astronauts in an orbiting space station genuinely weightless?

1.9. Raindrops are falling vertically at 1 m s^{-1} relative to the ground.
a. If you have a walking speed of 1.5 m s^{-1} at what angle to the vertical do the raindrops hit you?
b. Would you expect the Earth's motion in its orbit to have any effect on the observed positions of stars during the year? Explain any assumptions you have made.

1.10. The momentum of a photon is equal to E/c where E is its energy. The normal intensity of solar radiation near the top of the atmosphere is about 1400 W m^{-2}.
a. Calculate the force on a perfectly reflecting panel of area 100 m^2 in near Earth orbit if it is oriented at 90° to the incident sunlight.
b. How would your answer change if the surface was totally absorbing?
c. Comment on the feasibility of 'light-surfing' using solar reflectors as sails.

1.11. A space laboratory is freely falling from a great distance toward the surface of a planet. Are there any observations that could be made inside the laboratory to prove that it is not at rest in space?

1.12. A nucleus of mass $238m$ is moving through the laboratory at speed u in the x-direction. It undergoes an alpha decay (emits a helium nucleus of mass $4m$) at right angles to the direction of its original motion (in its own reference frame). The energy released in the decay is E.
a. Consider the decay in the centre of mass frame. Calculate the ratio of the final velocities v_α of the alpha particle and v_n of the nucleus produced by the decay in this frame.
b. Calculate the ratio of kinetic energies of these particles in the centre of mass frame.

c. Transform the linear momenta of the alpha particle and nucleus to the laboratory frame and use them to calculate the angle between each particle's path and the original direction of motion of the nucleus that decayed. Hence find the angle between the paths of the two particles after the decay.

1.13. A spacecraft is moving away from the Earth at speed v. Every T days it sends a brief radio message back to the Earth. After $10T$ days it reverses and returns to Earth at the same speed. What are the intervals between reception of messages during the outward and return journeys? (Assume that light has speed c relative to the Earth.)

1.14. The following problem has caused a fair amount of discussion in physics magazines. *'A truck has a vertical windscreen and is moving east at 25 m s^{-1}. A fly meanwhile is travelling west at 2 m s^{-1} along the same line. Since the fly will end up squashed against the windscreen it must change its velocity from 2 m s^{-1} due west to just less than 25 m s^{-1} due east. In doing this it must at some instant be at rest with respect to the road. At this instant the truck also is at rest.'*

a. Why is the final velocity 'just less than 25 m s^{-1} due east'?
b. Does the truck come momentarily to rest during the collision?
c. Draw a velocity time graph to show the motion of the fly and the truck.
d. Discuss the application of Newton's laws to this example.

1.15. Another problem with flies! Two cyclists are approaching each other along a straight road. Each has a road speed of 15 m s^{-1}. When they are 300 m apart a fly leaves the nose of one cyclist and flies to the nose of the other. On reaching the second cyclist it immediately flies back to the first and so on.

a. How far will the fly travel (relative to the air) before the cyclists pass if the fly's speed through the air is 3 m s^{-1}?
b. The two cyclists stop 15 m apart. The same fly buzzes back and forth between them. How long does it take to complete a round trip?
c. How long would it take if there was a 1 m s^{-1} east wind blowing directly from one cyclist to the other?
d. How long would it take if this wind changed direction and became a north wind?

1.16. A man of 75 kg is in a rowing boat of mass 225 kg. There are also oars (of negligible mass), a solar powered fan (of negligible mass), an umbrella (also of negligible mass) and ten 5 kg stones. He considers several possible means of propulsion. For each one say whether it would be successful and explain how Newton's laws of motion and the conservation of linear momentum and energy apply.

a. Rowing the boat.
b. Using the umbrella as a sail when the wind blows.
c. Using the umbrella as a paddle in the air to 'row' the boat when no wind blows.

d. Using the fan as an air propeller.
e. Throwing the stones overboard behind the boat.
f. Running from the front of the boat to the back and then returning slowly to the front.
g. As in the previous case but carrying a stone in each hand.

1.17. A terrestrial laser beam is directed at the Moon which is about 400 000 km away.
a. If the laser is rotated through 10 degrees in 0.1 seconds how fast does the point of the beam scan across the Moon's surface?
b. Is this an example of light travelling faster than light?

1.3 GRAVITY.

1.3.1 Explaining The Motion Of The Moon

Bodies close to the Earth's surface accelerate downwards when released. Applying the first law of motion Newton argued that this revealed the existence of a resultant force toward the Earth's centre. The third law implied that a force of equal magnitude but opposite direction would act on the Earth. He assumed this arose from a mutual attraction between the masses of the two bodies. He called this attraction 'gravity' and postulated that all massive particles exert attractive forces on all other massive particles. Mass enters mechanics in two apparently distinct ways, as the inertia of a body and as a measure of how strongly it is affected by or can produce gravitational forces. Both of these seem related to our naive intuition of mass as 'amount of substance', but there is no obvious link between them in classical mechanics.

Galileo contended that, in the absence of friction, all falling bodies accelerate at the same rate. This means that a larger mass, which has a greater reluctance to change its state of motion, must experience a larger attraction to the Earth in order to accelerate at the same rate as a body of smaller mass. In fact gravitational force and inertia must increase in direct proportion to one another, indicating that the inertial and gravitational effects of mass are also proportional and perhaps equivalent. This is a subtle point and one that would form the cornerstone of Einstein's General Theory of Relativity.

Newton's theory of gravity was not limited to terrestrial effects. The Moon is observed to follow an almost circular orbit around the Earth, changing its direction of motion continuously. According to the first law this continuous change of velocity is an acceleration towards the centre of the Earth (a centripetal acceleration) produced by a resultant force in the same direction. In fact, if there was no force the Moon would continue along a tangential path and fly off into space. Newton argued that this centripetal force has the same origin as the force on falling bodies close to the Earth's surface, it is produced by a gravitational attraction to the mass of the Earth. By calculating the Moon's centripetal acceleration and comparing it with that of objects falling close to the Earth he was able to show that the force exerted per kilogram of the Moon's mass is very much smaller than per kilogram of a terrestrial projectile. Rather than abandon his idea he assumed that gravitational forces diminish with distance and showed that an inverse square law (such that gravitational forces are reduced by a factor of 100 when we increase the separation of bodies by a factor of 10) could explain both motions. The distance of the Moon from the Earth is about 60 times the Earth's own radius. Newton calculated that the Moon's centripetal acceleration is about 0.0027 m s^{-2}, $1/60^2$ of the acceleration of free fall near the surface of the Earth.

1.3.2 Newton's Law Of Universal Gravitation.

All massive bodies attract one another with a gravitational force. The magnitude of this force between any two point masses separated by a distance r is proportional to the product of their masses and the inverse square of their separation. The constant of proportionality is called the universal gravitational constant G.

$$F = \frac{Gm_1m_2}{r^2} \qquad G = 6.67 \times 10^{-11}\ \mathrm{N\,m^2\,kg^{-2}}$$

Although Newton's law of gravitation is stated in terms of point masses it also applies for uniform spherical masses. This allows us to treat planets and stars like point objects. Newton's laws were soon applied to the solar system and had remarkable success in calculating trajectories. Kepler's three laws of planetary motion, for example, follow directly from the inverse square law.

At first planetary orbits were calculated by assuming the only significant interaction is that between the planet concerned and the Sun. Later it became necessary to allow for the perturbations caused by the gravitational influence of other planets. This always led to close agreement between observation and theory and such was the confidence in theory that tiny disagreements were often taken to imply the existence of as yet undiscovered bodies. In 1844 Uranus had deviated from its predicted position by 2 minutes of arc and two astronomers, J.C. Adams in Britain and J.J. Leverrier in France, calculated the position of a then undiscovered planet that would affect Uranus's path as it passed. On the basis of Leverrier's calculations J.G. Galle at the Berlin observatory found the new planet (Neptune) within half a degree of the predicted position after less than half an hour of searching! By the end of the nineteenth century another small deviation from Newtonian predictions was beginning to cause problems. The axes of Mercury's elliptical orbit seemed to be rotating. Even after allowing for a wobble in the Earth's axis that causes an apparent motion of Mercury, and adjusting the prediction for the non-spherical shape of the Sun, an advance of perihelion (closest approach to Sun) of about 43 seconds of arc per century remains. Nineteenth century astronomers suggested the existence of another undiscovered planet, Vulcan, but this remained undiscovered. Some even wondered whether Newton's theory might not be quite correct. We shall see how this effect is explained when we consider the effects of General Relativity.

1.4 MAXWELL'S ELECTROMAGNETISM.

1.4.1 The Concept Of Field.

fig 1.10 Michael Faraday
'The long and constant persuasion that all the forces of nature are mutually dependent, having one common origin, or rather being different manifestations of one fundamental power, has made me often think upon the possibility of establishing, by experiment, a connection between gravity and electricity.'
Artwork by Nick Adams ©1996

Electric and magnetic forces are observed to act on bodies separated by considerable distances. This idea of an 'action-at-a-distance' posed considerable problems of interpretation. Faraday found it helpful to describe magnetism by tracing out 'lines of force' in the space surrounding a magnet. The lines of force map directions in which a 'free north pole' (if such things exist) would move. The density of lines represent the strength of force exerted. A similar pictorial representation was used for electrical forces.

The significance of Faraday's representation is that it attributes properties to the spaces between charged particles. We describe this as an electromagnetic field. The lines of force bear a similar relation to this field as do contour lines to the topography of a landscape. The introduction of a field description means that the charges no longer experience action-at-a-distance, they are now acted upon by a direct local force from the field at the point they occupy. This model is appealing since it allows for a simple cause and effect explanation of electromagnetic effects (e.g. charge A creates a field in space and charge B, placed at some point in space, experiences a force from the field at that point). It also introduces a revolutionary new concept, the field itself. From Faraday's time on, attempts would be made to explain either the field in terms of particles or particles in terms of fields.

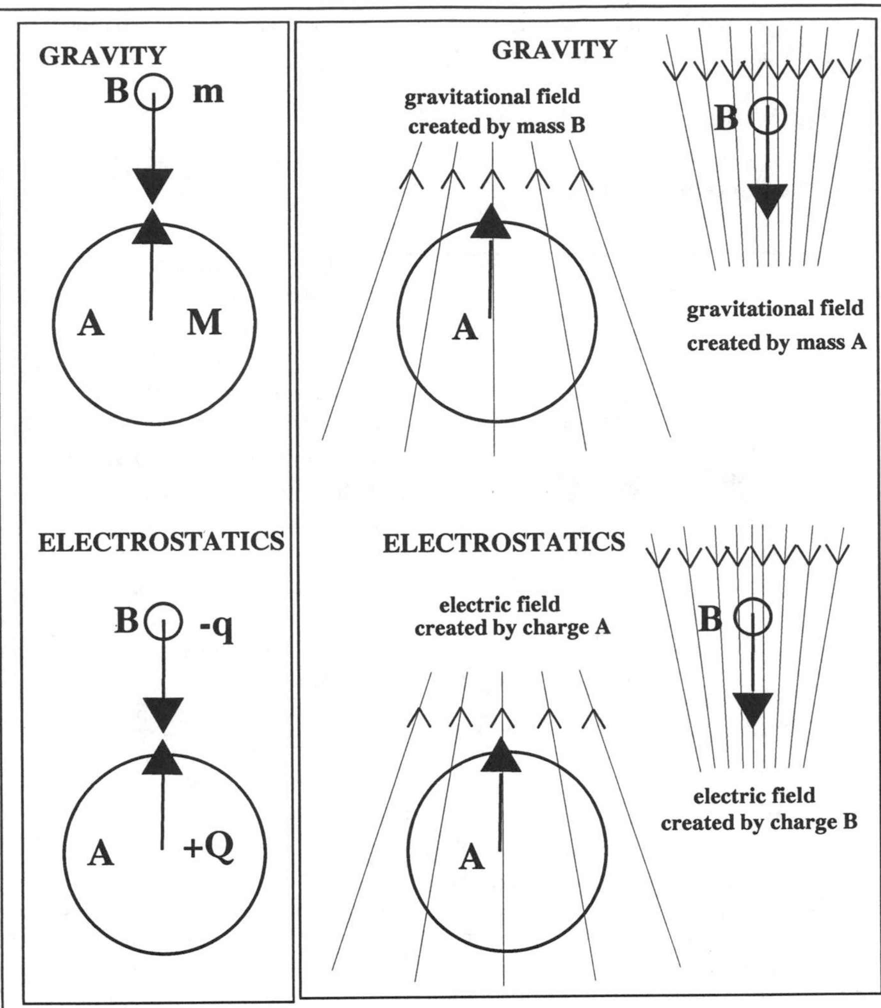

fig 1.11(a) Action-at-a-distance in gravity and electrostatics.

fig 1.11(b) Field theories of gravity and electrostatics.

In (a) forces act on distant objects with nothing between them to transfer the force. In (b) both bodies create a field and each experiences a local force from the field created by the other.

1.4.2 Maxwell's Equations.

For the field theory to be really useful it must account for all the known experimental laws of electromagnetism (e.g. Coulomb's Law for the force between charges). Conversely these laws should provide vital clues to the mathematical form such a field theory should take. Maxwell succeeded in generating a set of four equations which describe the behaviour of fields in all circumstances. They show how the fields depend on the distribution of charges and how they change as a result of changes in the fields adjacent to them. It is this connectivity in the field which allows a simple explanation of electromagnetic waves.

1.4.3 Field And Particle.

Whilst the Maxwell equations describe the fields themselves we also need to describe the forces exerted by these continuous fields on the discrete charges which reside on material bodies (e.g. electrons). In 1895 Lorentz provided an equation of motion for charged particles in the electromagnetic field. The Lorentz equation shows that the force produced by an electric field is parallel to the field and independent of the velocity of the charge whereas the magnetic force is only experienced by a moving charge and is always perpendicular to the velocity and field. (The dependence of magnetism on motion was one of the clues that led Einstein to Relativity.) Combining Lorentz's equation of motion with Maxwell's equations gives a complete theory within which we can predict future configurations of fields and charges if we know their present configuration. Treated in this way electromagnetism gives the forces between charged particles and Newton's laws then tell us how the trajectories of these particles evolve.

1.4.4 Light As An Electromagnetic Wave.

In 1845 Faraday showed that a magnetic field was capable of rotating the plane of polarization of light:

'Thus is established, I think for the first time, a true and direct dependence between light and the magnetic and electric forces....'
(Faraday, Philosophical Magazine, Nov. 1845)

In Maxwell's theory the electromagnetic field is capable of sustaining rapid vibrations of both the electric and magnetic fields. These electromagnetic waves travel at a speed determined purely by constants that arise form electrical and magnetic phenomena (the speed of electromagnetic waves in empty space is given by $c = 1/\sqrt{\varepsilon_o \mu_o}$ (where ε_o and μ_o are the permittivity and permeability of free space respectively). This speed is the same (allowing for experimental errors) as that obtained for the speed of light in terrestrial and astronomical measurements.

'The agreement of these results seems to show that light and magnetism are affectations of the same substance and that light is an electromagnetic disturbance propagated through the field according to electromagnetic laws.'
(Maxwell, Philosophical Transactions, 1864)

fig 1.12 James Clerk Maxwell (1831-1879)
'I have reason to believe that the magnetic and luminiferous media are identical.'
Artwork by Nick Adams © 1996

Imagine a stationary charge. It will be surrounded by a static field. If it is moved suddenly the field close to it will change. As the electric field changes it induces a corresponding change in the local magnetic field. These local disturbances of the electromagnetic field will spread to regions further from the original charge. The further a point is from the charge the later will the disturbance reach it. An electromagnetic pulse travels outwards at a finite speed rather like a ripple travelling out from the place where a raindrop has disturbed an otherwise smooth surface of a pond. Continuous electromagnetic waves are produced by continually oscillating charges. Visible light differs from say radio waves or X-rays only in the frequency of vibration that produces it (and consequently the length of the waves produced). They all travel at the same speed in a vacuum, the speed of electromagnetic waves derived from Maxwell's equations.

1.4.5 The Luminiferous Ether.

All mechanical waves are disturbances of some material medium (e.g. sound waves are disturbances of the air) so it was natural that Maxwell and others should seek a mechanical medium to support the electromagnetic fields and waves. This 'luminiferous ether' should be an all pervasive but tenuous medium in which the electric and magnetic fields exist as distortions or strains. These strains would be released when the ether exerted a force on a charge and caused it to move. The similarities between Maxwell's equations and equations for the distribution of stresses in an elastic medium reinforced these ideas and provided a suggestive analogy for further speculation, and the ether provided a medium in which electromagnetic energy could be stored like the elastic potential energy of a stretched spring. Moreover, if the ether filled space it might provide an accessible reference for absolute measurements. Soon the question of the ether became intimately tangled with that of absolute space and physicists began to wonder if it was possible to measure the Earth's velocity relative to the ether.

1.4.6 Energy.

Whenever electromagnetic waves are transmitted energy must be supplied to charges in the transmitter and we can detect and extract energy from the charges in the receiver. It is clear that energy has been transmitted from one to the other. Also, since the transmitting charges experience a 'recoil' force from the fields as they emit the waves, and the receiving charges experience a thrust as the waves arrive, momentum has also been transferred (momentum change being equal to the impulse of the radiation). In 1897 Lebedev managed to measure the tiny radiation pressure exerted on a metal surface as light falls on it.

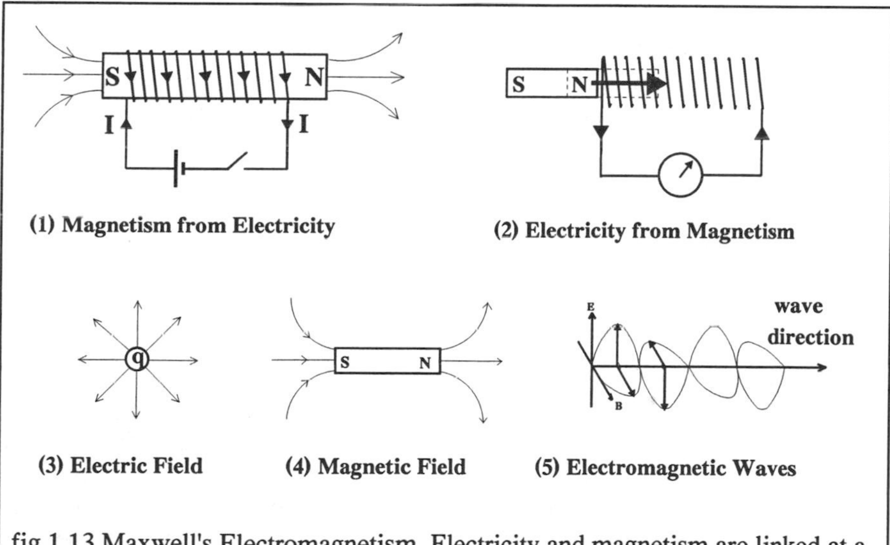

fig 1.13 Maxwell's Electromagnetism. Electricity and magnetism are linked at a fundamental level. The energy and momentum of electromagnetic waves links electromagnetism to mechanics. The investigation of this link led to relativity.

These observations raise more serious considerations. If the field can store and transmit energy and momentum then how are these manifested in the ether? Momentum is classically associated with mass in motion, leading several late nineteenth century theorists to suggest that the energy of an electromagnetic field might have an associated mass and inertia. Some went so far as to suggest that all mass is electromagnetic in origin.

1.5 THE WAVE MODEL OF LIGHT.

1.5.1 Waves Or Particles?

Whilst both the wave and the particle models of light were able to account for the phenomena of rectilinear propagation, reflection and refraction, only the wave theory could explain:

- Diffraction: The effect of light apparently bending around corners after passing an obstacle or aperture (first reported, posthumously, by Grimaldi in 1665).

- Interference: The addition or cancellation of light where waves from different sources overlap. (This was used by Young in 1801 to measure the wavelength of visible light.)

- Velocity: The speed of light in a denser medium (e.g. water) is less than that in a less dense medium (e.g. air or the vacuum). The particle model predicted it would increase. (Foucault measured the speed of light in air and water in 1862.)

Both relativity and quantum theory cause us to question some of the ideas of the classical wave theory of light so here we summarize some of the fundamental concepts.

1.5.2 Wave Velocity.

The speed of sound is measured with respect to the air that carries it. The speed of a water wave is relative to the body of water on which it moves. The speed of a wave on a string is with respect to the string. In all mechanical waves there is some clear medium that propagates the wave and whose properties determine its speed. It was assumed that light would behave in a similar way, and that its velocity would be measured relative to the ether and determined by the mechanical properties of the ether. Since the ether's rather tenuous properties had not shown up in any physical experiments it was hoped to detect it indirectly by observing the alteration of the velocity of light relative to the Earth due to the Earth's own motion through the ether. In the latter part of the nineteenth century several such experiments were carried out. We shall return to these in the context of the special theory of relativity later.

1.5.3 The Doppler Effect.

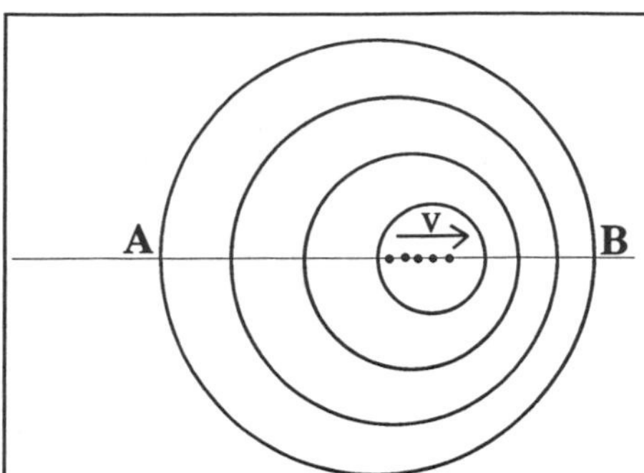

fig 1.14 The Doppler effect. The dots show successive source positions as it moves to the right. Waves at A have increased wavelength and reduced frequency whilst those at B have decreased wavelength and increased frequency compared to the source.

If waves are received from a moving source their frequency and wavelength will be shifted with respect to waves from the same source at rest. This is easy to understand. Imagine a source moving away at a constant velocity. Each wavefront (e.g. the crest of each emerging wave) will be emitted from a point beyond the release point of the wavefront preceding it. This will effectively 'stretch' the waves with respect to the receiver and reduce their frequency since successive waves are delayed because of the extra distance travelled. If the source approaches the shift is in the opposite sense, the wavelength is reduced and the frequency increased. With mechanical waves we must consider the motion of source and receiver with respect to the wave medium, by analogy we might expect the motion of electromagnetic wave sources relative to the ether to affect the Doppler shifts of light. All that matters in electromagnetism is the relative velocity of source and receiver, once again the idea of an absolute motion is unnecessary. The motion of the Earth in our Local Group of galaxies causes a Doppler shift in the background radiation received from space, it is blue shifted ahead of the Earth and red shifted behind it (see the colour image from the COBE satellite data). This shift must be allowed for when looking for variations in intensity.

1.5.4 Diffraction.

When a wave passes through an aperture there is a slight tendency for the edges of the wave to spread outwards into the region of geometric shadow. This diffraction effect becomes more significant as the size of the aperture decreases or the wavelength increases, becoming extreme as the aperture size is comparable to the wavelength, when emerging waves are effectively circular (or spherical in 3D). Similar effects occur when light waves encounter small solid obstacles (Babinet's principle). The two diagrams below (fig 1.15) show diffraction for plane waves incident on a linear aperture (a form of Fraunhofer diffraction). In practice the pattern of diffraction is more complicated than this because wavelets spreading from each point on the aperture superpose to create an interference pattern. The general approach to problems like this is based on a geometric construction suggested by Huygens.

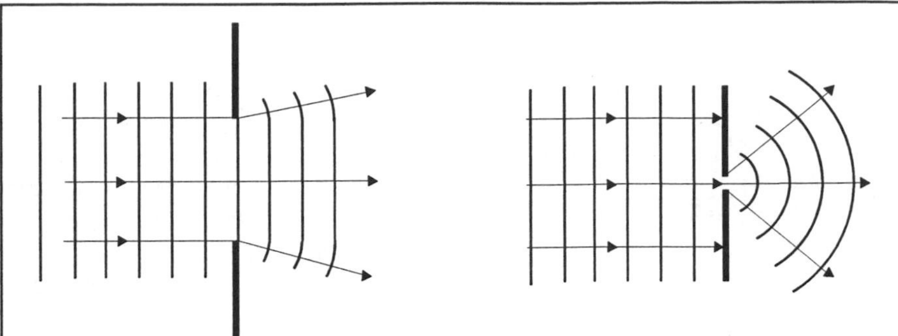

fig 1.15 Diffraction. Waves passing through an aperture spread into the region of geometric shadow. The effect becomes more significant as the ratio of wavelength to aperture size increases. Huygens explained diffraction as the superposition of secondary wavelets from each point on a wavefront.

1.5.5 The Principle Of Superposition.

Where waves from two or more sources overlap (superpose) the resultant disturbance at that point is the vector sum of disturbances due to each wave. For example, where the crests of two water waves pass over one another there is a moment when the water surface is lifted up into a very large crest equal in height to the sum of heights of the two individual crests. Similarly when light waves meet their electric and magnetic field vectors superpose; in some circumstances this can result in a pattern of maxima and minima or 'interference fringes'.

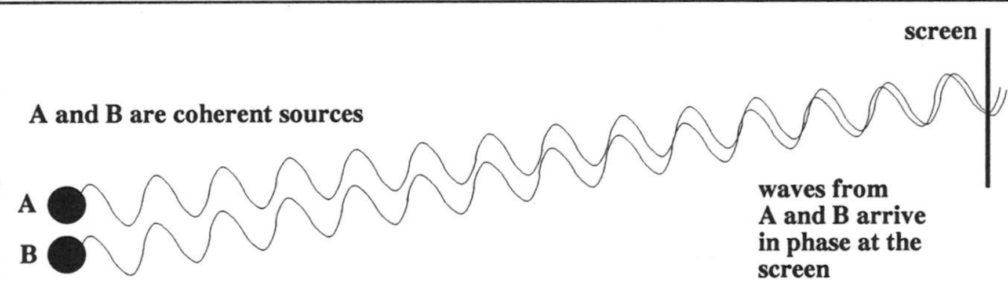

fig 1.16 Superposition. Waves from A and B pass through each other. As they do their combined effect depends on their relative phase. At the screen waves arrive in phase and add to produce constructive interference (giving maximum amplitude). Interferometers provide a very accurate way to measure very small displacements and play an important role in experimental tests of relativity.

1.5.6 Young's Double Slit Interference Experiment.

In Young's experiment light falling onto a small single slit is made to diffract onto a pair of double slits. This ensures that the double slits act as coherent sources, that is they produce light waves that have the same frequency, wavelength and amplitude, and which keep a constant phase relation (as the wave leaving the first slit reaches its amplitude so does that from the second). The double slits diffract light so that there is a significant region of overlap in which interference patterns can be observed.

If a screen were placed parallel to the barrier containing the two slits it would show a regular pattern of bright and dark interference fringes. This is the characteristic double slit interference pattern produced in Young's experiment. The maxima are caused by constructive interference at places where waves from both slits arrive in phase. These positions correspond to a path difference for the light of a whole number of wavelengths. Minima occur where the waves arrive π out of phase and destructive interference occurs (when the path difference is an odd number of half wavelengths). By considering the geometry of the experiment Young showed how the separation of maxima is related to the wavelength of light, the separation of the slits and the distance to the screen. He measured the experimental parameters and calculated a value for the wavelength of visible light (of the order of $0.5\ \mu m$).

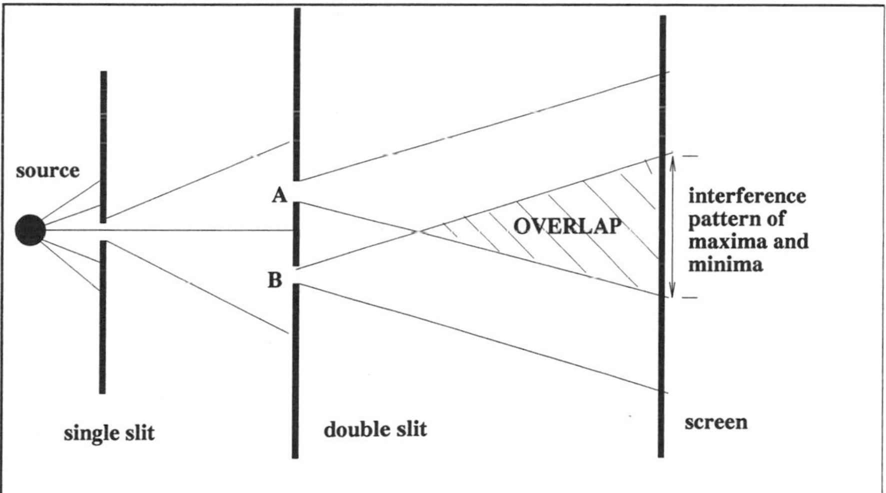

fig 1.17 Young's double slit experiment seemed to prove that light must be a wave phenomenon. It is hard to imagine how particles could interfere.

1.5.7 Interferometers.

Interference is often used to make measurements of small changes in lengths and has been involved in several of the tests of relativity and the search for gravity waves. Most of these experiments make use of an instrument based on a Michelson interferometer.

The principle of an interferometer is very simple. Incident light, usually from a laser (which produces a monochromatic, coherent and very intense source) is split into two beams. The beams are then directed along separate paths before being recombined. Where they superpose they produce interference effects. For example, if both paths are identical the beams superpose in phase and produce an intense central maximum. If one beam is delayed then the resulting path difference causes a phase difference when they superpose and the maxima and minima positions change. The basic layout of an interferometer is shown below. (The same principle is used on a much larger scale when the signals from radio telescopes are combined in long baseline interferometers. The advantage is that the resolving power of the pair of telescopes is much greater than that of a single telescope.)

Interferometers are also being used to test the fundamental principles of quantum theory. If light is so weak that only one photon at a time passes through the apparatus it might seem reasonable to suppose that each photon goes by a particular route and so no interference effects should be observed. However, the interference effects persist as long as it remains impossible to determine the actual route taken by the photon.

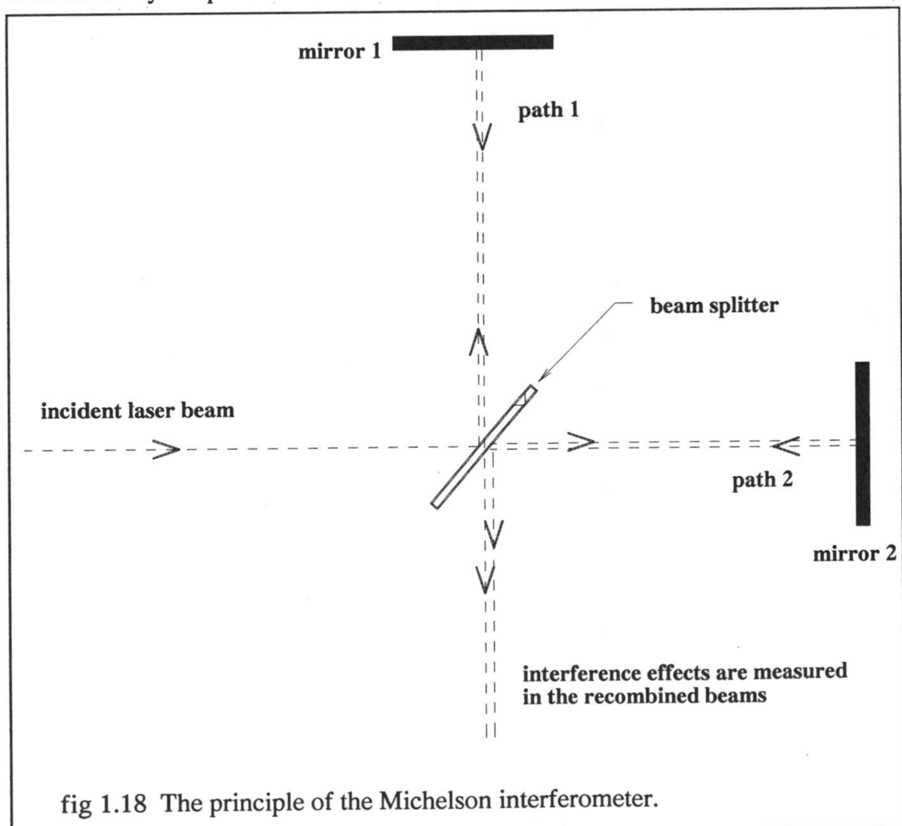

fig 1.18 The principle of the Michelson interferometer.

1.6 A SUMMARY OF CLASSICAL PHYSICS TO 1900.

- Matter consists of particles (ultimately atoms) which are rather like miniature billiard balls possessing properties of mass, size, position and motion, and may be charged positively or negatively. (The first subatomic particle, the electron, was discovered by J.J. Thomson in 1897. It was assumed to be a tiny charged mechanical particle.)
- Particles interact with one another. Their changes of motion under the influence of forces are described and determined by Newton's laws of motion and take place against a background of absolute space and absolute time.
- The forces between particles may arise in collision as they resist deformation (electromagnetic forces) or at a distance as a result of gravitational or electromagnetic interactions. Forces combine as vector quantities.
- The gravitational force is given by Newton's inverse square law whilst electromagnetic forces are given by the Lorentz equation in terms of field strengths derived from Maxwell's equations. Electrostatic forces obey a simple inverse square law like gravitation (Coulomb's law).
- Problems in classical physics often reduce to identifying the nature and form of resultant forces and calculating their effects from Newton's laws of motion.
- The spectrum of visible light is a particular range of frequencies in the electromagnetic spectrum. All electromagnetic waves are disturbances of the electromagnetic field that travel at the speed of light in a vacuum. These electromagnetic waves transmit energy and momentum between source and receiver and exert a radiation pressure on absorbing or reflecting surfaces.
- The small scale structure of matter is governed by electromagnetic forces, mainly through interatomic bonds. The large scale structure of the Universe is governed by the much weaker gravitational force. This dominates for large masses because atoms are neutral and so long-range electrostatic effects are negligible.
- The Earth and planets orbit the Sun because of gravitational attraction. The Universe is probably infinite in extent so that the average gravitational effect at any point cancels. A finite universe would tend to collapse.

1.7 PROBLEMS.

Developing The Ideas.

The problems below explore the theoretical and philosophical ideas about space time and motion from which relativity emerged. Some have clear-cut answers, but most are more open-ended and should be thought-provoking.

1.18. This problem is about the concept of inertia. Medieval mechanics was based on a concept of force suggested by Aristotle - forces are responsible for maintaining motion, or more simply, 'no force, no motion'. The classical concept of force was developed by Galileo and Descartes and finally stated as Newton's first law of motion. It can be paraphrased, 'no force, no change of motion'. Forces in classical physics are therefore responsible for changing the magnitude or direction of a body's motion.

a. Why did Aristotle's idea of force seem reasonable when applied to most of the motions experienced in everyday life?
b. Compare how Aristotle and Newton might explain the role of forces in causing a vehicle to travel along a horizontal road at a steady speed. If you asked the average non-physicist to explain the same example whose explanation do you think they would agree with?
c. Objections to Aristotle's idea of force were raised by several scientists. One of the famous examples involved the flight of an arrow. Why, if a force is necessary to maintain its motion does it continue to move after leaving the bow? Aristotle explained this by arguing that 'nature abhors a vacuum'. How might this help? Does it lead to any other problems (e.g. concerning motion in a vacuum, or why the arrow eventually comes to rest)?

1.19. Galileo, in his brilliant work, 'Dialogue Concerning the Two Chief World Systems', has Salviati and Simplicio discuss the motion of a free moving ball on a flat horizontal surface:

'SALV:....... what would happen if it were given an impulse in any direction?
SIMP: It would follow that it would move in that direction.
SALV: But with what sort of movement? One continuously accelerated, as on the downward plane, or increasingly retarded as on the upward one?
SIMP: I cannot see any cause for acceleration or deceleration, there being no slope upward or downward.
SALV: Exactly so. But if there is no cause for the ball's retardation there ought to be still less for its coming to rest; so how far do you have the ball continue to move?
SIMP: As far as the extension of the surface continued without rising or falling.
SALV: The if such a space were unbounded, the motion on it would likewise be boundless? That is, perpetual?
SIMP: It seems so to me ....'

a. Reading this seems to lead inevitably to the law of inertia (that is Newton's first law of motion). What has been assumed in the argument?
b. To what extent is it an idealization of experience rather than a description of it?
How valid is such a 'thought experiment' in establishing actual Laws of Nature?

1.20. Newton's laws assume the existence of an absolute space and time within which physical events take place. This question compares our experience of space and time with Newton's assumption. A measuring rod is used to measure the separation of two buildings.

a. What properties do we demand of an ideal measuring rod? To what extent are these realized in the devices we actually use to measure distance?

b. If the same rod were to be used to measure the separation of two different buildings elsewhere and at a later time what must we assume about the rod in order to compare the new measurement with the old measurement? How might we try to check this? Does your answer suggest any problems with the idea of an absolute length?

1.21. In his influential book, 'The Science of Mechanics', the Viennese philosopher/scientist Ernst Mach criticized Newton's assumption about absolute space. He pointed out that all measurements of length are carried out by comparing the extension of one object with the extension of another. So any definition of length produces a standard for comparison but not an absolute length.

a. If the linear dimensions of the universe and all it contains were to double what would be the effect on our measurements (consider both static and dynamic measurements)?

b. Mach levelled a similar criticism at Newton's conception of absolute time. Compare and discuss the comments of Newton and Mach in the quotations below:

NEWTON: 'Absolute, true and mathematical time, of itself and from its own nature, flows equably without relation to anything external, and by another name is called 'duration'; relative, apparent and common time is some sensible and external .... measure of duration by the means of motion, which is commonly used instead of true time, such as an hour, a day, a month, a year.'
(Principia)

MACH: 'It is utterly beyond our power to measure the changes of things by time. Quite the contrary, time is an abstraction, at which we arrive by means of the changes of things .... a motion may, with respect to another motion, be uniform. But the question whether a motion is in itself uniform is senseless. With just as little justice .... we may speak of an 'absolute time' - of a time independent of change. This absolute time can be measured by comparison with no motion; it has therefore neither a practical nor a scientific value; and no one is justified in saying he knows aught about it ....'
(Science of Mechanics)

1.22. Two space travellers carry out a number of standard physical experiments within their spacecraft whist travelling with a velocity v relative to one another. Their results for all the physical constants are in agreement. All the laws of physics are obeyed in all the experiments they undertake. However, as they pass one another they compare the lengths of their measuring rods and the rates of their clocks. These seem to disagree. How could we decide which of the two travellers has accurate measuring instruments?

1.23. This question is about frames of reference and the problem of detecting absolute motion. If absolute space exists we might expect it to have some absolute properties, such as shape, position or motion. Consider Einstein's thought experiment described below:

> *'The purpose of mechanics is to describe how bodies change their position in space with "time" .... it is not clear what is to be understood here by "position" and "space" .... I stand at the window of a railway carriage which is traveling normally, and drop a stone on the embankment without throwing it. Then, disregarding the influence of the air resistance, I see the stone falls to the Earth in a parabolic curve. I now ask: 'do the "positions" traversed by the stone lie "in reality" on a straight line or on a parabola? Moreover, what is meant here by motion "in space"? .... The stone traverses a straight line relative to a system of co-ordinates rigidly attached to the carriage, but relative to a system of co-ordinates rigidly attached to the ground it describes a parabola. With the aid of this example it is clearly seen that there is no such thing as an independently existing trajectory but only a trajectory relative to a particular body of reference.'*
> (Einstein, Relativity)

What conclusions do you draw about the geometry of space?

1.24. Imagine you are inside a closed compartment with no windows on a perfectly smooth silent train travelling at a high but constant velocity along a straight horizontal track.

a. Is there any experiment you could carry out *inside* the compartment to determine whether the train was at rest or in motion?

b. Could you detect the acceleration or deceleration of the train by any experiment carried out entirely within the compartment?

c. Would it be possible to distinguish the acceleration of the train from a tilting of the carriage (perhaps due to motion uphill)?

1.25. Galileo also wondered whether rest and motion could be distinguished. He asked where an object dropped from the top of a ship's mast would land (i) when the ship is tied to the quay and (ii) when it is sailing past the quay at constant velocity. (He assumed that the effect of air resistance could be neglected.)

a. Where will it land in each case? How does its path appear to observers on the boat and on the quay in each case?

b. In the light of your answers to (a) what can be concluded about the motion of the Earth through absolute space from the fact that objects dropped from the top of a tall building land near its base?

1.26. When we observe the behaviour of moving objects from an accelerated reference frame they appear to experience additional forces which cause them to follow apparently curved paths. In order to explain this we introduce 'inertial forces' which are sometimes dismissed as illusory since they do not arise when we describe the same motions from a non-accelerated, or inertial reference frame.

a. You are in a closed room somewhere in the Universe. You feel yourself apparently 'pressed' against the floor of the room. How can you account for this 'force'? On what assumptions does your answer depend?

b. Newton explained the motion of falling bodies in terms of a gravitational force of attraction to the Earth. Use Newton's laws of motion to reconstruct his possible line of reasoning. Is the gravitational force 'real'?

c. You are in a car which is negotiating a tight bend at high speed. You feel pushed toward the outside edge of the car. This apparent force is called a 'centrifugal' force. Is it 'real'? How can your experience be explained without the need to introduce such a centrifugal force?

d. It is possible to place a satellite in orbit above the Earth's equator such that its orbital period is exactly one day. It then remains geostationary (i.e. at rest relative to a point directly beneath it on the earth's surface) and can be used for communications. Conventional mechanics explains this motion in the following way. The satellite is placed at such an altitude that the Earth's gravitational attraction applies exactly the required centripetal force to keep it in circular motion at that radius with this period. What reference frame is assumed for this explanation?

e. If we adopt the rotating Earth itself as our reference frame then the satellite is stationary in this frame. What force must now be added in order to support the satellite? Can you suggest any possible source for this?

1.27. This question is about Mach's Principle, an idea which made a strong impression on the young Einstein. Examples like that of the geostationary satellite have prompted some physicists to consider the possibility that all inertial forces, and indeed inertia itself, might be induced by motion with respect to a reference frame defined by the 'fixed stars' or the overall mass distribution of the Universe. Think about Newton's bucket experiment, the surface of the water is only distorted when rotation relative to the fixed stars occurs. Think about Foucault's pendulum, somehow it 'senses' the orientation of its plane of oscillation with respect to those same fixed stars.

a. If the fixed stars really do act back on local matter, what is missing from our present laws of physics?

b. Consider a mass in an otherwise empty Universe. Would there be any way to determine whether it was at rest, in uniform motion or accelerating? Is there any physical distinction between these alternatives? (Your answer may well depend on what you understand by 'empty'.)

c. Now consider the same mass in this Universe. It now possesses inertia, a reluctance to change its state of motion. Is it possible that the reaction force we experience when we accelerate a body is in some way produced by its interaction with the mass of the rest of the Universe? Discuss this idea.

d. A Foucault pendulum suspended directly above the north pole would be observed to swing such that its plane of oscillation rotates once in 24 hours in the reverse sense to the Earth's rotation. How does this suggest that the rotation of the Earth is absolute and not relative?

e. How do the coriolis forces which deflect air masses as they move north and south from the equator originate?

f. Compare the ideas of inertia as they appear in Newtonian mechanics and in Mach's principle.

1.28. Here are some related conceptual questions. They do not all have clear cut answers and could well be used as the basis for discussion of the concepts of space, time and force.

a. Is motion possible without time? Is time possible without motion? Is time possible without change?

b. I think it was John Wheeler who said that time is simply that which prevents everything from happening at once. Respond to this statement.

c. Paraphrasing one of Aristotle's arguments: the past has gone, the future has not yet happened, the present moment has no duration so time is an illusion! Any comments?

d. If the future is predictable in every detail, that is if it is totally determined by the present, does it in some sense already exist? And by the same token does the past 'continue' to exist?

e. A stationary charge sets up an electric field in space. A moving charge creates both an electric and a magnetic field. Consider two observers watching a moving charge. The first sees the charge pass at velocity v, the second moves with it so that the charge appears stationary to him. Does the particle produce an electric field alone or both an electric and a magnetic field? What does your answer tell you about the nature of electromagnetic fields?

f. A FACT: If you were to build a vertical scaffold the right height (approximately 36 000 km) at the equator, climb up it, and step off into space you would remain stationary above a fixed point on the Earth's surface below. What holds you up? What would happen if you stepped off from a point just above or below this position?

g. Does space exist? In other words, is an empty universe any different from the absence of a universe? If your answer to this is 'No' then what we call space is presumably brought about by the existence of matter. How can this be? If your answer is 'Yes' then what properties does space possess?

h. Is all space uniformly packed? If it wasn't, how would we know? Imagine a telephone box containing space at 10 times 'normal' density, what might you experience on entering? If the density of space varied from point to point in the universe, how would this affect our perception of mechanics?

i. Does the term 'density' make any sense when applied to space as in the last question? What assumptions are hidden in this terminology?

j. People often discuss the rate of flow of time, what meaning can be attached to such a term? What problems does this kind of description lead to?
k. Give a clear definition of 'Time Travel'.
l. It has been suggested that the fundamental particles might be some kind of 'knot' in the geometric fabric of space and time. What physical predictions might be made on the basis of such a theory?

1.29. As a young man Einstein was intrigued by the question, 'What would a light wave look like if we could catch up with it?'
a. What would it look like on the basis of classical physics?
b. Is there any reason to worry about this conclusion?

2
PROBLEMS WITH SPACE AND TIME

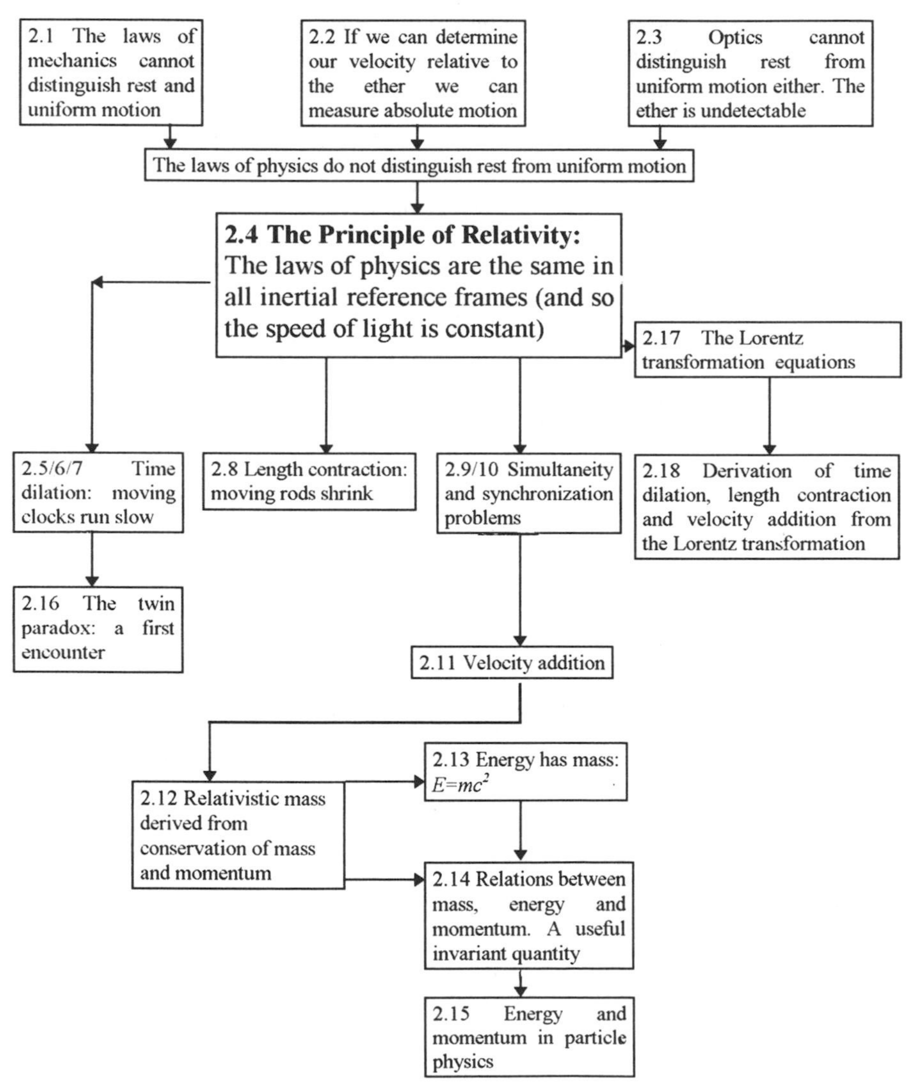

2.1 PROBLEMS OF MOTION AND CHANGE.

'Nature has been defined as a "principle of motion and change", and it is the subject of our inquiry. We must therefore see that we understand the meaning of "motion"; for if it were unknown, the meaning of "nature" too would be unknown.' (Aristotle, Physics, Book III Chapter 1)

2.1.1 Where Is Absolute Rest?

fig 2.1 Galileo Galilei (1564-1642)
'My purpose is to set forth a very new science dealing with a very ancient subject. There is, in nature, perhaps nothing older than motion, concerning which the books written by philosophers are neither few nor small; nevertheless I have discovered by experiment some properties of which are worth knowing and which have not hitherto been either observed or demonstrated.'
Artwork by Nick Adams © 1996

Newton asserted the existence of an absolute space and time in which classical physics took place, but it was clear from his own laws of motion that no mechanical experiment would be able to identify this reference frame. Galileo drew attention to this even before Newton in his 'Dialogue Concerning Two New Sciences' in 1638:

'Shut yourself up with some friend....below decks on some large ship and have with you there some flies, butterflies, and other small flying animals. Have a large bowl of water with some fish in it; hang up a bottle that empties drop by drop into a wide vessel beneath it. With the ship standing still observe how the little creatures fly with equal speed to all sides of the cabin. The fish swim indifferently in all directions; the drops fall to the vessel beneath; and in throwing something to your friend, you need throw it no more strongly in one direction than another, the distances being equal; jumping with your feet together, you pass equal spaces in every direction. When you have observed all these things carefully, have the ship proceed with any speed you like, so long as the motion is not fluctuating this way and that. You will discover not the least change in all the effects named, nor could you tell from any of them whether the ship was moving or standing still.'

The laws of mechanics are completely indifferent to whether a system is truly at rest or is moving with a constant speed in a straight line. This statement is generally referred to as Galilean Relativity. In this sense all such uniformly moving reference frames are equivalent and it makes no sense to select one as a special case and call it absolute rest. As far as mechanics is concerned there is no absolute rest frame.

However, mechanics *can* distinguish between uniform motion and acceleration. This is obvious to anyone who has tried to drink coffee in a moving car - as long as the car is travelling in a straight line on a smooth road at constant speed there is no problem, but any sudden acceleration, deceleration, bumping or turning makes it very difficult not to spill it. It is as if new forces act in the accelerating reference frame causing it to slop about. These are called *inertial forces* since they arise when we view the inertial motion of an object from a non-inertial reference frame. On a global scale the coriolis forces that deflect winds in the weather system are inertial forces. Special relativity is restricted in that it deals only with physics as seen from the point of view of an inertial reference frame.

2.1.2 What Is Absolute Time?

For Newton all changes in the mechanical universe must be measured in absolute time but, as Ernst Mach pointed out:

'It is utterly beyond our power to measure the changes of things by time. Quite the contrary, time is an abstraction, at which we arrive by means of the changes of things; made because we are not restricted to any one definite measure, all being interconnected.... A motion may, with respect to another motion, be uniform. But the question whether a motion is itself uniform is senseless. With just as little justice, also we may speak of an "absolute time" - of a time independent of change. This absolute time can be measured by comparison with no motion; it has therefore neither a practical nor a scientific value; and no one is justified in saying he knows aught about it. It is an idle metaphysical conception....'
(Mach, Science of Mechanics, 1895)

Newton was well aware that all practical measurements of time would be by comparison with objects undergoing some kind of cyclical change (e.g. a simple pendulum, or the rotation of the Earth) but he felt that the observed agreement of these devices would be sufficient to imply that they all actually provide measurements of a single absolute time to which we have no direct experimental access. Mach pointed out that there was no empirical evidence for such an opinion and realized that adopting it when there was no reason to do so may well lead to problems in the future.

Mach's view that time should be defined in terms of the *operations* used to measure it was influential in the approach Einstein adopted to the question of time in Relativity. By making time a property of the universe itself rather than something external to it Mach opened up a new route of approach to many thorny philosophical problems. For Einstein it meant that we must start not with absolute time but with time measurements on clocks subject to the laws of physics.

2.1.3 Does The Earth Move?

The Copernican revolution seemed to have established once and for all that the Earth moves through space. However, if we apply the ideas of Galilean relativity we see that there is no mechanical experiment that can be carried out on the Earth to determine whether or not it has a velocity relative to absolute space. We are quite used to referring all our observations of the external world to the frame of reference defined by the Earth rather than that defined by the Sun or 'fixed stars'. What Copernicus really achieved was to show that there is a particular choice of reference frame, that is one at rest with respect to the Sun, from which the laws of planetary motion are simplified. This makes no judgment about the motion of the Earth relative to absolute space. The very question of such a motion may be meaningless. The motion or lack of motion of the Earth may only be judged relative to other objects, not relative to absolute space.

2.1.4 Inertial Reference Frames.

The inertia of a body is a reluctance to change its state of motion, the intrinsic property that is responsible for all the objects on Galileo's ship maintaining the same uniform velocity as the ship. If this were not the case then his passengers would have been able to detect the motion of the ship by observing the anomalous way that objects inside the cabin moved relative to it. Uniformly moving reference frames (e.g. those considered at 'rest' or moving with constant velocity in a straight line) are called inertial reference frames and play a very important role in relativity theory. Strictly speaking special relativity deals only with physics viewed from inertial reference frames.

The simplest way to test whether you are in an inertial reference frame is to hold a small mass at arm's length and release it. It should remain at rest. If given a small impulse it will move off at constant speed in a straight line. This is the kind of mechanics we see in operation in films shot inside an orbiting spacecraft. It is *not* what happens on Earth. However, we are so familiar with the almost constant gravitational field close to the Earth's surface that often we treat the Earth as if it were an inertial reference frame with gravity added. A similar approach is often taken when electric or magnetic fields are involved. Having said this, it is interesting that Einstein managed to incorporate gravity into relativity only by realizing *how to transform it away* by choosing a freely falling reference frame. This led him to the equivalence principle and showed how intimately gravity and acceleration are linked. The crucial clue was an observation made by Galileo that all objects fall with same acceleration, this is discussed in more detail in chapter 4.

2.1.5 The Operational Approach.

In classical physics a measurement of length or time would be expected to yield a unique objective result corresponding (within the limits of accuracy and repeatability of the measurement process) to the absolute length or time between the events concerned. If there is no absolute background for these measurements then the value obtained can only be relative. Thus the operation by which the measurements are carried out becomes all-important. The time between events must be related to a particular clock in a particular reference frame and the distance between objects must be related to a particular measuring rod in a particular frame of reference. Strangely it seems we may only approach the nature of space and time through the behaviour of the devices used to measure them.

2.2 THE PROBLEM OF THE ETHER.

2.2.1 Introduction.

No mechanical experiment can distinguish between rest and uniform straight line motion. However, the possibility remains that an optical experiment might be able to do so. Let us return below decks on Galileo's ship and consider two new experiments. The first is an attempt to measure the speed of sound and the second to measure the speed of light.

Sound is a mechanical wave that travels at approximately 330 m s^{-1} relative to the air. Since the air itself moves with the cabin our experimenter will measure exactly the same value regardless of whether the ship is at rest or in motion on the sea. It will make no difference in which direction of travel the wave speed is measured. This is of course consistent with the assumption of Galilean relativity.

Light is an electromagnetic wave that travels at a velocity of about 3.0×10^8 m s^{-1}. The classical assumption was that this velocity is relative to the ether that pervades all space but is not in any special way associated with the particular frame in which we make our measurement. Thus a motion of the ship relative to the ether ought to reveal itself by modifying the measured value of the speed of light. If, for example, the velocity of light (c) were to be measured in the same direction as the velocity of the boat (v) through the ether then the result of the measurement would be the difference of these velocities ($c - v$). This is the same line of reasoning that leads us to expect that when a car traveling at 100 km h^{-1} overtakes one travelling at 80 km h^{-1} it will draw away at a relative speed of 20 km h^{-1}. It appears that, at least in principle, optical experiments *are* capable of distinguishing between rest and motion through the ether. Since the ether is assumed to fill all of space this may be equivalent to locating Newton's absolute rest frame.

This argument was the basis of many nineteenth century attempts to determine the Earth's 'absolute motion' with respect to the ether. However, as is often the case, it was not practically possible to carry out the experiments exactly as described above, and the simple clear-cut result that such an experiment might have given

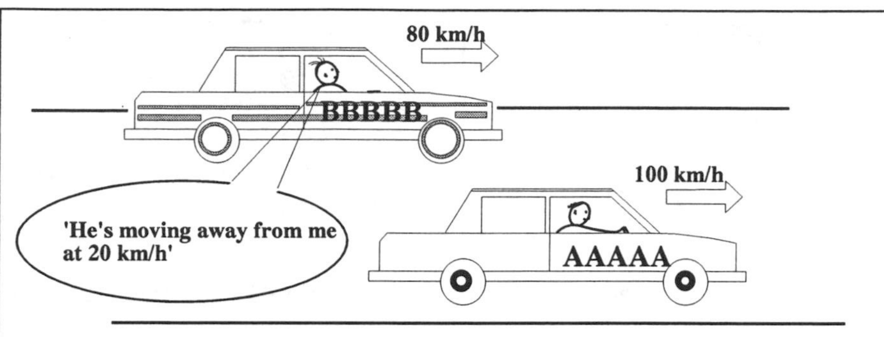

fig 2.2a In classical physics, when a car moving at 100km/h along a road passes a car moving at 80 km/h it draws away at a relative velocity of 20 km/h relative to outside observers or the car drivers.

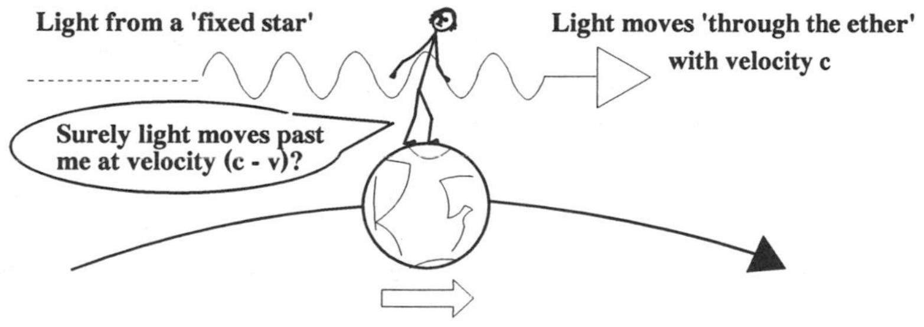

fig 2.2b If the same logic is applied to optics we should expect the velocity of light to vary according to the motion of the Earth through the hypothetical 'ether'.

was rather longer in arriving. Historically the most significant experiment was carried out by Michelson and Morley in 1887 but it is not clear to what extent Einstein was influenced by this. We shall consider the Michelson-Morley experiment in some detail shortly but there were a number of other observations that are worth discussing first.

2.2.2 Stellar Aberration.

If the Earth moves relative to the ether then light from distant stars will appear to be displaced due to this motion. This is for a reason similar to the one that causes us to tip an umbrella forwards when we walk through the rain. If we move horizontally and the rain falls vertically then the resultant path of the rain with respect to us is at an angle to the vertical which increases the faster we move. If we were to run around in a circle then we would need to point the umbrella continually ahead of us in order to give best protection.

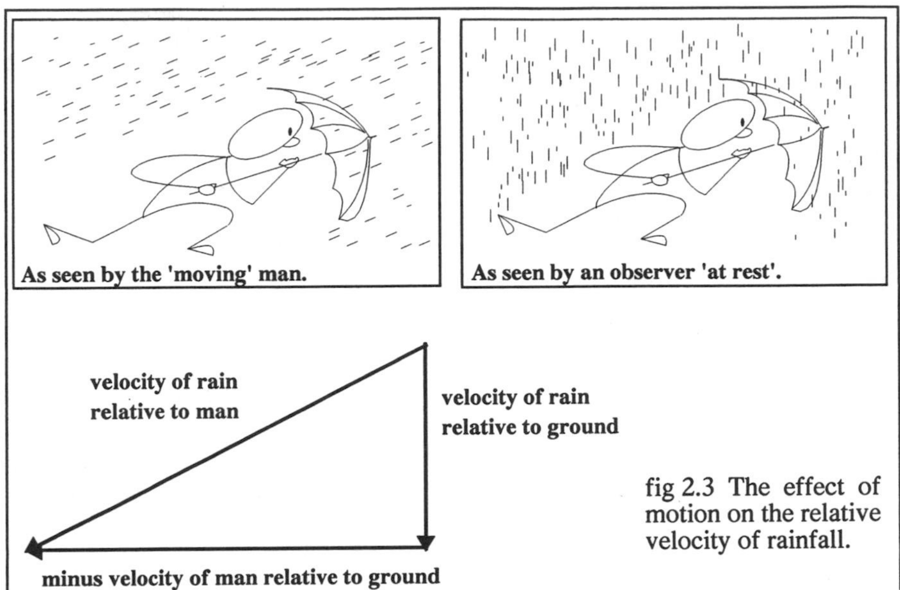

fig 2.3 The effect of motion on the relative velocity of rainfall.

The Earth moves in an almost circular orbit around the Sun and the light from distant stars behaves in a similar way. This was first discovered by Bradley in 1728 who noted an elliptical motion of the stars against the heavens associated with the Earth's motion about the Sun (he was trying to measure distances to stars using parallax with the Earth's orbital diameter as a baseline). Young gave a simple explanation of stellar aberration in terms of the wave theory of light in 1799. He suggested that aberration was caused by the movement of the telescope relative to the ether while the light travels down the telescope tube. Young's analysis suggested that *'....the luminiferous ether pervades the substance of all material bodies with little or no resistance....'* Since the angular displacement could be calculated from the ratio of the Earth's orbital speed and the speed of light this seemed to provide powerful support for the ether theory.

Example: Calculate the stellar aberration for a star (a) along a line perpendicular to the plane of the Earth's orbit (ecliptic) and (b) along a line making an angle of 25° to the ecliptic.

(a) At all points in Earth's orbit its orbital velocity (30 km s^{-1}) is perpendicular to the incident light (ignoring the small effect of parallax) so the telescope must be tilted in the direction of the Earth's motion through an angle:

$$\tan\alpha = \frac{v}{c} \qquad \text{but } v << c \qquad \text{so} \qquad \tan\alpha \approx \alpha$$

$$\alpha \approx \frac{v}{c} = \frac{3.0\times10^{4}\,m\,s^{-1}}{3.0\times10^{8}\,m\,s^{-1}} = 10^{-4}\,rad = \underline{\text{ 21 seconds of arc}}$$

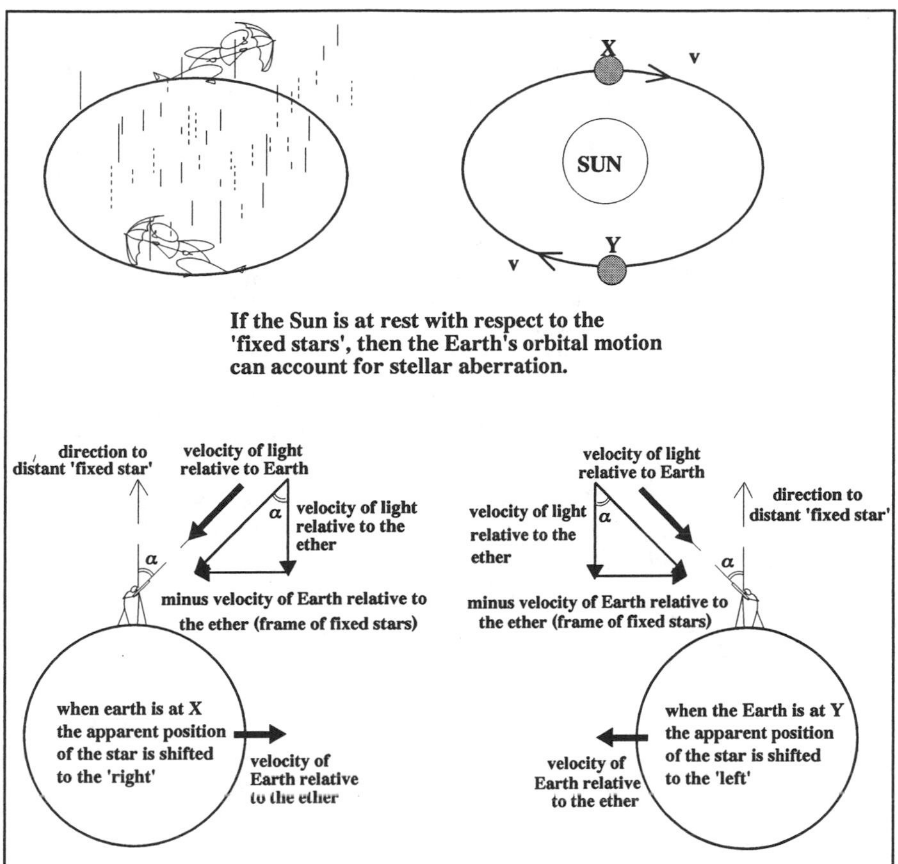

fig 2.4 Stellar aberration. The diagrams above assume the existence of an all-pervading 'ether' in which the Sun and distant stars are at rest, through which the Earth moves with a varying velocity. The case shown is for a star whose direction is perpendicular to the plane of the Earth's orbit. in this case the angle of aberration has a sine equal to the ratio v/c This is very much exagerated in this diagram. In fact stellar aberration is about 21 seconds of arc for such a star.

(b) There are two points on the Earth's orbit (along the axis parallel to the projection of the line to the star onto the ecliptic) where the Earth's orbital velocity is perpendicular to the incident light. Here the aberration α will be as calculated in (a) above, 21 seconds. The minimum aberration will occur at the positions where the orbital velocity is parallel to this projection. The component of the Earth's velocity perpendicular to the incident light is then just $v\sin\theta$ and the aberration is:

$$\tan\alpha' = \frac{v\sin\theta}{c} \quad \text{but} \quad v \ll c \quad \text{so} \quad \alpha' \approx \frac{v\sin\theta}{c}$$

$$\alpha' \approx \frac{v\sin\theta}{c} = \frac{3.0\times10^{4}\,m\,s^{-1}\,\sin 25^{o}}{3.0\times10^{8}\,m\,s^{-1}} = 4.2\times10^{-5}\,rad = \underline{8.7 \text{ seconds}}$$

The annual 'motion' of the star is *elliptical*. Although this approach works it is *incorrect*. The relativistic treatment of stellar aberration is given later.

2.2.3 Arago's Experiment.

The refraction of light as it crosses a boundary between media depends on the ratio of its velocities in those media. Arago suggested that this should cause the position at which a terrestrial telescope focuses an image of a distant star to change depending on whether the Earth approaches or recedes from the light emitted by that star. It is easiest to visualize this effect if we imagine the star at rest in the ether and the Earth moving directly towards or away from it. While the Earth approaches the star at velocity v the relative velocity of light to the Earth increases to $(c + v)$. The light entering the telescope is not deviated as much as if the Earth were at rest in the ether and the image is formed further from the objective lens. If Arago's reasoning was correct the deviation should depend on the ratio of the Earth's velocity through the ether to the velocity of light through the ether (v/c) and so the adjustment to focus during the Earth's orbit would be a measure of the Earth's velocity through the ether. He carried out an experiment to test this in 1810 but the expected deviation *did not appear*. Terrestrial telescopes formed images as if the Earth were at rest in the ether.

A related experiment carried out by Airy in 1871 also failed to detect the effect of the Earth's motion through the ether. Airy formed an image of a distant star and then observed the same star through a telescope filled with water. Since the speed of light in water is less than in air, light should take longer to travel the length of the tube and so the amount of stellar aberration was expected to increase. It didn't.

Arago's 'null result' could not be accounted for by Young's hypothesis of a stationary ether, so Fresnel, in 1818, suggested that the light might be partially dragged along inside material bodies. Fresnel's drag hypothesis would conceal the effect of a body's motion in the ether. He calculated a drag coefficient which depends on the refractive index of a material and is just the right size to conceal the effect of motion relative to the ether in experiments like Arago's or Airy's. In 1851 Fizeau reinforced the drag hypothesis by 'racing' two light beams through moving water such that one beam moved against the flow and the other moved with it. The beams were recombined and formed interference fringes. When the flow rate was changed or reversed the fringes moved. Detailed analysis showed that the light moving against the flow was delayed by just the amount predicted by the Fresnel drag coefficient (but that was not the correct explanation - we return to this when we look at relativistic velocity addition).

2.2.4 The Ether Wind.

By 1880 there was a reasonable amount of empirical evidence to support the idea of a stationary ether and a partial drag of light by matter in motion. However, it should still be possible to detect the change in velocity of light due to the Earth's motion since this would effectively create an ether wind through the laboratory. This idea is discussed in 'The Evolution of Physics' by A. Einstein and L. Infield:

'Again we return to our moving room with two observers, one inside and one outside. The outside observer will represent the standard CS [coordinate system, or reference frame], designated by the ether-sea. It is the distinguished CS in which the velocity of light always has the same standard value. All light sources, whether moving or at rest in the calm ether-sea, propagate light with the same velocity. The room and its observer move through the ether. Imagine that a light in the centre of the room is flashed on and off and, furthermore, that the walls of the room are transparent so that the observers, both inside and outside, can measure the velocity of the light. If we ask the two observers what results they expect to obtain, their answers would run something like this:
The outside observer: My CS is designated by the ether-sea. Light in my CS always has the standard value. I need not care whether or not the source of light or other bodies are moving, for they never carry my ether-sea with them. My CS is distinguished from all others and the velocity of light must have its standard value in this CS, independent of the direction of the light beam or the motion of its source.
The inside observer: My room moves through the ether-sea. One of the walls runs away from the light and the other approaches it. If my room travelled, relative to the ether-sea, with the velocity of light, then the light emitted from the centre of the room would never reach the wall running away with the velocity of light. If the room travelled with a velocity smaller than that of light, then a wave sent from the centre of the room would reach one of the walls before the other. The wall moving towards the light wave would be reached before the one retreating from the light wave. Therefore, although the source of light is rigidly connected with my CS, the velocity of light will not be the same in all directions. It will be smaller in the direction of motion relative to the ether-sea as the wall runs away, and greater in the opposite direction as the wall moves towards the wave and meets it sooner.
Thus, only in the one CS distinguished by the ether-sea should the velocity of light be equal in all directions. For other CS moving relatively to the ether-sea it should depend on the direction in which we are measuring.'

The most famous and ingenious experiment designed to detect this motion was carried out by Michelson and Morley in 1887.

2.3 THE MICHELSON-MORLEY EXPERIMENT.

2.3.1 Introduction.

Fresnel's drag coefficient for air is so small that we can neglect it in a terrestrial laboratory. It should therefore be possible to determine the velocity of the Earth's

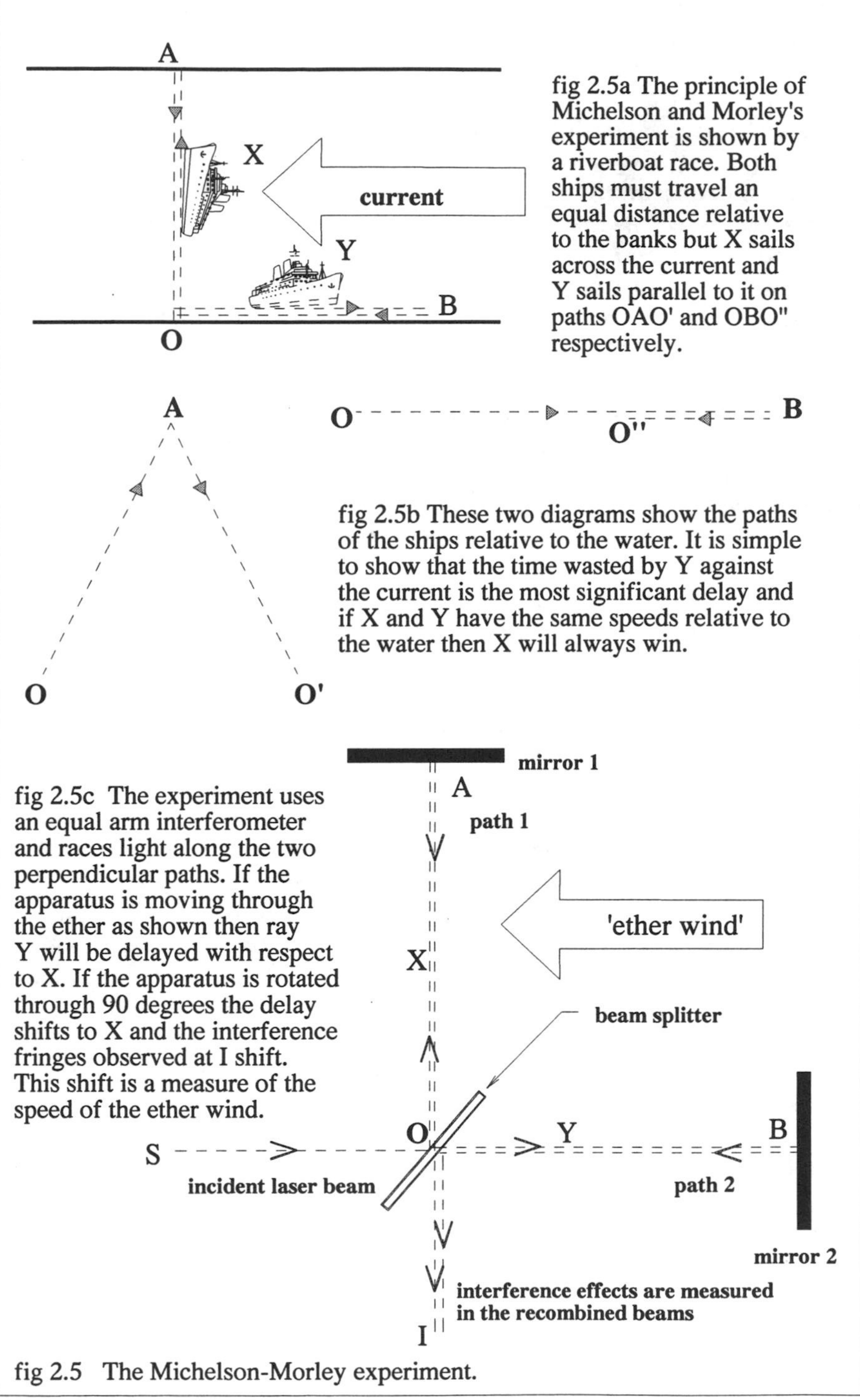

fig 2.5a The principle of Michelson and Morley's experiment is shown by a riverboat race. Both ships must travel an equal distance relative to the banks but X sails across the current and Y sails parallel to it on paths OAO' and OBO" respectively.

fig 2.5b These two diagrams show the paths of the ships relative to the water. It is simple to show that the time wasted by Y against the current is the most significant delay and if X and Y have the same speeds relative to the water then X will always win.

fig 2.5c The experiment uses an equal arm interferometer and races light along the two perpendicular paths. If the apparatus is moving through the ether as shown then ray Y will be delayed with respect to X. If the apparatus is rotated through 90 degrees the delay shifts to X and the interference fringes observed at I shift. This shift is a measure of the speed of the ether wind.

fig 2.5 The Michelson-Morley experiment.

motion through the ether by measuring deviations from the theoretical value of the speed of light. However, as Maxwell pointed out:

'All methods .... by which it is practicable to determine the velocity of light from terrestrial experiments depend on the measurement of the time required for the double journey from one station to the other and back again, and the increase of this time on account of a relative velocity of the ether equal to that of the earth would be only about one hundred millionth part of the whole time of transmission, and would therefore be insensible ....'

fig 2.6 Albert Michelson (1852-1931)
Michelson won the Nobel Prize in 1907 for his increasingly accurate measurements of the speed of light and for his part in establishing that this speed is independent of the motion of the source or observer.
Artwork by Nick Adams © 1996

The problem is that light travels in both directions in all terrestrial measurements of the speed of light (c). Thus, even if we could be sure that we orientated our apparatus along the line in which the Earth moves through the ether (at velocity v), the delay when travelling against the ether wind (at $(c - v)$ relative to the laboratory) will be partially compensated for by the additional speed (of $c + v$) when it is travelling with the wind.

Maxwell's remarks were taken as a challenge by Michelson who set about devising experimental arrangements which might detect such a small change. He proposed an experimental arrangement in which a light beam is split in two and each half is then 'raced' over paths of equal length in perpendicular directions. If the apparatus is at rest in the ether then the two rays should return together. If it is in motion through the ether such that one path is parallel and the other perpendicular to the motion then the parallel beam is delayed with respect to the other perpendicular beam. The amount of delay depends on the velocity v and is a measure of the ether wind speed in the laboratory. In other words if the delay can be measured it will tell us the Earth's speed relative to the ether.

Michelson and Morley carried out their famous experiment in 1887. The two rays were compared after their journeys by bringing them together and causing them to interfere with one another. A slight delay in either beam would result in a shift of the interference fringes. Since the actual direction of the Earth's motion through the ether was unknown they mounted their entire apparatus on a large stone slab floated in mercury and looked for a movement of the fringes when they rotated it through ninety degrees. If, in a particular orientation path 1 took longer than path 2 then, after rotation path 2 should take longer. The shift in fringes was then used to calculate the speed of the ether wind.

The result of the experiment was surprising: allowing for experimental errors there was *no shift in the position of the interference fringes*. This most famous of all null-results has been repeated in many later and more accurate versions of the experiment.

2.3.2 The Significance Of The Null Result.

The immediate and obvious conclusion would be that the Earth is actually at rest in the ether, but this would contradict the evidence of stellar aberration and also seems to go set the clock back before the Copernican revolution. An alternative could be, as Stokes suggested, that the Earth drags some of the ether close to its surface along with it, rather like the layers of a viscous fluid that adhere to the surface of an object moving through it, but this too runs into problems in explaining aberration and disagrees with some other optical experiments carried out by Michelson.

A more radical hypothesis was proposed independently by Fitzgerald and Lorentz. Perhaps bodies in motion undergo a contraction of length in the direction of motion of exactly the amount required to make the two alternative paths in the Michelson-Morley experiment equal. That is to say, although the path parallel to the ether wind should take slightly longer to traverse at the velocity $c - v$, its length is reduced by just the right amount to ensure that the race between light beams always ends in a dead heat! Although the Lorentz-Fitzgerald contraction was fairly generally adopted as a practical explanation of the null result it left a feeling of unease among the physics community. It seemed too much like a convenient 'fudge factor' explaining away rather than accounting for the difficult results.

Meanwhile Poincaré had become increasingly convinced that it is impossible to detect the absolute motion of the Earth *by any experimental method whatever.* In a prophetic paper delivered in 1904 (the year before Einstein's own famous paper on relativity) he stated what he described as a 'Principle of Relativity':

'.... the laws of physical phenomena must be the same for a "fixed" observer or for an observer who has a uniform motion of translation relative to him: so that we have not, and cannot possibly have, any means of discerning whether we are, or are not, carried along in such a motion.'

In many respects this anticipates Einstein, but for Poincaré it was the result of compensatory effects such as the Lorentz-Fitzgerald contraction whereas Einstein's approach led to a new understanding of the nature of space and time.

2.4 THE PRINCIPLE OF RELATIVITY

2.4.1 Laws Of Physics Are Universal.

fig 2.7 Albert Einstein (1879 1955)
'I am anxious to draw attention to the fact that this theory is not speculative in origin; it owes its invention entirely to the desire to make physical theory fit observed facts as well as possible ... The abandonment of certain notions connected with space, time and motion hitherto treated as fundamental must not be regarded as arbitrary, but only as conditioned by observed facts.'
Artwork by Nick Adams

Einstein's first relativity paper was entitled 'On the Electrodynamics of Moving Bodies', and was published in volume 17 of Annalen der Physik in 1905. (Einstein had two other papers published in the same journal, one of which, on the photoelectric effect, later earned him the Nobel Prize!) Rather than solve existing problems by introducing new effects in the ether Einstein proposed a universal Principle of Relativity which made the need for an ether disappear:

'The phenomena of electrodynamics as well as those of mechanics possess no properties corresponding to the idea of absolute rest. They suggest rather that .... the same laws of electrodynamics and optics will be valid for all frames of reference in which the equations of mechanics hold good. We will raise this conjecture (the purport of which will hereafter be called the "Principle of Relativity") to the status of a postulate, and also introduce another postulate, which is only apparently irreconcilable with the former, namely that light is propagated in empty space with a definite velocity c which is independent of the state of motion of the emitting body .... the introduction of a "luminiferous ether" will prove to be superfluous inasmuch as the view here to be developed will not require an "absolutely stationary space".'
(Einstein, 1905 Paper)

The simplicity of Einstein's principle is striking. Galileo showed that mechanics could not distinguish rest from uniform motion or Galilean Relativity:

- *The laws of physics have the same form in all inertial reference frames.*

Einstein 'merely' extended the principle to include electromagnetism. Thus, if we accept the speed of light as a constant arising from a law of physics then the principle of relativity can be stated as simply:

- *The laws of physics are the same in all inertial reference frames.*

Special relativity is the working out of the consequences of this very general principle. It is special because the principle is restricted to *inertial reference frames* and does not tell us how to deal with reference frames subject to acceleration or gravitational fields. In special relativity the laws of physics are said to be covariant (that is they have exactly the same form) with respect to all inertial reference frames.

2.4.2 The Speed Of Light As An Invariant.

The velocity of light plays a special role in physics. Since it is a universal constant it can be used as a common reference value for the comparison of measurements made by observers in different reference frames. Quantities such as c which have the same value in all inertial reference frames are called 'invariants' and are of great importance in relativity. As a velocity it relates measurements in space to measurements in time and so can be used to compare distances and time intervals measured by differently moving observers. This is why, in what follows, we shall make use of 'light clocks' and use light beams to synchronize our experiments. It is important to appreciate that the time measured by a light clock is exactly the same as the time measured by *any other type of clock* in the same inertial reference frame, including pendulums, digital watches and biological clocks! If this were not the case then it would be possible to determine the absolute motion of a light clock, and hence of its reference frame, by noting its deviation from some other motion designated as a standard.

It follows that we could use any type of clock at all to measure time since they can all be calibrated against a light clock in their rest frame. This may seem a sensible thing to do since digital electronic clocks are far more familiar than the hypothetical light clock, but there is a major problem. Electronic clocks and watches depend on the fundamental laws of physics in a far more complicated way than the light clock and so make it much harder to see how to apply the rule of covariance. The light clock depends directly on the speed of light, a value derived directly from Maxwell's electromagnetic theory and so having the same value in all inertial frames; it is easy to deduce how a light clock must appear to an inertial observer.

Einstein proposed the law of constancy of the speed of light as a second principle separate from the principle of relativity, but it is not really an additional postulate. It must be constant if the laws of electromagnetism are to be covariant. Also the term 'speed of light' is misleading. All electromagnetic waves propagate at this speed, but so do all other influences carried by particles of zero rest mass (like photons). The speed c plays a fundamental role in physics which is not limited to optics or electromagnetism, it is quite simply the most important velocity in the universe (in some formulations of relativity it is the *only* velocity). Light happens to travel at this speed and the historical name is unfortunate. This special velocity will be our point of reference as we begin to explore the consequences of relativity.

Although the speed of light in a vacuum is a limit for all material bodies, there are situations where particles can move faster than light in a medium. One particularly important example is Cerenkov radiation, emitted when a charged particle passes through a transparent medium faster than the speed of light in that medium. This is discussed in the caption to figure 2.8. Another example involves the neutrinos from supernovae. These particles interact so weakly with matter that they beat light to us despite the near vacuum of space. Neutrinos emitted from supernova 1987A arrived about 20 hours before the light from the explosion.

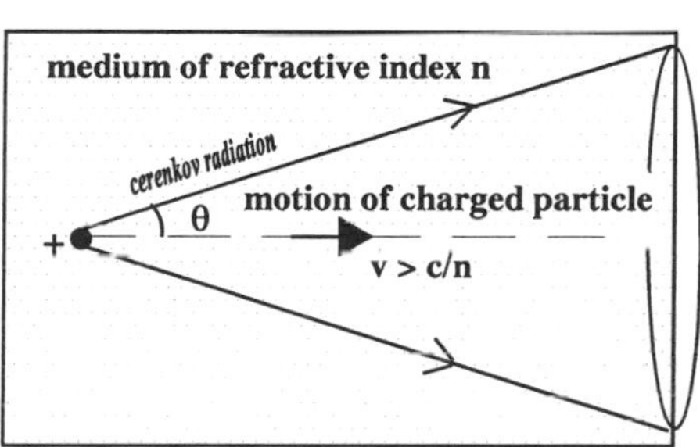

fig 2.8 Cerenkov radiation

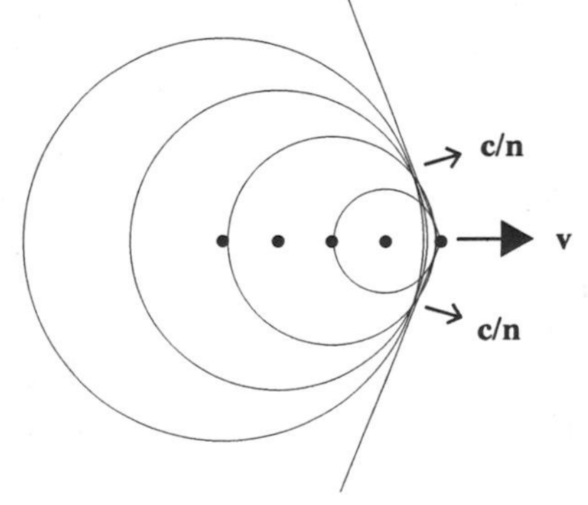

Faster than light? Yes.... the velocity of light *in a medium* is c/n where n is the refractive index. Light is slowed down because of its electromagnetic interactions with atoms in the medium. High energy particles can travel at any speed up to c so they sometimes exceed the speed of light in a particular medium. When this happens, the moving charge emits Cerenkov radiation. Waves leaving nearby atoms as the charge passes reinforce in a direction at angle θ to the velocity of the moving charge. This angle is given by:

$$\cos\theta = \frac{c}{nv}$$

The angle increases with particle velocity and although the Cerenkov effect is weak it can be used in detectors designed to respond to particles moving only at or above a particular velocity. This selectivity makes Cerenkov detectors very useful in high energy physics.

2.4.3 The Light Clock.

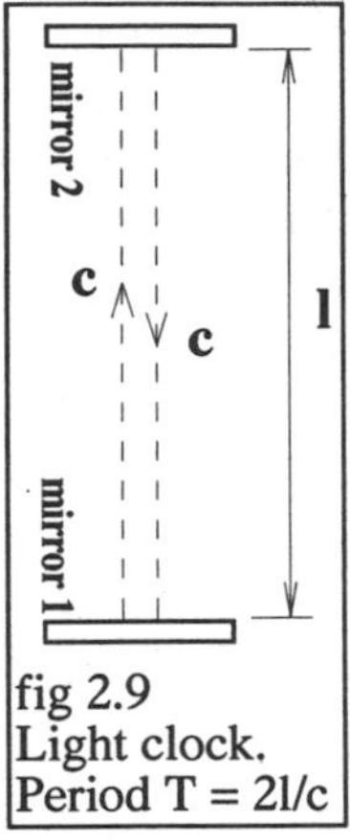

fig 2.9
Light clock.
Period T = 2l/c

We shall use a very simple optical arrangement to measure time. It consists of two parallel mirrors and a pulse of light that bounces backwards and forwards between them. One return trip marks one 'tick' of our light clock. if the mirrors are separated by 1.5 m and light travels at 3×10^8 ms^{-1} then the time between two ticks is 10^{-8} s. In general, for a mirror separation l the time between ticks will be $T = 2l/c$. The beauty of using such a light clock is that all inertial observers see the same velocity of light *relative to their own inertial reference frame*. They are therefore able to compare the time intervals between the ticks of a moving clock with the time intervals between ticks on their own clock by a simple calculation based on the observed distances travelled by the light in each clock while their own clock ticks once. The time between ticks on a clock in its own rest frame is called the *proper time*.

2.5 TIME DILATION.

2.5.1 Explaining The Michelson-Morley Result.

The explanation of null-results in ether drift experiments like the Michelson-Morley experiment is trivial once we adopt the principle of relativity. Since the apparatus is in an inertial reference frame light will be transmitted at velocity c in all directions. Neither route will introduce a delay however the apparatus is orientated and no change in the positions of interference fringes will occur.

This is all very well until we ask how experiments carried out in one inertial frame appear to an observer in another inertial frame in relative motion. For example, if both we and a moving experimenter measure the velocity of the same light beam as it passes us we must both find its relative speed to be c. The effect of our relative motion cannot affect the outcome. This is only possible if our perceptions of space and time are *different* (if we adopted a common background of absolute space and time we would find that one observer gets c and the other gets either $c + v$ or $c - v$). Of course this example is exactly that of the ether experiments, our failure to detect a change of velocity of light due to relative motion is equivalent to failing to detect the presence of the 'ether wind'.

The principle of relativity has removed the distinction between rest and motion and made the concepts of absolute space and absolute time redundant. There is a price to pay, however: if there are no absolutes then the measure of space and time must be changed by motion. How then are we to compare a moving observer's measurements with our own?

Let us assume that both ourselves and the moving observer are equipped with identical light clocks which, when placed side by side with no relative motion, keep time perfectly with one another. What we cannot infer from this is that they will remain in step when one moves at velocity v with respect to the other. To compare them in this way we must first choose a reference frame in which to make the comparison and then apply the principle of relativity to it in order to deduce the consequences. In the examples below 'applying the principle of relativity' will mean no more than making sure the speed of all light beams relative to any particular observer is c.

We shall label our observers A(rthur) and B(etty) and consider how each of them views their own and their counterpart's light clock.

2.5.2 Equivalent Reference Frames.

A and B both consider everything to be normal in their own reference frames. Their own light clocks will tick at the expected rate with respect to any other physical timer they happen to have (e.g. masses on springs or atomic oscillators) and time will seem to pass normally. Moreover, if they look out at phenomena in the universe they will see the same laws of physics producing the same values for invariants everywhere. As far as physics is concerned there is nothing to distinguish A's experiences from B's. Their reference frames are equivalent in all respects.

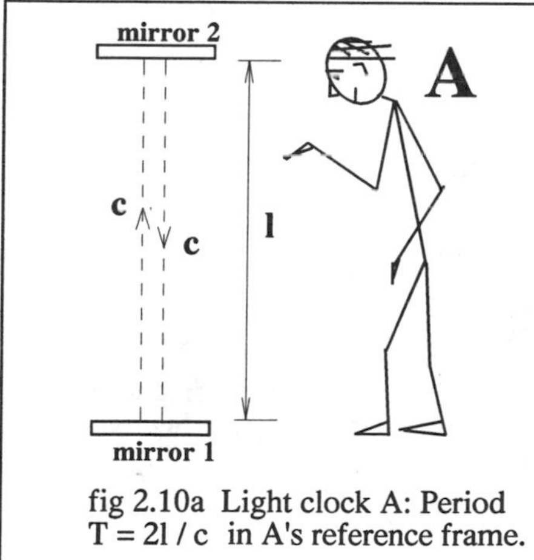

fig 2.10a Light clock A: Period T = 2l / c in A's reference frame.

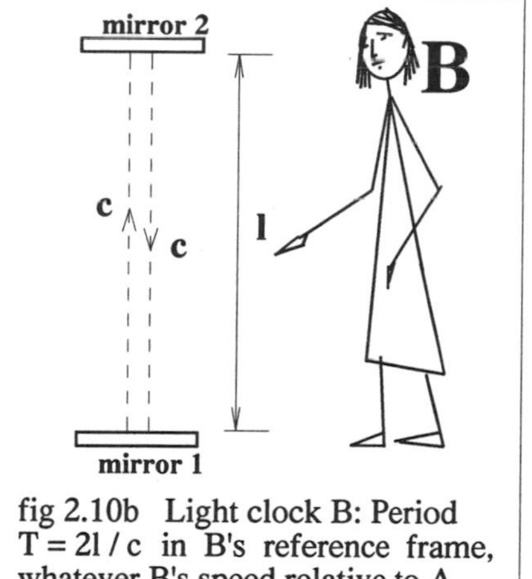

fig 2.10b Light clock B: Period T = 2l / c in B's reference frame, whatever B's speed relative to A.

2.5.3 The Light Clock Comparison According To A.

B's relative motion will carry her past A at a velocity v. If A considers light pulses in B's light clock he will come to the conclusion that the light paths are extended in the direction of B's motion. This is easy to see: a pulse leaving B's lower mirror takes time to reach the upper mirror, by which time it will have moved on. Where A's own clock sends light vertically up and down the light paths in B's clock will be diagonal and hence longer. Since A sees light in his own and in B's clock traveling at velocity *c* relative to his own reference frame but also sees B's light paths as longer than those in his own clock he concludes that B's clock runs slow. That is, his own clock ticks faster than B's. Since all other changes in B's frame proceed at their normal rate relative to B's clock (including the rate at which B ages), A sees *time* pass more slowly in the 'moving' frame than in his own 'rest' frame. A concludes that time runs slow in a moving reference frame and that B ages more slowly than he does! This is the famous 'time dilation' effect: moving clocks run slow.

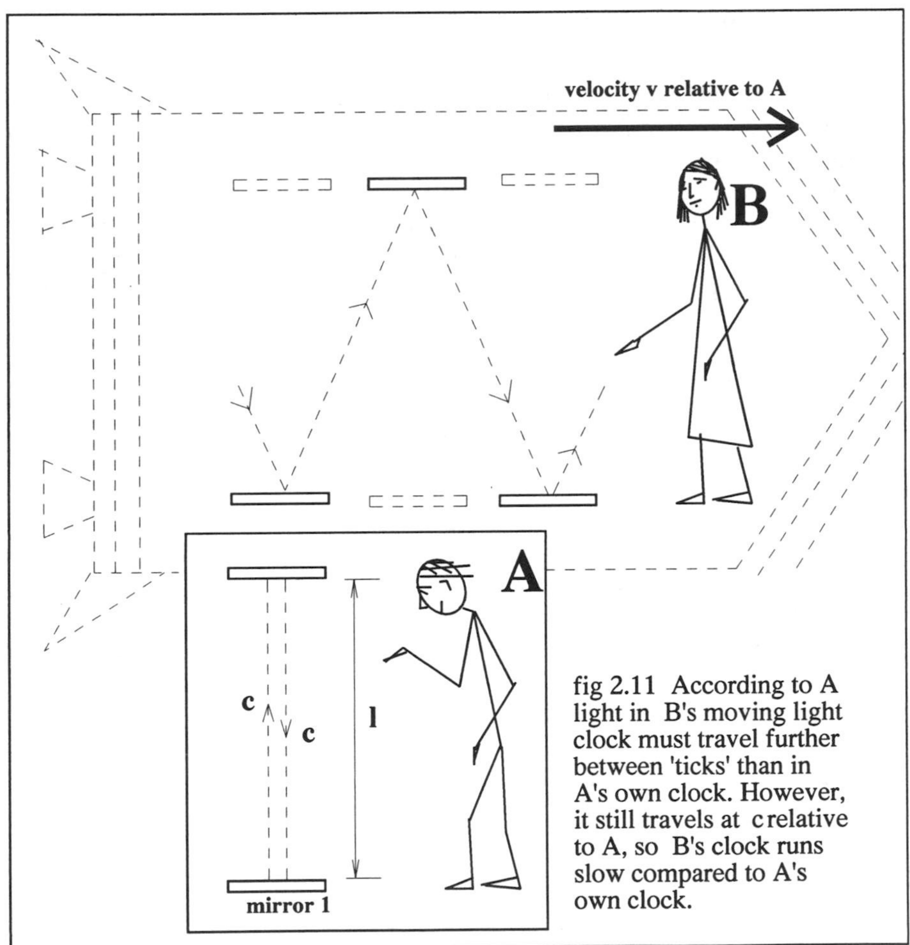

fig 2.11 According to A light in B's moving light clock must travel further between 'ticks' than in A's own clock. However, it still travels at c relative to A, so B's clock runs slow compared to A's own clock.

2.5.4 The Comparison Of Light Clocks According To B.

Time dilation in itself is quite startling, but the situation is more subtle than we have shown so far. How does time in A's frame appear to B? Since both A and B occupy inertial reference frames we are free to take either frame as a 'rest' frame and consider the other to be in motion. According to B, A's laboratory moves past at velocity v to the left. It is obvious that B will now conclude that light pulses in A's clock travel further than those in her own clock (though once again at the same velocity, c) and so take longer to complete a cycle. Time dilation is a reciprocal effect: A sees time slow down in B's frame and B sees time slow down in A's frame. Whichever way we look at it, the 'moving' clock always runs slow.

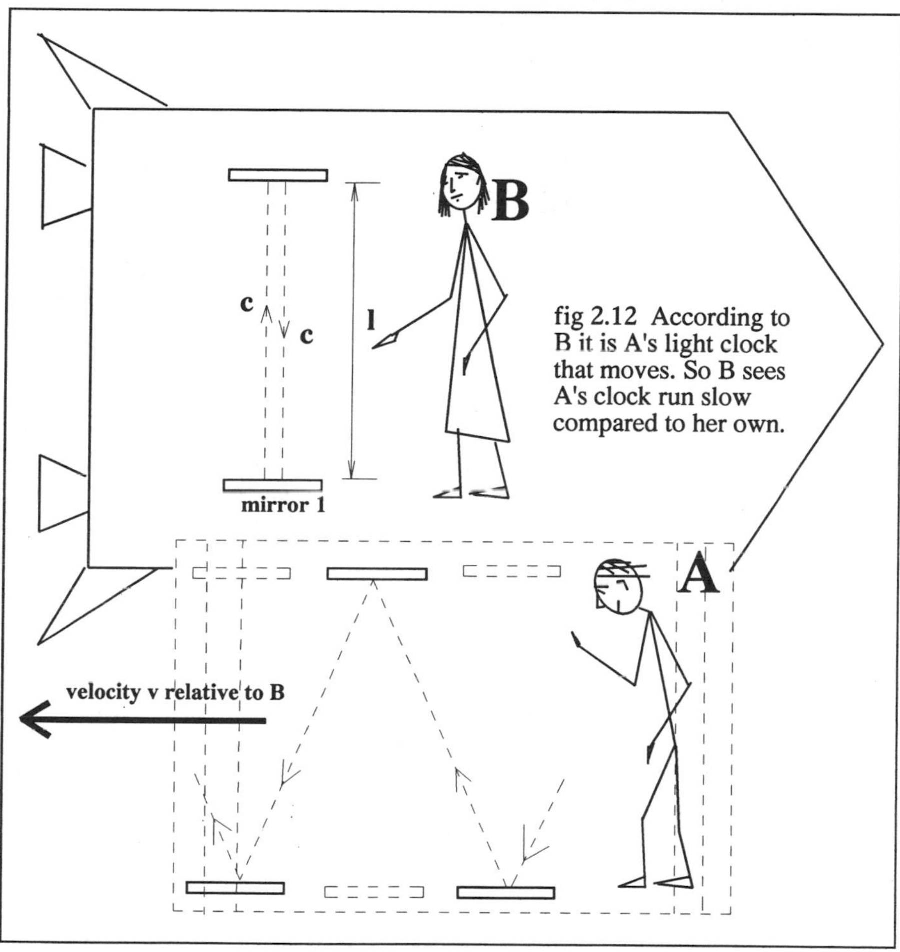

fig 2.12 According to B it is A's light clock that moves. So B sees A's clock run slow compared to her own.

Surely there is a paradox here? How is it possible for A's clock to be slower than B's and yet for B's to be slower than A's? We shall face squarely up to problems of this sort later when we consider the twin paradox, but for now an analogy with visual perception might be helpful to bear in mind. Imagine standing some distance away from a friend and then raising your hand at arm's length in front of your face. It is easy to block out their entire image so that they appear smaller than your hand.

By the same token they are able to block you out with their hand. Assuming your hands are roughly the same size this could lead to the conclusion that you are smaller than your friend and your friend is smaller than you. There is obviously no great mystery or even a hint of a paradox here, we are very familiar with the effects of perspective.

2.6 THE TIME DILATION FORMULA.

2.6.1 Comparing Light Clocks.

A simple geometric comparison between light clocks in uniform relative motion leads to a formula that will allow us to calculate the rate at which a 'moving' clock ticks. To do this we must compare the time for light to complete one round trip in the 'moving' clock with the time for light to complete a similar journey in the 'stationary' clock. Before we start we must bear two things in mind. Firstly that 'stationary' merely identifies a reference frame from which we intend to make our measurements, that is stationary with respect to us as the observers. The moving frame is moving with respect to us and our measuring apparatus. Secondly, having identified a reference frame from which to make our observations we must remember that all our measured values are on measuring devices fixed in this frame. When we conclude that a moving clock is running slow what we really mean is that if a time t passes between ticks of our own clock then our clock measures a time t' that is greater than t between ticks on the moving clock. An observer moving with that clock would certainly not think that his or her clock was running slow (in fact, as we have already seen, their conclusion would be that it was *our* clock that ticked slowly).

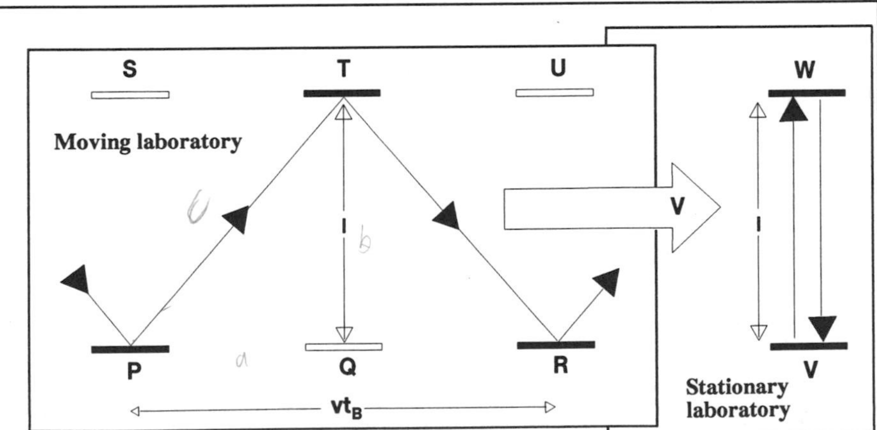

fig 2.13 Light rays in a moving clock are extended with respect to a 'stationary' observer. The diagram above is drawn from the point of view of an observer at rest with respect to the light clock on the right.

We return to observers A and B in the previous section and adopt A's point of view (see fig 2.13). B moves past at uniform velocity v and B's clock runs slow. We shall assume that A and B carry identical light clocks whose mirror separation is l metres and we know that the light will travel at c relative to us in both our own and the moving reference frame. Let the time between ticks (i.e. the round trip time for light) on A's clock be t_A and the time between ticks on B's clock *as measured on A's clock* be t_B.

For the stationary clock $$t_A = \frac{2l}{c}$$

For the moving clock $$PT = \frac{1}{2}ct_B \quad \text{and} \quad PQ = \frac{1}{2}vt_B$$

by Pythagoras $$PT^2 - PQ^2 = TQ^2$$

therefore $$\frac{1}{4}c^2t_B^2 - \frac{1}{4}v^2t_B^2 = l^2$$

rearranging we obtain $$t_B^2\left(1 - \frac{v^2}{c^2}\right) = \frac{4l^2}{c^2}$$

$$t_B = \frac{2l}{c} \cdot \frac{1}{\sqrt{1 - \frac{v^2}{c^2}}}$$

therefore $$t_B = \frac{t_A}{\sqrt{1 - \frac{v^2}{c^2}}}$$

or $$t_B = \gamma t_A \tag{2.1}$$

where $$\gamma = \frac{1}{\sqrt{1 - \frac{v^2}{c^2}}} \tag{2.2}$$

The ratio v/c occurs so often and is so significant that it is usually written as β and then:

$$\gamma = \frac{1}{\sqrt{1 - \beta^2}} = \left(1 - \beta^2\right)^{-½} \tag{2.3}$$

Since $t_B > t_A$ more than one tick of A's clock will elapse during a single tick of B's clock. If $v << c$ then the denominator of the fraction is close to one and $t_A = t_B$ to a very good approximation. The two clock rates are then indistinguishable. As v increases and approaches c so B's clock is observed to run slower and slower. If v could reach the speed of light then the denominator would be zero and A would have to wait an infinite time for B's clock to tick once. In other words A would see time in B's frame come to a complete standstill! We shall soon see that it is not possible for B to reach the speed of light with respect to A, but there is no theoretical reason why B's speed cannot come arbitrarily close to c (there are of course many practical problems involved in attempting to accelerate a large object to such a speed, but particle accelerators routinely produce velocities in excess of $0.99c$).

It is important to emphasize once again, and at the risk of becoming rather repetitive, that for B herself, at rest with respect to her own clock, life proceeds *as normal*. The clock rate agrees with any other measure she might choose to determine time flow in her own reference frame, and the light in her clock moves at velocity c. The symmetry of the situation dictates that if she were to measure the ticks of A's clock relative to her own she would derive the same time dilation factor but the subscripts A and B would be interchanged in the equation since now everything is measured on her clock rather than A's and she would conclude that A's clock runs slow. This is how relativity avoids a genuine paradox over time dilation. A's clock tells him that B's time is running slow. B's clock tells her that A's time is running slow but there is no independent clock belonging to some observer C who can agree with both A and B. The reason for this is simple, C must make measurements from a particular reference frame and if that happens to coincide with A then C will agree with A's measurements. If it is B's frame then C agrees with B. From any other frame C would disagree with *both* A and B (if C moves with respect to both A and B then he must conclude that both their clocks run slow with respect to his own).

Example: Astronomers have discovered that gamma-ray bursts from a group of distant galaxies last on average twice as long as bursts from similar galaxies closer to Earth. Assuming the nearby galaxies are moving slowly relative to Earth estimate the recession velocity of the distant galaxies (most distant galaxies are moving away from us because of the expansion of the Universe).

Let average burst duration be T_o in the rest frame of a galaxy.
The average duration of bursts received on earth is T'.

$$T' = \gamma T_o = 2T_o$$

$$\gamma = 2 = \frac{1}{\sqrt{1-\beta^2}} \quad \text{so} \quad 1-\beta^2 = \frac{1}{4} \quad \text{giving} \quad \beta = \sqrt{\frac{3}{4}} = 0.87$$

recession velocity $\underline{v = 0.87c}$

The gamma-rays are also red-shifted because of the relative motion.

Example: A star is moving away from the Earth at velocity v in the x-direction. The light received from the star contains a characteristic spectral line of frequency f' in the star's reference frame. What is the frequency of this spectral line when it is received on Earth?

The relativistic Doppler effect depends on two things: the frequency is reduced because each wavefront is emitted from a point further from the Earth than the previous one (thereby delayed slightly) and reduced further because time 'runs slow' in the 'moving' reference frame (relative to Earth time).

Let f', T' be the frequency and period of the radiation in the rest frame of the star.

Let f and T be the frequency and period of the radiation as it is received on Earth.

In the Earth frame a time $t = \gamma T'$ passes during one period of the source.

During this time the star moves an extra distance $\Delta x = \gamma v T'$ from Earth.

$T =$ time of one period + delay due to extra distance (both in the Earth's frame)

$$T = \gamma T' + \frac{\gamma v T'}{c} = \frac{T'(1+\beta)}{\sqrt{1-\beta^2}} = \frac{T'(1+\beta)}{(1+\beta)^{1/2}(1-\beta)^{1/2}} = \frac{T'(1+\beta)^{1/2}}{(1-\beta)^{1/2}}$$

$$f = \frac{1}{T} \quad \text{and} \quad f' = \frac{1}{T'} \quad \text{so:}$$

$$\underline{f = f'\frac{(1-\beta)^{1/2}}{(1+\beta)^{1/2}}} = \underline{f'\left\{\frac{1-\beta}{1+\beta}\right\}^{1/2}} \qquad (2.4)$$

If the star had been approaching rather than receding the signs would swap from top to bottom.

Example: The H and K absorption lines in the hydrogen spectrum from a distant galaxy are red-shifted from $\lambda = 394$ nm to $\lambda' = 500$ nm. Assuming this is due to the recession of the galaxy calculate its velocity relative to the Earth.

$$r = \frac{\lambda'}{\lambda} = \frac{f}{f'} = \frac{(1-\beta)^{1/2}}{(1+\beta)^{1/2}} = 1.27 \quad \text{(where the primes refer to the frame of the galaxy)}$$

rearranging this gives: $\beta = \dfrac{r^2-1}{r^2+1} = \dfrac{1.27^2-1}{1.27^2+1} = 0.235$

The galaxy is moving away from the Earth at a velocity $\underline{v = 0.235c}$

2.7 TESTING TIME DILATION.

2.7.1 Radioactive Clocks.

To make an experimental test of time dilation we need to compare the times elapsed on two similar clocks moving relative to one another. One way to do this is to synchronize two atomic clocks and then send one of them on a high speed jet ride before bringing them back together. This has actually been carried out. In 1971 two American physicists, Hafele and Keating synchronized the times on four identical cesium atomic clocks with standard clocks at the US Naval Observatory in Washington DC. They then flew two of the clocks eastwards around the world and two westwards, comparing them with the standard clock on their return. The clocks travelling east lost about 59 nanoseconds whilst the clocks travelling west gained 273 nanoseconds. There are two distinct relativistic phenomena which contribute to these results. One is an effect of general relativity that makes clocks run slower in stronger gravitational fields (in this case the clocks that remained in Washington), and the time dilation due to motion that we discussed in the last couple of sections. Allowing for both effects and for the actual geographical course followed by the aircraft the measured time differences were in good agreement with the values calculated from special and general relativity.

Nature provides us with another form of clock that we can use to test time dilation. This is the intrinsic clock associated with an unstable particle or nucleus. Although it is never possible to predict when an individual particle will decay, it is possible to measure the half-life for a large population of such particles. This is the time during which half the population will decay and is independent of the original number in the sample. (However, if that original number is too small then statistical fluctuations caused by the random nature of radioactivity may cause significant deviations from the expected half-life, so experiments based upon half-lives ought to involve large numbers of particles.) If we could measure the half-life of a particular type of particle in the laboratory and compare this value with the half-life of similar particles moving rapidly relative to the laboratory then we could put the time dilation formula to test.

2.7.2 Cosmic Rays And Muons.

Muons are a form of elementary particle produced in some high energy nuclear reactions. They are also produced in large numbers when high energy cosmic rays collide with gas nuclei in the upper atmosphere (at an altitude of about 60 km). Those produced in the atmosphere travel downwards with velocities close to the speed of light.

Muons produced in the laboratory with low velocities are observed to decay with a half-life of about 1.5 μs. Presumably this is also the half-life of muons produced at the top of the atmosphere. A quick calculation tells us that it will take approximately 6×10^4 m/3×10^8 m s^{-1} = 2×10^{-6} s or 200 μs for a muon to travel down through the atmosphere to the surface. An Earthbound observer would reckon this to be about 133 times the half-life of the muon. Therefore, from a laboratory on Earth we would expect the flux of muons at the surface to be about $1/2^{133}$ times the flux in the upper atmosphere. This would mean about one in 10^{40} muons make it down to our laboratory at sea-level.

If this simple calculation was to be borne out by experiment then we should be unable to detect *any* significant muon flux at the surface. However, it is discovered that muons do reach the surface in significant numbers and that the flux at sea-level is about 1/8 of the flux at high altitude! This implies that, for the muons themselves, only three half-lives have elapsed ($2^3 = 8$). A mere 4.5 μs passes in the muon reference frame while 200 μs passes for us. Time in the muon frame, seen from Earth, passes at 3/133 or a little over 2% of the rate at which it passes on Earth. This time dilation factor $\gamma \approx 44$ can be substituted into the equation to derive the muon velocity. It gives a velocity of about 0.9997*c*.

In 1963 Frisch and Smith carried out a very similar experiment in which they compared the rates of arrival of muons in laboratories at the top and bottom of Mt. Washington. They selected muons of a particular velocity by counting only those which were brought to rest by a particular thickness of absorbing material. Knowing their velocity they could calculate the expected time dilation factor and compare this with the value calculated from the proportion that arrive at the lower laboratory. Using a velocity of 0.994*c* they obtained a time dilation factor of about 9 that was in good agreement with their experimental ratio.

Another classic experiment was carried out at CERN in 1966. Muons produced in high energy collisions were deflected into circular 'storage rings' by strong magnetic fields. The muon velocities were 0.997*c*. The half-life of muons in the storage ring was measured to be about 12 times that of 'stationary' muons in the laboratory, a value within 2% of the relativistic prediction.

2.7.3 Making Use Of Time Dilation.

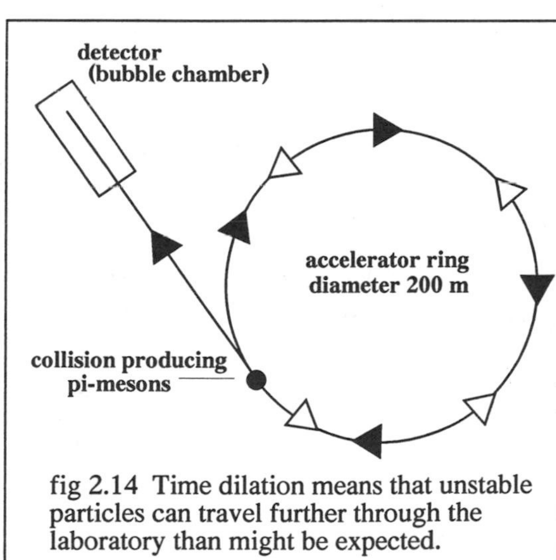

fig 2.14 Time dilation means that unstable particles can travel further through the laboratory than might be expected.

There is one branch of physics in which very high speed particles are produced every day. This is called high energy physics and involves the big accelerators at laboratories like CERN and Fermilab. The basic idea behind many of these experiments is very simple: fast moving particles undergo violent collisions and showers of new, sometimes exotic, particles are produced. Various detectors are set up to register the products of these collisions.

Unfortunately many of the particles produced have very short half-lives and so are short lived. If time dilation did not occur life would be very much harder for high energy physicists - they would have to place their detectors close enough to the original collision so that not too many half-lives elapsed before the particles entered the detector (or else there would be no particles left to detect). In one experiment beams of high energy pi-mesons, which have a half-life of about 17 ns, are produced. In 17 ns a particle travelling at the speed of light would only move 5 m, so pi-meson detectors might be expected to be placed at about this distance from the collision point. In fact they are detected in a bubble chamber about 120 m away! The pi-mesons are travelling at such high speeds (i.e. close to *c*) that far less time elapses between production and detection in their reference frame than in the laboratory so they still arrive in the chamber in significant numbers.

The images in plate 2.1 show tracks from the ALEPH detector at CERN following the collision of an electron and positron in the LEP ring. The tracks of short-lived particles are captured by an electronic vertex detector close to the interaction point. In this case the B and D meson tracks are about 3 times the length one would expect on the basis of their proper lifetimes.

Example The orbital velocity of Mercury is about 50 km s^{-1} whilst that of the Earth is about 30 km s^{-1}. How much time would a clock on Mercury 'lose' relative to a clock on Earth during one terrestrial hour when the planets are moving parallel but in opposite directions?

The moving clock (i.e. the one on Mercury) runs slow relative to the one on Earth by a factor γ. If a time T passes on the Earth clock the Mercury based clock will record a time T/γ.

$$\gamma = \sqrt{1-\beta^2} = \sqrt{1-\left(\frac{8.0\times10^4\ \text{m s}^{-1}}{3.0\times10^8\ \text{m s}^{-1}}\right)^2} = \sqrt{1-1.67\times10^{-8}} = 0.999999986$$

Time lost by Mercury clock = 0.000000014 hours = $\underline{50\mu\text{s}}$

N.B. When $\beta << 1$ γ and $\frac{1}{\gamma}$ can be expanded using the binomial expansion:

$$\gamma = (1-\beta^2)^{-\frac{1}{2}} \approx 1+\frac{1}{2}\beta^2 \qquad \text{and} \qquad \gamma^{-1} = (1-\beta^2)^{\frac{1}{2}} \approx 1-\frac{1}{2}\beta^2$$

Example A beam of particles traveling at 99% of the speed of light passes through two detectors 60 m apart. The flux at the second detector is one quarter of the flux at the first, what is the half-life of the particle in its rest frame?

Time taken to travel 60 m in the laboratory: $T = \dfrac{60\ \text{m}}{0.99\times3.0\times10^8\ \text{m s}^{-1}} = 2.02\times10^{-7}\ \text{s}$

Time elapsed in the frame of the particles: $T' = \dfrac{T}{\gamma}$

$$\gamma = \frac{1}{\sqrt{1-\beta^2}} = 7.09 \qquad \text{so} \qquad T' = \frac{2.02\times10^{-7}\,s}{7.09} = 2.85\times10^{-8}\ \text{s}$$

Two half lives pass in this time so: $\underline{T_{\frac{1}{2}} = 1.42\times10^{-8}\ \text{s}}$

2.7.4 Light from the Beginning of Time.

It is claimed that the Hubble Space Telescope is recording images of galaxies as they were just a short time after the Big Bang. At first sight this seems rather strange. If all matter was together at the Big Bang and moved away from it at sub-light speeds then surely it (and us) was overtaken by light from the early universe long ago? This can be backed up by a classical argument. Assume a time T has passed since the Big Bang and that we are receiving light from a galaxy receding from us at velocity v (for the sake of this calculation we assume v has remained constant since the Big Bang). What is the earliest time T' that the light could have been emitted if it is arriving at the Earth now?

Time since Big Bang = time of light emission + time of flight to Earth

$$T = T' + \frac{vT'}{c}$$

giving $$T' = \frac{T}{\left(1 + \frac{v}{c}\right)}$$

If $v \approx c$ the earliest time of emission is: $$T' \approx \frac{T}{2}$$

fig 2.15 The Hubble Deep Field view.
Photograph credit: R.Villard, NASA/HST.

This suggests that the earliest light we could receive after 15 billion years would have been emitted about 7.5 billion years after the Big Bang. However, we have already recorded light emitted much earlier than this (radiation from distant quasars was also emitted earlier than this).

The calculation above is wrong because it does not allow for time dilation. If the source of light is moving rapidly away from us it ages more slowly than we do, that is for every t years passed in the Earth's reference frame only t/γ passes in the source frame. If this is included the conclusion is rather different.

This is another example where the effects of relativity allow us to get at information that would, in a classical or Newtonian world, be completely inaccessible. The Deep Field photographs from the Hubble Space Telescope are quite awe inspiring, revealing not just what is out there in the depths of space, but what happened in the distant past as well.

Let T_e be the time at which light is emitted by the distant galaxy in the Earth's frame and let T' be the time it is emitted in the frame of the galaxy:

Time since BB = time of light emission + time of flight (both relative to Earth)

$$T = T_e + \frac{vT_e}{c}$$

Time of emission in frame of source is $T' = \dfrac{T_e}{\gamma}$ so:

$$T = \gamma T' + \frac{\gamma T' v}{c}$$

$$T' = \frac{T}{\gamma\,(1+\beta)}$$

As $v \to c$ $\gamma \to \infty$ so $T' \to 0$

In principle we can see back to any time after the Big Bang. (This is not quite true - before about one million years after the Big Bang photons interacted so frequently with matter that the universe was opaque to electromagnetic radiation). Also we have ignored the fact that the expansion velocity of the Universe has decreased since the Big Bang.

2.8 LENGTH CONTRACTION.

2.8.1 A Short Cut For Muons.

The muon experiments make a convincing experimental case for time dilation, only three muon half-lives pass in their frame as they move from the top to the bottom of the atmosphere whilst 133 half-lives would pass for muons at rest in the laboratory between the same events. However, if we look at this from the reference frame of the moving particles there seems to be a problem. This is as good a reference frame as any other to measure time intervals in so we must agree (since we are riding on the muon!) that only 4.5 μs passes. We also agree that the relative velocity of approach of the Earth is close to *c*. The problem is that in 4.5 μs travelling at 3×10^8 m s^{-1} we would move a mere 1.35 km! And yet, at the end of this time we crash into some physicist's detector at sea level: how can this be? The answer must be that relative to muons the Earth's atmosphere is only 1.35 km deep. In order for both inertial reference frames to yield consistent results, that is results in accord with the principle of relativity, time dilation *and* length contraction must both occur. It is also clear that far from being independent effects they are intimately related; in fact what appears 'timelike' to one observer appears 'spacelike' to another.

The existence of this length contraction is encouraging for the future of human space exploration. One of the most daunting aspects of the Universe is its vast scale and the apparent insignificance of our local solar system compared to this. Apart from the Sun our nearest star is Alpha Centauri, a mere 4.3 light years away. Galaxies are separated by distances measured in tens of millions of light years! If, as relativity implies, it is impossible to travel at a speed greater than or equal to c then there is no hope of any individual embarking on a journey to a distant galaxy. But this of course is exactly the dilemma faced by muons created at the top of the atmosphere. They survive their journeys because the journey distance contracts. The faster the object the greater the contraction of distance and the shorter the journey time. As far as a journey to a distant galaxy is concerned, the distance of say ten million light years is a value relative to the Earth. An observer moving at high velocity relative to the Earth would measure a greatly reduced distance and so calculate a much shorter journey time. In principle the distance and time can be contracted to as little as we like. At a speed barely different from c we might contract the space between Earth and a distant galaxy to as little as 1 m and arrive in a few nanoseconds! (Of course there is no way we would survive the accelerations and decelerations that would be required to do this, not to mention the fuel bill!) There is also a serious side effect to such journeys. The observers (and our loved ones) who remain behind on Earth will see our journey lasting a little over ten million years. In other words they are not going to be there to welcome us back. A few other things may also have changed in our absence.... In this sense high speed return journeys are a way of travelling into the future. The traveller lives through a short time whilst the stay-at-homes live through a far longer period. The traveller has moved into their future.

It is worth mentioning what happens to light which does, obviously, travel at speed c. If we could follow a photon from its creation to its absorption by matter we would see that it is totally unchanged. The time dilation factor has blown up to become infinite so clocks in the light frame have stopped completely. As far as length contraction is concerned this means the length of a light ray according to the light is always zero and it arrives at its destination no time after leaving its source! Remember this is what would happen if we could travel with the light at velocity c. Unfortunately we cannot do this (or perhaps fortunately, it might be rather disconcerting being everywhere at once!).

2.8.2 The Length Contraction Factor.

The muon experiment discussed above shows that the length contraction factor must be numerically equal to the time dilation factor. However, whereas the time between ticks on a moving clock is increased (when measured on our rest clock) the length of a moving object is reduced (in its direction of motion) when compared to a similar object at rest. To see that this is the case you could imagine the observer in the muon reference frame to carry a 1 km measuring rod that he uses to measure the lengths of similar rods placed vertically at rest in the atmosphere. Since about 60 of these rods stretch from sea-level to the top of the atmosphere the muon observer concludes that they are all much shorter than his own measuring rod which fits only 1.35 times into the same distance. The contraction factor is about 1/44. (Remember, according to this observer his rod is at rest and the atmosphere and its measuring rods are in high speed motion.) We can simply write down a formula for the length contraction by using the reciprocal of the time dilation factor (γ) derived earlier:

$$l = l_o\sqrt{1-\frac{v^2}{c^2}} = \frac{l_o}{\gamma} \tag{2.5}$$

Here l is the length of a moving rod and l_o the length of a stationary rod measured from the reference frame in which it is at rest. l_o is called the 'proper length' of the rod.

Example Our galaxy is 10^5 light years across. How long does it take to cross in a space craft traveling at $0.995c$?

The width of the galaxy is length contracted in the spacecraft frame, so the distance travelled becomes:

$$l' = \frac{l}{\gamma} = l\sqrt{1-\beta^2} = 10^5 \text{ light years} \times \sqrt{1 - 0.995^2} = 10^4 \text{ light years}$$

$$t' = \frac{l'}{v} = \frac{10^4 \text{ light years}}{0.995c} = 10^4 \text{ years.}$$

This is the time that passes for the travellers. Of course just over 10^5 years will pass in their absence on Earth.

2.8.3 Length Contraction And Light Clocks.

Here is another simple argument for length contraction. The logic used is similar to that employed when we derived the time dilation formula. Once again the power of the argument resides in its use of symmetry. Two identical laboratories are set up with a standard measuring rod lying along the floor and a light clock perpendicular to it placed at its centre. The laboratories have a relative velocity v along a line parallel to the rods. Each observer notes that it takes exactly one 'tick' of their own light clock for the moving rod to pass them. If the time for one tick in each clock's rest frame is T_o then both observers agree that the moving rods have length $l = vT_o$.

The interesting question to ask is how this length compares with the length of a rod at rest relative to the observers. Take A's point of view, he can use B's motion to measure the length of the rod in his own reference frame. He agrees that B's clock ticks once as B passes but the light path in B's clock is extended so B's time runs slow. The time taken in A's frame while B's clock ticks once and B moves along the length of A's rod is $T = \gamma T_o$. The length l_o of A's rod according to A is therefore:

$$l_o = v\gamma T_o = \gamma l$$

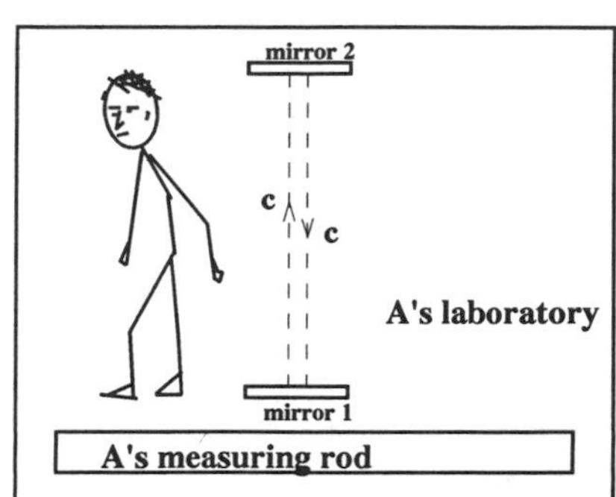

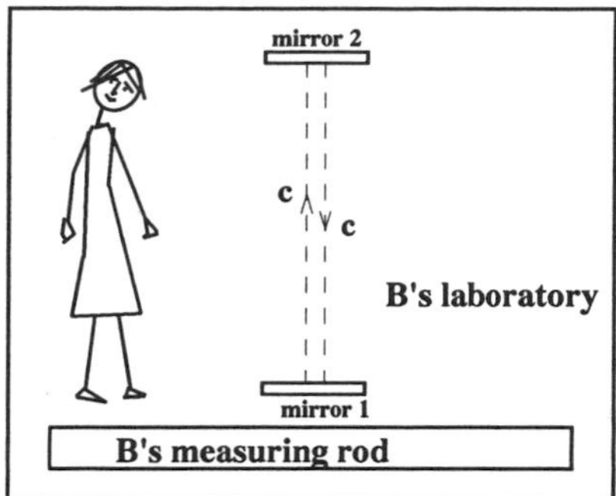

fig 2.16a A and B are in identical laboratories. Both have a light clock and a measuring rod. Both reference frames are inertial so the laws of physics are the same for both observers.

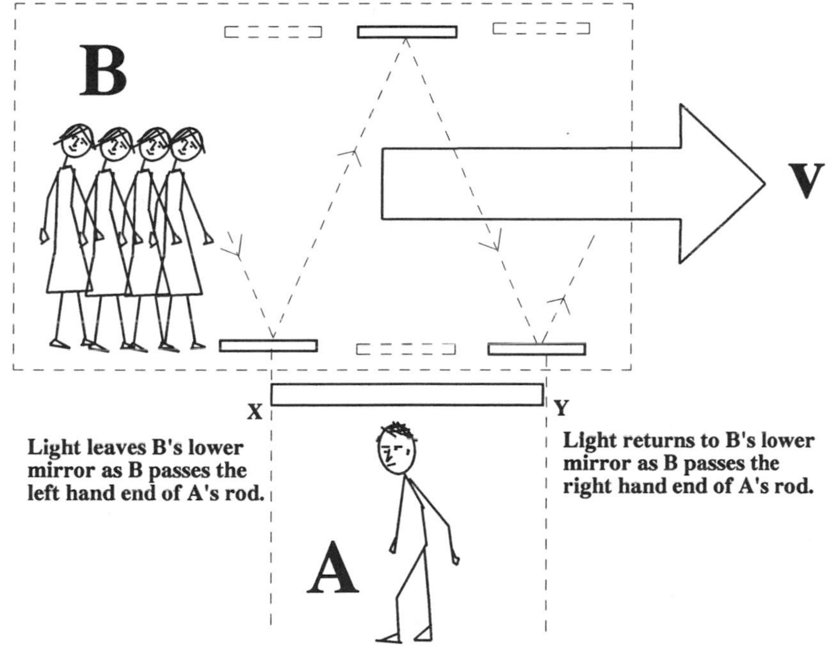

fig 2.16b From A's point of view B's light clock runs slow, so A's clock records a longer time for B to move from X to Y than B's clock. If both observers then calculate the length of A's rod from relative velocity times time they will disagree about the length. A will find the rod is longer than B measures it to be. However, the rod is moving relative to B so this 'length contraction' affects moving rods. The situation is completely symmetric and A will think that the rod carried in B's laboratory has contracted by the same factor.

where l is the length of the moving rod. The symmetry of the situation results in identical conclusions about rest and moving rods in B's reference frame. The conclusion is inescapable, moving rods contract in their direction of motion:

$$l = \frac{l_o}{\gamma} \quad \text{or} \quad l = l_o\sqrt{1 - \frac{v^2}{c^2}} = l_o\sqrt{1-\beta^2} \tag{2.5}$$

This is exactly the Lorentz-Fitzgerald contraction which was offered as an ad hoc explanation of the Michelson-Morley result. Here it arises naturally from the principle of relativity. Later (in section 2.18) we shall derive the length contraction formula from the Lorentz transformation equations.

2.8.4 Some Comments On Length Contraction.

- Just like time dilation this is a reciprocal effect. A sees B's rod contract, B sees A's rod contract. There is no paradox here because, in order to make a genuine comparison of the rod lengths we should have to put them both into the same reference frame, that is bring rod A to rest in B's frame or bring rod B to rest in A's frame. In either case this would remove the very effect we are discussing which arises only when there is relative motion between the two systems.
- Each observer sees no change in the lengths of objects in their own reference frame. It is the scale of length as judged by an external moving observer that has changed, and then only in the direction of motion.
- Since length contraction only affects dimensions parallel to the direction of motion the shape of a moving object changes too. For example, if we passed the Sun at high enough speed, its diameter parallel to our direction of motion would contract: it would appear as an ellipse with its major axis perpendicular to the direction of our relative motion.
- As mentioned previously, we do not have to measure the length of a rod to demonstrate length contraction. If there is a star one hundred light years from Earth then, merely by travelling rapidly towards it, the journey length contracts and we do not have so far to travel. For example, at $0.99995c$ ($\gamma = 100$) the journey distance would contract to just one light year and the journey time (for the traveller) would be a shade over a year. Of course, a little over one hundred years would pass on Earth before the traveller completed his outward bound journey.
- The length of a rod measured in its rest frame is called its *proper length*. Similarly, the time between two events measured on a clock which coincides with both events is called the *proper time*. The measured length of a rod never exceeds its proper length and the measured time between two events is never less than the proper time.

Example Rigel is 900 light years from Earth. What is the minimum Earth time a spacecraft could take to make a return trip to Rigel? (b) How fast should the spacecraft travel to complete the journey in just 30 astronaut years? (c) What is the distance from Earth to Rigel as measured by astronauts on their 30 year round trip?

(a) Just over 1800 years - if the journey is completed at a velocity close to the speed of light.

(b) To complete the round trip in 30 years requires a time dilation factor of 1800/30 or 60.

$$\gamma = 60 = \frac{1}{\sqrt{1-\beta^2}} \quad \text{giving} \quad \beta = 0.99986 \quad v = 0.99986c$$

(c) The distance l' measured by the astronauts undergoes a length contraction relative to the distance l measured in the Earth's reference frame:

$$l' = \frac{l}{\gamma} = \frac{900 \text{ light years}}{60} = 15 \text{ light years}$$

The same result follows simply by noting that the astronauts travel for 30 years at almost exactly the speed of light, so the round trip distance is 30 light years.

2.9 THE RELATIVITY OF SIMULTANEITY.

2.9.1 The Meaning Of Simultaneity.

'What do we mean by "two simultaneous events in one coordinate system"? Intuitively everyone seems to know the meaning of this sentence. But let us make up our minds to be cautious and try to give rigorous definitions, as we know how dangerous it is to overestimate intuition.'
(A. Einstein and L. Infield, The Evolution of Physics)

It is one of the assumptions of classical physics that time passes at the same rate for all observers and that therefore it makes sense to ask whether or not two distinct events take place at the same time. For example, if two stars become supernovae in different regions of the universe we expect a definite answer to the question as to whether star A exploded before star B or vice versa. This question of simultaneity is complicated by relativity.

Even in classical physics it is possible that star A might *appear* to explode before star B if A is closer to us, even though, when we correct our observations for the time of flight of light, we discover that B exploded before A or that the explosions were simultaneous. Classically two events are simultaneous (regardless of how they appear to us) if, *after allowing for the time of flight of light*, they are calculated to have occurred at the same moment (that is both events have the same time co-ordinate). Since classical physics is based on the assumption of a single absolute time we are effectively locating the events at some value on this time scale. It is as if there were one absolute clock and if two events occur at the same time on this clock then they are absolutely simultaneous.

This is where relativity introduces a complication. The velocity of light is a constant for all observers in inertial reference frames, so a 'moving' observer will disagree with a 'stationary' observer as to the time that should be allowed for the light paths from the events to the observers. They will disagree about the corrections that must be made for time of flight of light and so will not be able to agree on the times of the events. There is no way around this since all inertial reference frames are equivalent and hence have equal authority when it comes to judging times or time orders of events. We cannot appeal to a particular observer and say that they are in a better position or state of motion to make a judgment, there are no privileged observers and we must conclude that the order of events in the universe is relative!

2.9.2 An Operational Definition Of Simultaneity.

We can illustrate the problem by considering an operational approach to the synchronization of a pair of distant clocks. This might be carried out in the following way:

- Set two identical clocks to read 'zero', but do not start them.
- Attach a light detector to one side of each clock so that when a light beam strikes the detector the clock starts.
- Now separate the clocks and place them at rest some distance apart.
- Locate the point midway between the clocks and place a small flashlight there.
- Operate the flashlight so that rays travel out to each detector.
- Both clocks begin at the same instant and so are synchronized.

This seems a fairly straightforward and apparently foolproof procedure. Two distant clocks will then measure the same time. In principle this process could be extended without limit until a uniform time frame was established throughout the universe. We could then attempt to answer the question about the simultaneity of different events (i.e. two supernovae) by comparing the occurrence of the events with the time indicated on local clocks. If a clock close to A when it explodes shows the same time as one close to B when it explodes then the events were simultaneous. It may take time to gather this information, but now this 'time of flight' has been eliminated from the judgment since the value communicated is determined by the local clock and will not change even if it is written down in a notebook which is then transported back to Earth on a slow spacecraft. Surely we

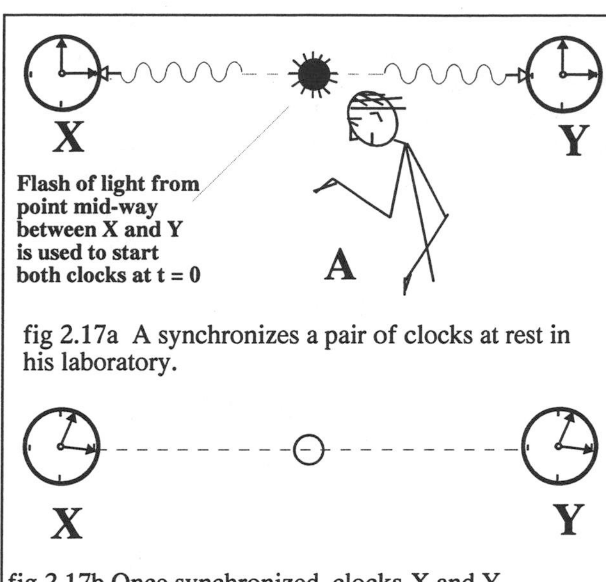

fig 2.17a A synchronizes a pair of clocks at rest in his laboratory.

fig 2.17b Once synchronized, clocks X and Y continue to agree in A's reference frame.

have now established an unambiguous criterion for simultaneity and a method by which (in principle at least) we can test for it.

2.9.3 Simultaneity And Relative Motion.

There is one point we should be very careful about. Our framework of synchronized clocks are at rest with respect to each other and so necessarily occupy a unique inertial reference frame. How does this synchronization process appear to another observer who moves relative to the entire framework? For simplicity we shall consider the straightforward synchronization of a pair of clocks whose separation lies parallel to the direction of relative motion of the two observers.

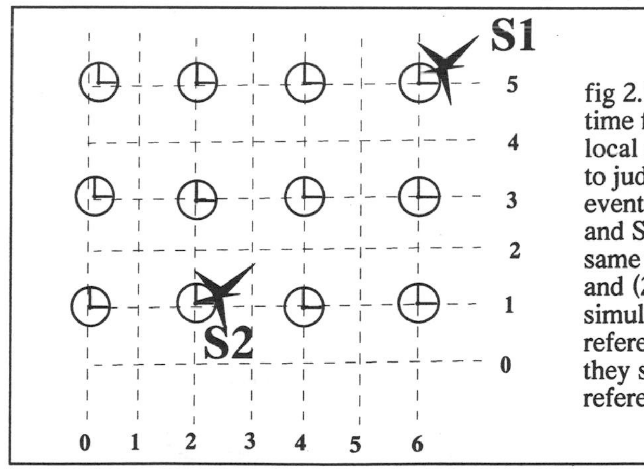

fig 2.18 Once a common time frame is established local clocks can be used to judge the time order of events. If supernovae S1 and S2 explode at the same time on clocks (6,5) and (2,1) they explode simultaneously in A's reference frame. But are they simultaneous in all reference frames?

Let A synchronize a pair of clocks according to the instructions given previously. B passes at velocity v. First of all she will disagree about the separation of A's clocks. This is length contraction but will have no effect on her judgement of whether the clocks are synchronized since she agrees that A has placed the flashlight at the mid-point between the two clocks. B will also see A's clocks run slow due to time dilation, but once again this will have no direct effect on the judgment of synchronization since it affects both clocks equally (it will, however, affect her judgment of synchronization error as we shall see later). However, when light leaves the flashlight it will travel at velocity c in all directions relative to B (just as it does for A). If B moves to the right then she will see light travelling in this direction heading toward an *approaching* clock whilst that heading to the left will be chasing a clock which *moves away* from it at velocity v. Although light was emitted from a central point, B will see it arrive at the right hand clock *before* it reaches the left hand clock and conclude that the clocks are not synchronized. It is clear from this argument that the synchronization error will become greater if the clock separation or the relative velocity of A and B is increased.

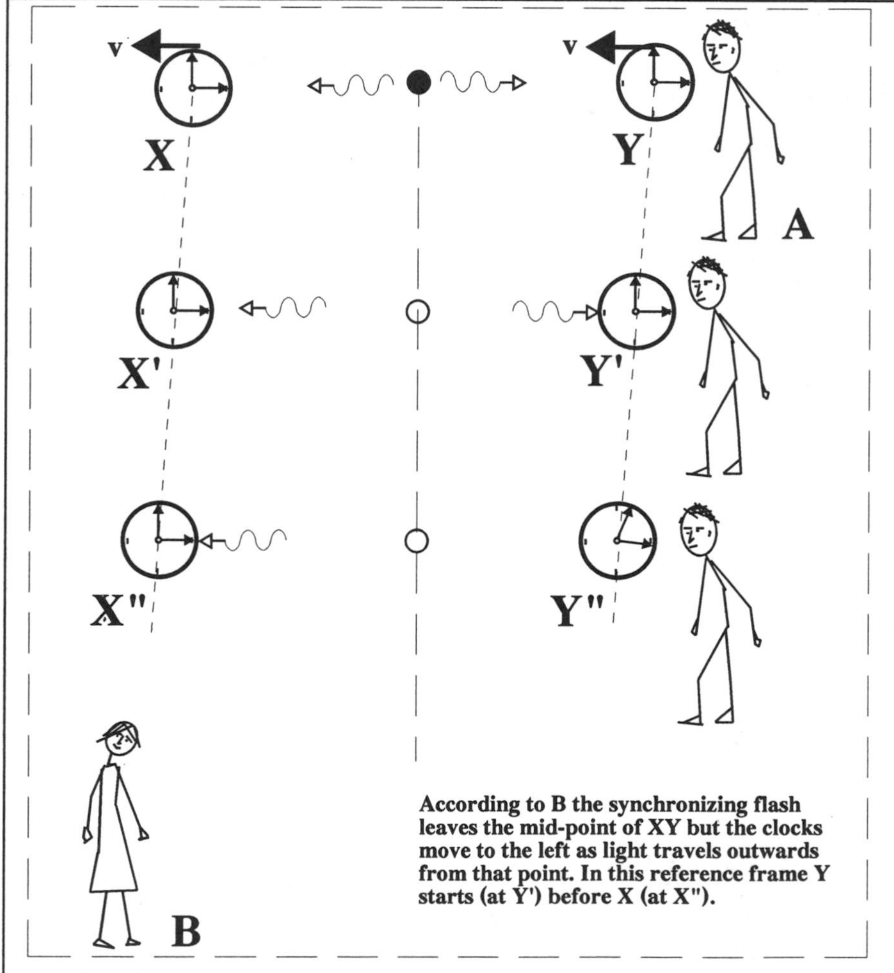

fig 2.19 The synchronization of A's clocks as seen from B's reference frame. Although the clocks start simultaneously in A's frame, Y starts before X in B's frame. Simultaneity is relative.

The only time A and B will agree on simultaneity is if two events occur at the same point in space since there is then no synchronization error.

If a system of clocks were to be set up and synchronized throughout the universe according to the original plan and then two supernovae were judged to be simultaneous against this reference frame then a second observer, moving to the right as in our example, would believe that the star further to the right exploded first. Furthermore a third observer, moving to the left, would believe that the star to the left exploded first! All three judgments are equally valid. The statement that two events are simultaneous only applies to a particular inertial reference frame.

2.10 SYNCHRONIZATION ERRORS AND CORRECTIONS.

2.10.1 Calculating The 'Errors'.

We have seen that the synchronization of clocks is only valid within a single inertial reference frame. Any observer moving relative to this frame will detect errors between clocks separated along the line of his motion. This gives an additional source of disagreement over derived quantities such as the velocity of a projectile viewed from each frame and must be taken into account when we combine velocities measured with respect to different reference frames (relativistic velocity addition).

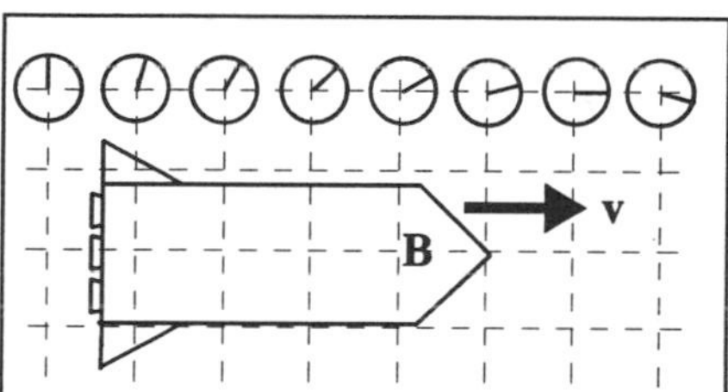

fig 2.20 Clocks synchronized in A's frame have synchronization errors relative to B. Clocks ahead of B's motion appear advanced, and the advance depends on their position.

To determine the size of these 'synchronization errors' we must analyse the procedure by which distant clocks are synchronized in a 'moving' reference frame. Once again B moves past A at relative velocity v. B places two identical zeroed clocks a distance $\Delta x'$ apart in her laboratory. A sees the entire laboratory and clocks moving in the positive x direction at velocity v.

B locates the mid-point between the clocks and emits a light pulse from this point to both clocks simultaneously. Light detectors on the clocks respond to the arrival of the pulse and start them. They are now synchronized in B's reference frame. A however, as we discussed previously, will see the left hand clock start first since this one (from his point of view) is approaching the light source whereas the other clock moves away from it. If t_X and t_Y are the times at which clocks X and Y start *as observed by A* then we can identify *three* steps in the synchronization process in figure 2.21 below:

- Event 1: Light emitted from centre of B's laboratory.
- Event 2: Clock X starts.
- Event 3: Clock Y starts.

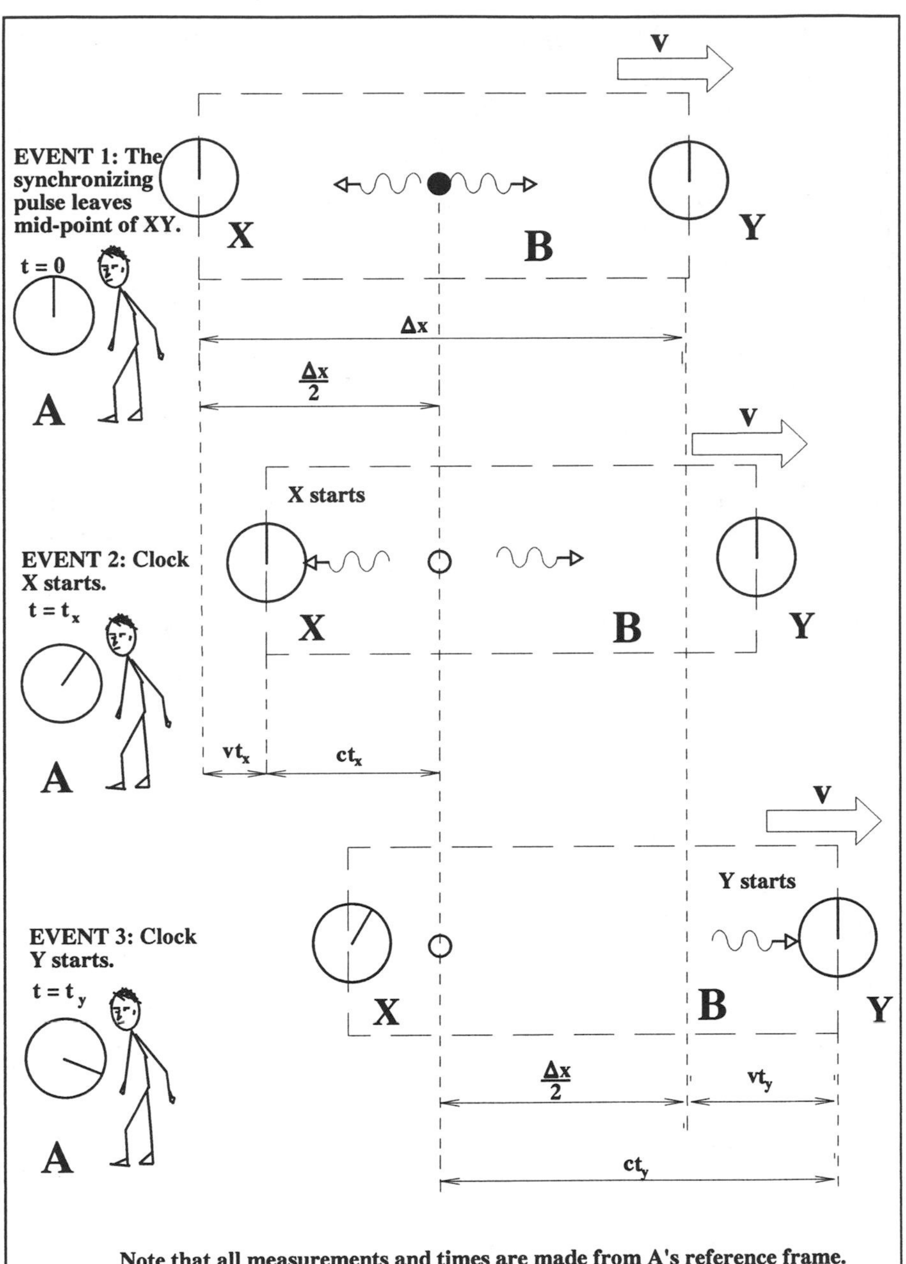

fig 2.21 A observes B as B synchronizes her clocks. A concludes that clock X starts before clock Y even though the clocks start simultaneously in B's reference frame.

Using distances and times of flight of the synchronization pulses in A's reference frame we can derive an expression for the observed synchronization error $(t_Y - t_X)$.

From diagram 2:	$\frac{\Delta x}{2} = vt_x + ct_x$	
From diagram 3:	$ct_y = \frac{\Delta x}{2} + vt_y$	
Therefore:	$t_x = \frac{\Delta x}{2(c+v)}$	
	$t_y = \frac{\Delta x}{2(c-v)}$	
Synchronization error	$\Delta t = \frac{v\Delta x}{(c^2 - v^2)}$	(2.6)

2.10.2 Simultaneity And Length Contraction.

When we considered length contraction it seemed rather strange that two observers moving relative to one another could both conclude that lengths measured in the other's frame of reference had undergone a contraction. We can make some sense of this if we consider the process by which they would measure lengths and how they would see such measurements carried out by the other observer. Consider how A might attempt to measure the length of a rod that is at rest in B's laboratory. Remember that B's laboratory moves past A at constant velocity v to the right. One method A might use would be to note the positions of the ends of B's rod in A's own reference frame *at the same instant*. To judge when this instant occurs A needs to make two observations:

- *The position of the left hand end of the rod at time t_A.*
- *The position of the right hand end of the rod at time t_A.*

This needs two synchronized clocks in A's reference frame, one opposite each end of the rod at the time the measurement is made. However, B, observing what A does, will conclude that A's clocks are not synchronized. B sees the right hand clock start first. Thus B will think that A recorded the position of the right hand end of the moving rod at some time t_B, and then delayed a short while (the synchronization error), during which time the rod continued moving to the right, and then recorded the position of the left hand end of the rod. According to B, A has recorded the position of the left hand end *late* and therefore too close to the right hand end and so measures a shorter length than the rest or proper length of the rod in B's frame. In other words B is reconciled to A's conclusion that moving rods contract - B believes that A's measurement process is flawed.

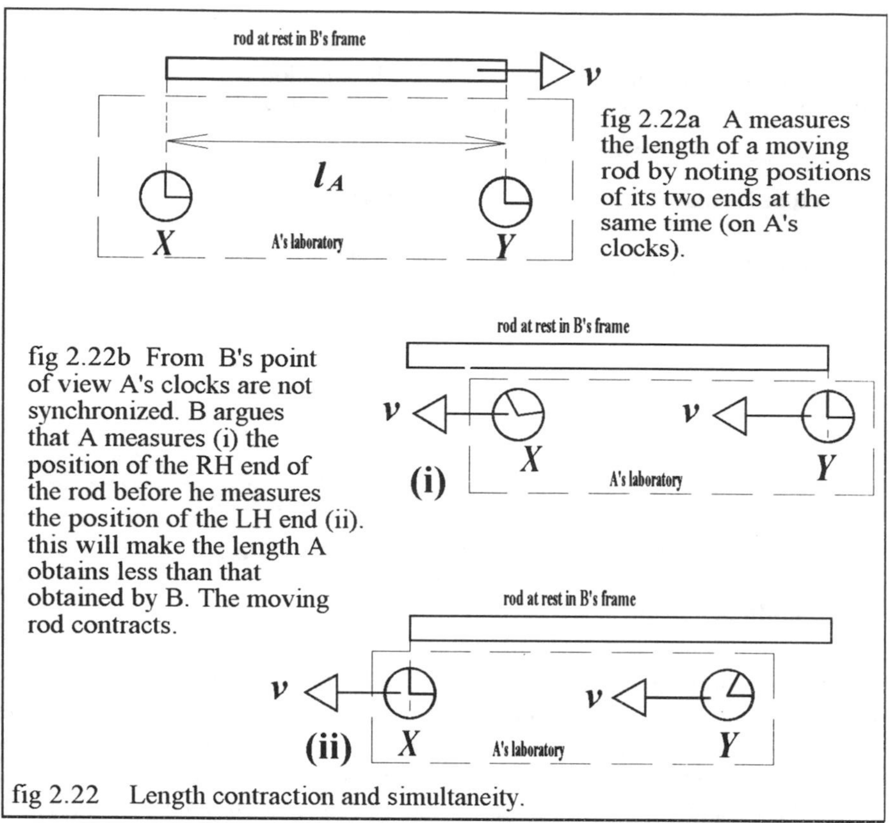

fig 2.22a A measures the length of a moving rod by noting positions of its two ends at the same time (on A's clocks).

fig 2.22b From B's point of view A's clocks are not synchronized. B argues that A measures (i) the position of the RH end of the rod before he measures the position of the LH end (ii). this will make the length A obtains less than that obtained by B. The moving rod contracts.

fig 2.22 Length contraction and simultaneity.

Of course, A is also able to criticize B's measurement process. If B measures the rod's length using a light beam A will argue that the return time of flight of the light measured by B is too long because B has failed to allow for the extra time of flight of the light due to the motion of B's apparatus relative to A. Once again there is no way to settle this argument, each observer is in an inertial reference frame and the laws of physics are the same in all inertial reference frames.

2.11 RELATIVISTIC VELOCITY ADDITION.

2.11.1 Classical Velocity Addition.

Observer A is standing on an embankment watching as a train passes at 20 m s^{-1}. Inside the train observer B, who has just purchased a coffee from the buffet car, is walking along the train at a tentative 1 m s^{-1} in the same direction as the train is moving. Classically A would reason that, in one second, B moves 1 m to the right relative to the train and the train itself moves 20 m to the right relative to A, so B moves a total of 21 m to the right relative to A. In other words A concludes that B has a velocity of 21 m s^{-1} relative to the embankment. He has applied the classical rule for velocity addition:

$$\textit{velocity of B relative to A} = \textit{velocity of B relative to train} + \textit{velocity of train relative to A}$$

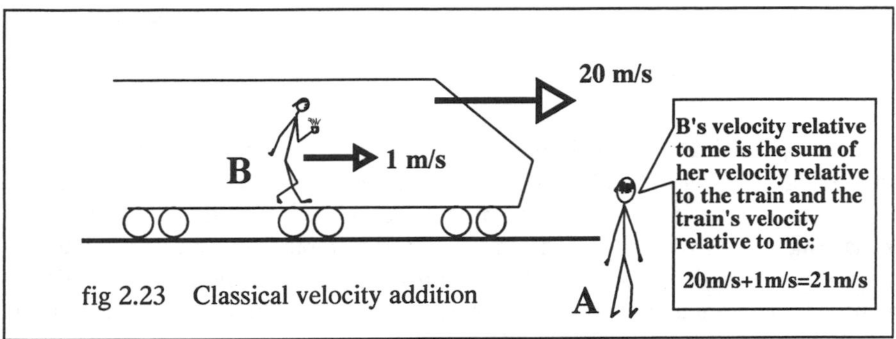

fig 2.23 Classical velocity addition

2.11.2 The Speed Of Light.

Imagine that, instead of walking along the train, B had shone a torch along it. The velocity of the train relative to A is 20 m s^{-1}, so classically we expect the velocity of light relative to A to be c + 20 m s^{-1}. This would violate the principle of relativity. We must therefore abandon the classical rule for velocity addition. The speed of light is the same for A as for B so adding 20 m s^{-1} to it must yield the velocity of light! In fact, adding any velocity to the velocity of light must still produce the same value, c.

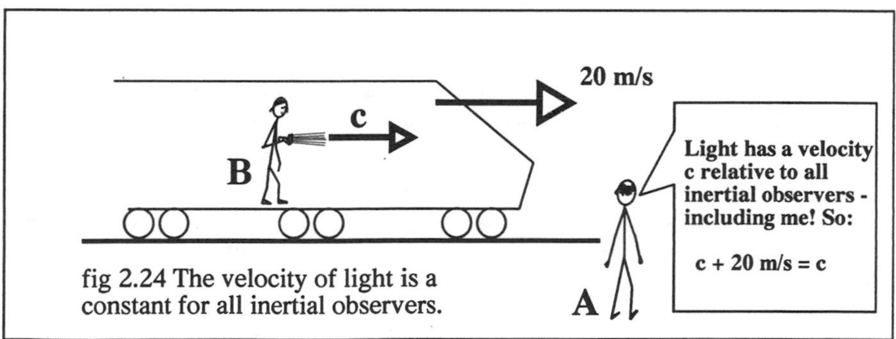

fig 2.24 The velocity of light is a constant for all inertial observers.

2.11.3 Adding Velocities.

What do we really mean by the addition of velocities? From the examples above we can see that the process involves the addition of a velocity measured *relative to A* (i.e. the velocity of the train) to a velocity measured *relative to the train* (i.e. the light's velocity) in order to produce a resultant velocity *relative to A*. The problem is that any measurement of velocity involves a determination of distance travelled and time taken and the two velocities we are adding have been defined in *different inertial reference frames*. A's clocks and measuring rods disagree with B's in three respects:

- *Time dilation;*
- *Length contraction;*
- *Synchronization.*

It is not surprising that the simple addition of velocities obtained partly with A's instruments and partly with B's instruments will result in errors. In order to calculate the velocity of an object relative to A we must use A's clocks and measuring rods throughout. To do this we need to replace the velocity relative to the train measured on clocks and rods *moving with the train,* with the velocity relative to the train *as it would be measured by A.* When the effects of each of the above corrections are carefully taken into account A will discover that B's measurements of the velocity of objects in B's own reference frame are always overestimates when compared with the values for the same relative velocities obtained by A using instruments at rest in A's own reference frame. Thus the result of adding velocities relativistically will always be less than the result obtained classically (and will never exceed the velocity of light).

2.11.4 The Velocity Addition Formula.

It is possible to derive this result very simply from the Lorentz transformation equations (see later) but here we approach velocity addition as a problem of combining velocities that happen to have been measured in different reference frames.

B measures a projectile's speed to be u in her laboratory. A who sees B's laboratory moving past at speed v (in the same direction) also observes the projectile. The question is, how does the relative velocity w of the projectile with respect to A derive from the two velocities u and v? To answer this question we must consider the procedure adopted by B in measuring the speed u and adjust her values taking account of time dilation, length contraction and synchronization errors. The outcome will be a value u_A that is the relative velocity of the projectile to B as measured by A.

B's PROCEDURE:

- Synchronize two clocks and place them a measured distance $\Delta x'$ apart.
- Record the times t'_x and t'_y that the projectile passes each clock.
- Calculate the speed of the projectile through the laboratory by dividing the distance between the clocks by the time difference.

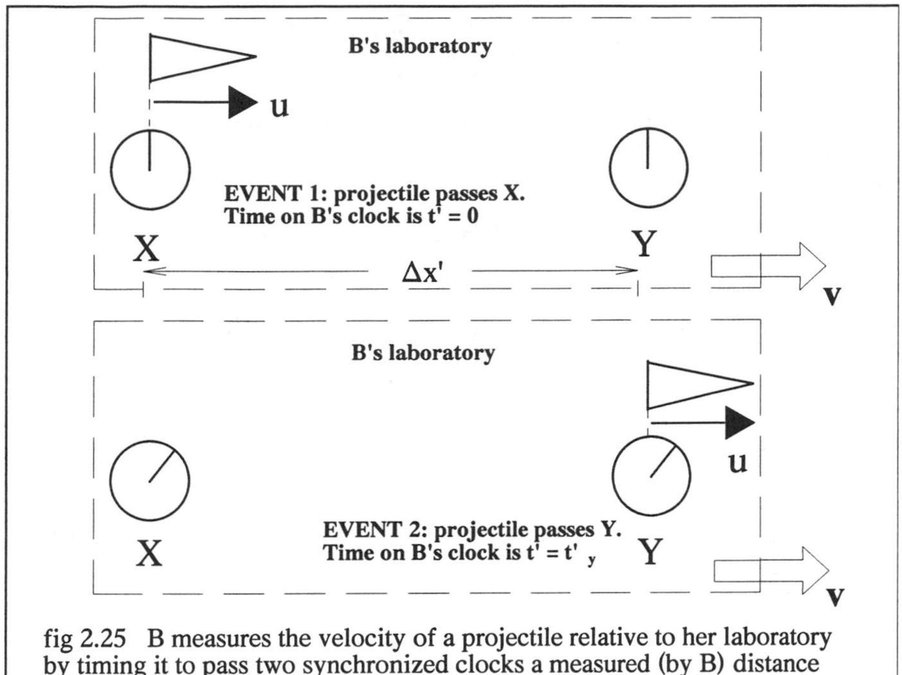

fig 2.25 B measures the velocity of a projectile relative to her laboratory by timing it to pass two synchronized clocks a measured (by B) distance apart. However, B's laboratory is moving relative to A.

A's INTERPRETATION:

A disagrees with B's measurement of the relative velocity u between B and the projectile. According to A this relative velocity is really u_A, a value A can determine by correcting B's measurements.

$$u = \frac{\Delta x'}{\Delta t'} = \frac{\Delta x'}{t_y - t_x} \qquad (1) \qquad \text{and} \qquad u_A = \frac{\Delta x}{\Delta t} \qquad (2)$$

$$\Delta x = \frac{\Delta x'}{\gamma}$$ because A sees the separation of B's clocks contract

A also sees clock Y start late because of a synchronization error τ:

$$\tau = \frac{v\Delta x}{\left(c^2 - v^2\right)}$$

and both of B's clocks appear to A to run slow because of time dilation. The overall effect of these two corrections on the time Δt observed by A between B's two measurements is:

$$\Delta t = \gamma\Delta t' + \frac{v\Delta x}{(c^2 - v^2)} = \gamma\Delta t' + \frac{\gamma^2 v\Delta x}{c^2} = \gamma\left(\Delta t' + \frac{v\Delta x'}{c^2}\right)$$

Using equations (1) and (2) we can now derive an expression for u_A

$$u_A = \frac{\Delta x}{\Delta t} = \frac{\Delta x'}{\gamma\left(\gamma\Delta t' + \frac{v\Delta x}{(c^2 - v^2)}\right)} = \frac{\Delta x'}{\gamma\left(\gamma\Delta t' + \frac{\Delta x'}{\gamma(c^2 - v^2)}\right)}$$

It is useful to note that $(c^2 - v^2) = \frac{c^2}{\gamma^2}$ then we can simplify to:

$$u_A = \frac{\Delta x'}{\gamma^2\left(1 + \frac{v\Delta x'}{c^2}\right)} = \frac{u}{\gamma^2\left(1 + \frac{vu}{c^2}\right)}$$

Now we simply *add* v and u_A as in classical physics to obtain the relative velocity w of the projectile to A:

$$w = v + u_A$$

$$w = v + \frac{u}{\gamma^2\left(1 + \frac{vu}{c^2}\right)}$$

which simplifies to give the relativistic velocity addition formula:

$$w = \frac{v + u}{1 + \frac{uv}{c^2}} \tag{2.7}$$

2.11.5 Comments.

- This is not the usual (or the most elegant) derivation of the velocity addition formula but it dispels any idea that we cannot add velocities in relativity. The final step above is the simple addition of two velocities. What relativity will not allow us to do is to add velocities measured on instruments located in different reference frames. ($v + u$) was not the correct value for the relative velocity w of the projectile with respect to A since v was measured in A's reference frame and u in B's. However, when we 'correct' u for the effects of relative motion on B's measuring apparatus we obtain u_A. It *is* then true that $w = v + u_A$.

- If both u and v are small compared to c, as is the case with the velocities of common experience, then uv/c^2 is negligible and the relativistic expression reduces to that of classical physics. For example, in the case of the man walking at 1 m s^{-1} relative to a train moving at 20 m s^{-1} the relativistic formula gives a velocity relative to the bank of 20.9999999999999998 m s^{-1} - indistinguishable (to put it mildly) from the classical prediction of 21 m s^{-1}.

- If either u or v is equal to c then $w = c$. This ensures, in keeping with the principle of relativity, and regardless of the velocity of a source or observer, that light has velocity c relative to all inertial reference frames.

- If both u and v are equal to c (as in the case of light beams travelling in opposite directions) then $w = c$.

- Imagine a situation in which two similar rockets travel in opposite directions away from a space station. If both move at 0.75c then classical addition of velocities would expect a velocity of separation of 1.5c. This *is indeed* the relative velocity of separation that would be measured by an observer situated on the space station but does not in any way violate the principle of relativity since neither rocket has an actual speed greater than c with respect to this observer. Also, if we travel in one of the rockets and calculate the relative velocity of the other rocket using the relativistic velocity addition formula (or even if we devise some way of measuring this separation speed) we obtain a value 0.96c, once again less than the speed of light and in agreement with the principle of relativity.

fig 2.26 A and B have equal velocities of 0.75c in opposite directions relative to some inertial reference frame. However, their velocity relative to one another is only 0.96c.

2.11.6 The Velocity Of Light As A Limit.

One of the consequences of the velocity addition formula is that it is not possible to take a body travelling at less than the speed of light and accelerate it to a velocity greater than the speed of light. In fact it is not possible for a massive body even to reach the speed of light (although it may come arbitrarily close to this). To see why this is we consider a body that has constant acceleration in its own reference frame. This means it will be continually adding small increments to its existing velocity. However, as this velocity approaches the speed of light the effect of each small addition becomes less significant. (This is because the term uv/c^2 has become larger in the velocity addition formula.) It is a law of diminishing returns and no increment can take it up to the speed of light since the effect of adding any velocity to a velocity less than c always produces a result less than c. It is rather like an

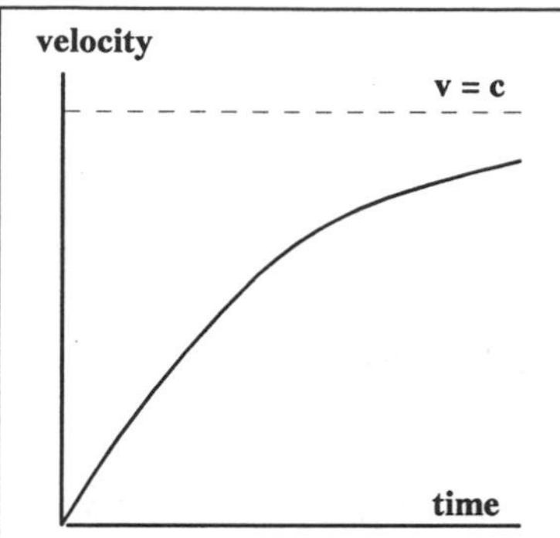

fig 2.27 An object with a constant acceleration in its own frame has a diminishing acceleration relative to an external inertial frame. Its velocity with respect to this frame is asymptotic to c.

open door which we attempt to close by pushing it halfway closed, then half of the remaining distance, half again etc.... It will take an infinite number of steps to close it, so if each step takes a finite time it will never be fully closed. However, it will approach closure as an asymptotic state. An observer watching the accelerating body would see its velocity increase toward the limiting speed c as in the graph on the left.

Example: Fresnel suggested that the movement of material bodies through the ether might drag the ether along with it and affect the speed of light relative to an external laboratory. In 1861 Fizeau tested this idea and found that the speed of light in water relative to a laboratory *was* affected by the velocity of flow of the water. If the ether is dragged along at the velocity of the medium the velocity of light relative to the laboratory would have this flow velocity, v added to it so that the velocity relative to the laboratory becomes

$$c_L = \frac{c}{n} + v \quad \text{where } n \text{ is the refractive index of the medium.}$$

In 1886 Michelson and Morley carried out an accurate measurement of this effect and found that the velocity relative to the laboratory is about:

$$c_L = \frac{c}{n} + 0.435v$$

Show that this result is consistent with relativity and does not need an ether hypothesis.

If the flow of water is parallel to the direction of the light beam the velocity c_L relative to the laboratory is found from the relativistic addition of flow velocity v and velocity of light relative to the medium c' $(= c/n)$:

$$c_L = \frac{c'+v}{1+\frac{vc'}{c^2}} = (c'+v)\left(1+\frac{vc'}{c^2}\right)^{-1}$$

since v is very small compared to c we can expand the second term using the binomial theorem:

$$c_L \approx (c'+v)\left(1-\frac{vc'}{c^2}\ldots\ldots\ldots\ldots\right)$$

$$c_L \approx c'+v-\frac{vc'^2}{c^2}+\frac{v^2c'}{c^2}\ldots\ldots\ldots\ldots \text{ but } \frac{v^2}{c^2}\approx 0 \text{ so}$$

$$c_L \approx c'+v-\frac{vc'^2}{c^2}=c'+v-\frac{v}{n^2}=c'+v\left(1-\frac{1}{n^2}\right)$$

$n = 1.33$ for water so $\quad c_L = c'+v\left(1-\frac{1}{1.33^2}\right) = \underline{c' - 0.435v}$

2.12 RELATIVISTIC MASS.

2.12.1 Inertia.

The inertia of a body is its reluctance to change its state of motion and is usually identified with its mass. Classically this is a value dependent upon the 'amount of matter' present, where 'amount of matter' is understood to relate in some way to the number of particles (protons, neutrons and electrons) in the body.

However, when a body accelerates uniformly in its own reference frame it is observed by an external observer to have a diminishing acceleration such that its velocity approaches but can never reach the speed of light. This is apparent in particle accelerators where ever greater energies are given to the particles but no particle quite reaches the speed c. The external observer will conclude that the body's 'reluctance to change its state of motion' increases as its velocity increases. *Inertia* increases with velocity. Relativity leads to the conclusion that the inertia, or mass, of a body depends both on the 'amount of matter' and the velocity at which that matter is moving.

- *Total mass increases with velocity.*

The mass of a body measured in its own rest frame is its *rest mass*, its total or *relativistic mass* when moving is always greater than this.

Linear momentum is defined in classical physics as the product of mass and velocity. In relativity this definition is retained, but the mass that must be used is the relativistic mass discussed above.

There is another aspect to this. Work done is equal to force times distance moved in the direction of the applied force. If the inertia of a body increases with velocity it implies that the force we apply produces a diminishing acceleration. What has happened to all the work done by this force if the body does not significantly increase its speed? Has it gone to some new form of internal energy? Can we identify the increased inertia with the extra energy? Does energy have mass? A relativistic approach will lead to new expressions for mass energy and momentum and a new interpretation of their relationship to one another.

2.12.2 Conservation Laws.

In classical physics the most important dynamic quantities are those which are conserved during interactions. For example, when two bodies collide with one another the total mass, energy and momentum before the collision are equal to the total after collision (of course some kinetic energy will be converted to other forms in an inelastic collision but no energy is created or destroyed). The principle of relativity demands that the laws of physics are the same in all inertial reference frames, so the laws of conservation of mass, energy and momentum must be true in all inertial reference frames. (Another way of looking at this is to say that we must define these quantities in such a way that they are conserved in all frames.)

Below we apply conservation laws to an inelastic collision viewed from two different inertial reference frames (one of which is the centre of mass frame for the collision). This leads to an equation for the dependence of mass on velocity and the idea that energy itself has mass. We shall then consider a thought experiment devised by Einstein that leads to the mass-energy relation.

2.12.3 Mass And Velocity.

To derive the correct formula for the variation of mass with velocity we assume:

- *Conservation of relativistic mass in all inertial reference frames.*
- *Conservation of linear momentum in all inertial reference frames.*

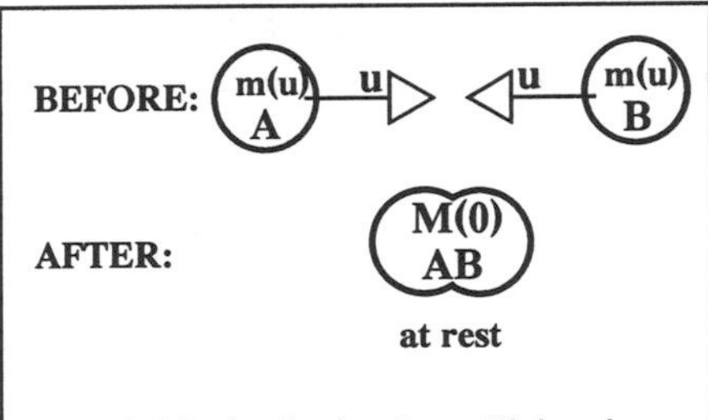

fig 2.28 An inelastic collision in the centre of mass frame.

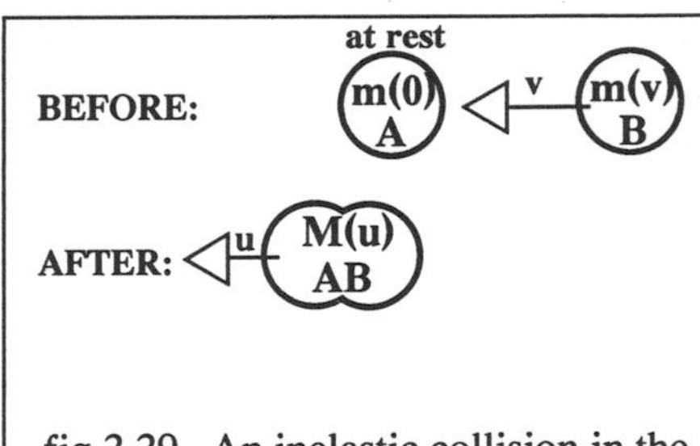

fig 2.29 An inelastic collision in the rest frame of mass A.

Consider an inelastic collision in which two identical bodies, each of rest mass $m(0)$, collide and stick together to form a composite body. We shall apply the conservation laws above in each of two inertial reference frames, one in which the final composite body will be at rest (e.g. the centre of mass frame) and another in which one of the two colliding bodies is at rest before the collision. ($m(u)$ etc.... represents the relativistic mass of the body when moving at velocity u. It is an expression for $m(u)$ that we seek.)

In the CM frame $M(0) = 2m(u)$ (1) (conservation law 1)

In the moving frame $M(u) = m(0) + m(v)$ (2) (conservation law 1)

$M(u)u = m(v)v$ (3) (conservation law 2)

The moving reference frame has velocity u relative to the CM frame so the velocity v of mass B in this frame will be the relativistic sum of u and u:

$$v = \frac{2u}{\left(1+\frac{u^2}{c^2}\right)} \qquad (4)$$

This will be useful later.

We are trying to find how m or M varies with velocity u, so we begin by eliminating $m(v)$ from equations (2) and (3). This leaves:

$$M(u) = \frac{m(0)}{\left(1-\frac{u}{v}\right)} \qquad (5)$$

Dividing (5) by (1):

$$\frac{M(u)}{M(0)} = \frac{m(0)}{m(u)} \cdot \frac{1}{2\left(1-\frac{u}{v}\right)} \qquad (6)$$

If the variation of mass with velocity is to be universal we must have:

$$\frac{M(u)}{M(0)} = \frac{m(u)}{m(0)}$$

so that (6) becomes:

$$\left(\frac{m(u)}{m(0)}\right)^2 = \frac{1}{2\left(1-\frac{u}{v}\right)}$$

we now substitute for v from (4), simplify and take the square root to obtain an expression for relativistic mass which is consistent with the conservation laws:

$$\frac{m(u)}{m(0)} = \frac{1}{\sqrt{1-\frac{u^2}{c^2}}}$$

or

$$\underline{m(u) = \gamma\, m(0)} \qquad (2.8)$$

2.12.4 Comments.

- If $v<<c$ then $m(v)$ is effectively equal to the rest mass $m(0)$. This is the classical limit and justifies our assumption of constant mass at low velocity. When we quote the mass of an electron as 9.1×10^{-31} kg we are referring to its rest mass.

- As the velocity of a body increases so does the ratio v/c and the relativistic mass becomes significantly greater than the rest mass. The relativistic mass of a body travelling at about 99% of the speed of light is roughly seven times its rest mass. Electrons emerging from a 1 GeV accelerator have masses about 2000 times their rest mass, comparable to the mass of a hydrogen atom!

- As v approaches c the mass of the particle approaches infinity. This huge increase in inertia makes it impossible to accelerate bodies of non-zero rest mass up to the velocity of light.

- We are in a sense placing rather too much emphasis on the concept of mass here. This is understandable since it plays such a central role in classical mechanics. In relativity there are several different ways in which mass and inertia can be defined. Our use of 'relativistic mass' (referred to as 'momental mass' by some authors) makes the laws of conservation of energy and mass equivalent statements. Once we have developed a proper relativistic concept of energy we will be able to tackle dynamic problems simply by considering the conservation of energy and momentum (and these too are closely related, as we shall see when we consider a spacetime representation in the next chapter).

2.12.5 Testing The Expression For Relativistic Mass.

It is interesting that some of the pre-relativistic theories also predicted an increase of mass with velocity and evidence that this did in fact occur was available before Einstein's 1905 paper. In 1897 J.J. Thomson carried out an experiment in which he measured the ratio of charge (e) to mass (m) of the electron. This value e/m is called the specific charge of the electron and, when combined with the electronic charge (determined initially from electrolysis and later directly in Millikan's oil drop experiment) enabled the mass of the electron to be calculated.

In Thomson's experiments the electrons were accelerated through an applied voltage (V) which generated velocities of about $0.1c$. The electrons then entered a region of crossed electric field (E) and magnetic field (B) oriented so that the electric and magnetic forces on the electrons opposed one another. Since the electric force is independent of the electron speed whereas the magnetic force depends upon it there is only one ratio of electric to magnetic field strengths that will allow electrons of a particular speed to pass undeflected. e/m can then be calculated from this ratio and the accelerating voltage.

In 1902 and 1906 W. Kaufmann carried out similar experiments to look for the variation of mass with electron velocity. Electrons traveling at $0.1c$ would only be expected to have a relativistic mass 0.5% greater than their rest mass, a variation that would have been undetectable in Kaufmann's experiment. He needed a source of much faster electrons and so used the beta rays emitted from radioactive sources. The nuclei of beta-emitters are unstable because they are neutron-rich. In order to become more stable a neutron in the nucleus decays to a proton, which remains in the nucleus, and emits an electron and an antineutrino. The energy released in the decay is shared between the two particles so that electrons from a particular nuclide may have any energy from a small value up to some maximum. This results in a range of velocities up to a significant fraction of c.

Kaufmann selected the more rapid electrons (velocity selection can also be achieved using crossed electric and magnetic fields) and then used Thomson's method to measure *e/m*. If mass increased with velocity then the value of *e/m* should fall. Kaufmann's results are shown in the diagram below.

These results show that *e/m* does indeed decrease with velocity. If we assume that *e* is the same value for all observers then Kaufmann's results provide strong evidence for the increase of mass with velocity.

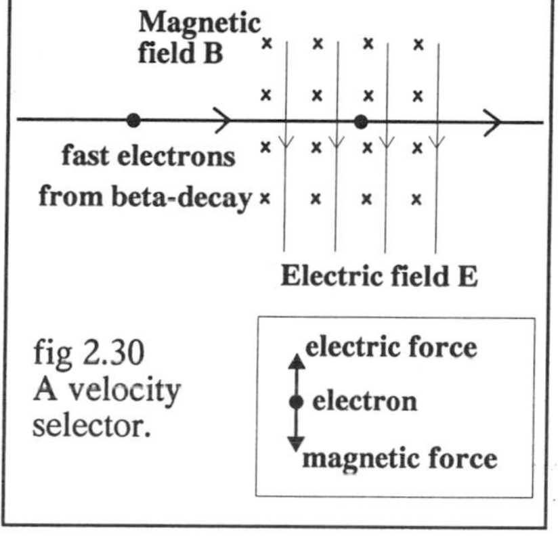

fig 2.30 A velocity selector.

electron velocity v **(m s^{-1})**	2.4×10^8	2.5×10^8	2.6×10^8	2.7×10^8	2.9×10^8
specific charge e/m **(C kg^{-1})**	1.3×10^{11}	1.2×10^{11}	1.0×10^{11}	0.8×10^{11}	0.6×10^{11}
electron mass m/m_0	1.3	1.5	1.8	2.3	2.8

fig 2.31 Kaufmann's results show that electron mass increases with velocity.

Many other more sensitive experiments confirm this conclusion (e.g. Bucherer in 1908), but by far the most compelling evidence comes from modern experiments in high energy particle physics and fusion research. Accelerators like CERN 'store' particle beams in rings by deflecting them with strong magnetic fields. The field strength needed to confine a particle in a ring of radius r depends on the mass of the particle and values are determined by the relativistic mass, not the rest mass.

2.12.6 Deuteron Fusion Reactions.

BEFORE
fast deuteron
deuteron at rest
proton
AFTER
triton

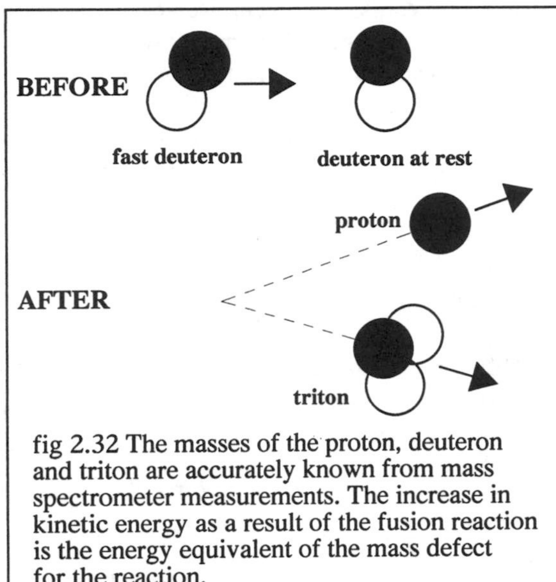

fig 2.32 The masses of the proton, deuteron and triton are accurately known from mass spectrometer measurements. The increase in kinetic energy as a result of the fusion reaction is the energy equivalent of the mass defect for the reaction.

The conservation of mass is closely linked to the other conservation laws we have discussed and these can be used to predict the mass of particles produced in nuclear reactions. One very accurate test of these conservation laws is provided by the fusion of two deuterons to produce a nucleus of tritium and a proton. The relativistic conservation laws can be used to predict the rest mass of the tritium nucleus and this value can be compared with the measured mass using a mass spectrometer. The calculated value is found to agree with the measured value to the limits imposed by the uncertainties in the nuclear measurements themselves (about 2 parts in a million).

2.13 MASS AND ENERGY.

2.13.1 An Inelastic Collision.

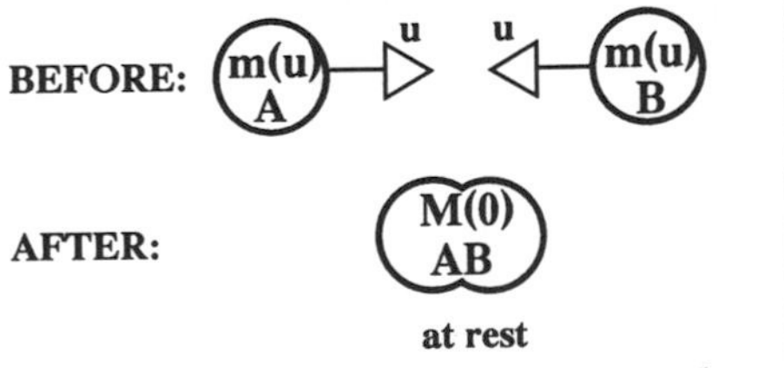

BEFORE:
m(u)
A
u
u
m(u)
B
AFTER:
M(0)
AB
at rest

fig 2.33 Both A and B have increased masses because of their motion. This means that M(0) must be greater than 2m(0), that is it has a greater rest mass than the sum of the rest masses of the two bodies that form it.

Consider the inelastic collision discussed previously in which two identical bodies moving with equal speeds in opposite directions collide and coalesce to form a single composite body. Since A and B (each of rest mass $m(0)$) are both moving at speed u relative to the centre of mass frame they will both have increased mass $m(u)$. If we apply the law of conservation of relativistic mass in this reference frame we can see that the composite body AB has a mass of $2m(u)$. This is obviously greater than twice the rest mass, $2m(0)$ of the composite body created by joining A and B together at rest. But AB is at rest, so how do we account for its enhanced mass?

There is one aspect of the collision we have not yet considered - that is how the kinetic energy of the colliding bodies is dissipated. If we assume energy conservation then this energy must be converted to some internal form within the composite body AB. Therefore AB differs from a static combination of A and B in two respects: it has greater internal energy and it has greater mass. In all other respects it is the same body that would be formed if A and B combined at rest. It is hard to avoid the conclusion that the extra mass is directly related to the extra energy content of the body.

There is one other point we could make here. The composite body AB has an enhanced rest mass *because* of the extra internal energy that it possesses. It would seem reasonable to suppose that, if energy has mass, there may be an energy equivalent of the rest mass of particles as well. This would make the symmetry between mass and energy complete.

2.13.2 Radiation Pressure.

When radiation is absorbed or reflected by a surface it exerts a pressure on that surface. This property follows from classical electromagnetism and was predicted by Maxwell and others in the nineteenth century. (The possibility of a radiation pressure was suggested by Kepler as early as the seventeenth century to explain the observation, a century before, that comets' tails always pointed away from the Sun.) Its measurement was complicated by the larger forces exerted by the random thermal motions of molecules as they impact upon surfaces. The familiar Crookes radiometer is an example of these thermal forces. The small but significant force caused by radiation pressure was first demonstrated by Lebedev and by Nichols and Hull in 1901-2. (The photon theory developed after Einstein's 1905 paper on the photoelectric effect allowed radiation in a cavity to be treated like a photon gas with radiation pressure explained by the rate of change of photon momentum at the cavity walls.)

The essential result (see appendix 2), which follows from either classical electromagnetism or photon theory, is that the radiation pressure R on an absorbing surface is equal to the intensity of radiation I divided by the speed of light:

$$R = \frac{I}{c} \tag{2.9}$$

This leads to a force on area A of $F = RA = IA/c = P/c$ where P is the power absorbed by the surface. (This force will obviously change if the radiation is partially or completely reflected.)

The impulse applied to the surface if it absorbs radiation of constant intensity for a time t is:

$$Ft = \frac{Pt}{c} = \frac{E}{c}$$

where E is the total energy absorbed. But impulse is equal to change of momentum so the momentum associated with radiation of energy E is simply the momentum lost when the radiation is absorbed, in other words:

$$p = \frac{E}{c} \quad \text{or} \quad E = pc \tag{2.10}$$

This is a fundamental and extremely important relation for electromagnetic radiation and can be used to derive the mass-energy relation directly. It implies that radiation transmits linear momentum between source and absorber and so presumably also transfers mass. It can also be applied to individual photons. If we use the usual definition of momentum as mv then in this case the velocity is c and we have:

$$mc = \frac{E}{c} \quad \text{or} \quad m = \frac{E}{c^2} \quad \text{or} \quad E = mc^2 \tag{2.11}$$

It follows from this that all sources of radiation will experience a recoil in order to conserve momentum and all absorbers will experience a force as they absorb radiation. These can be regarded as two ends of an interaction mediated by photons with the photons themselves acting as a temporary store of energy and momentum between emission and absorption.

2.13.3 Einstein's Thought Experiment.

Einstein suggested a simple thought experiment that shows that energy has mass and also allows us to determine the relation between the two quantities. It is based on the idea of radiation pressure and is really little more than an elaboration of the argument above, but it gives a very clear mechanical argument for mass-energy equivalence.

Consider a long cylindrical spacecraft in empty space. The craft contains two bodies, A and B, placed at opposite ends of the cylinder. A and B are identical in every respect except that A possesses an additional quantity of energy that it will later transmit as a pulse of electromagnetic radiation to B. The following sequence of events takes place:

- *A transmits a pulse of radiation to B. As the radiation is emitted A experiences a recoil from the radiation pressure and the whole tube recoils to the right.*
- *The radiation travels from A to B during which time the cylinder travels to the right at a uniform velocity.*
- *B absorbs the radiation and experiences an impulse to the left that is equal in magnitude to the impulse previously given to A and so brings the whole system to rest. The net effect has been a translation of the whole spacecraft and its*

contents to the right and the transmission of a quantity of energy from A to B (that is to the left).

- *If A and B are of equal mass they can now be exchanged by the operation of forces internal to the craft without any further displacement. The system is returned to its initial state but is displaced a distance x to the right. Repetition of this cycle of events could propel the tube any distance to the right at any velocity lower than c.*

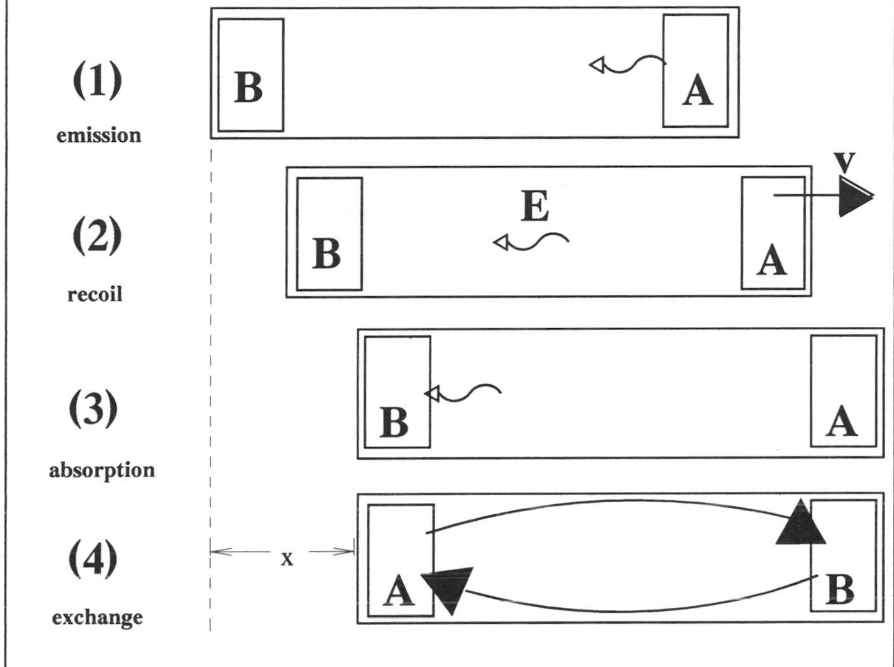

fig 2.34 Einstein's thought experiment. A cylindrical spacecraft in deep space contains two similar bodies that exchange energy by radiation. If step (4) is possible this would allow the body to repeat the process and accelerate from rest without the influence of an external resultant force.

This clearly violates Newton's first law of motion since an isolated system has changed its state of motion without the action of any external resultant force. If instead of exchanging radiation the two bodies had exchanged a massive object then we should have reached a different solution. We can see this if we consider an example with twin boys sitting at opposite ends of a small boat playing catch with a medicine ball. As A throws the ball to B he experiences an impulse to the right that causes him and the boat to move that way. When the ball is caught by B it applies an equal and opposite impulse to B and stops the boat. If A and B now carefully swap places B, who carries the ball, will apply a slightly greater force to the boat in order to propel himself forwards than A will. The boat receives a small resultant impulse to the left that will return it to its initial position when the swapping over is completed. This sequence of events produces no net displacement however often it is repeated. Throughout the entire process the centre of mass has remained stationary.

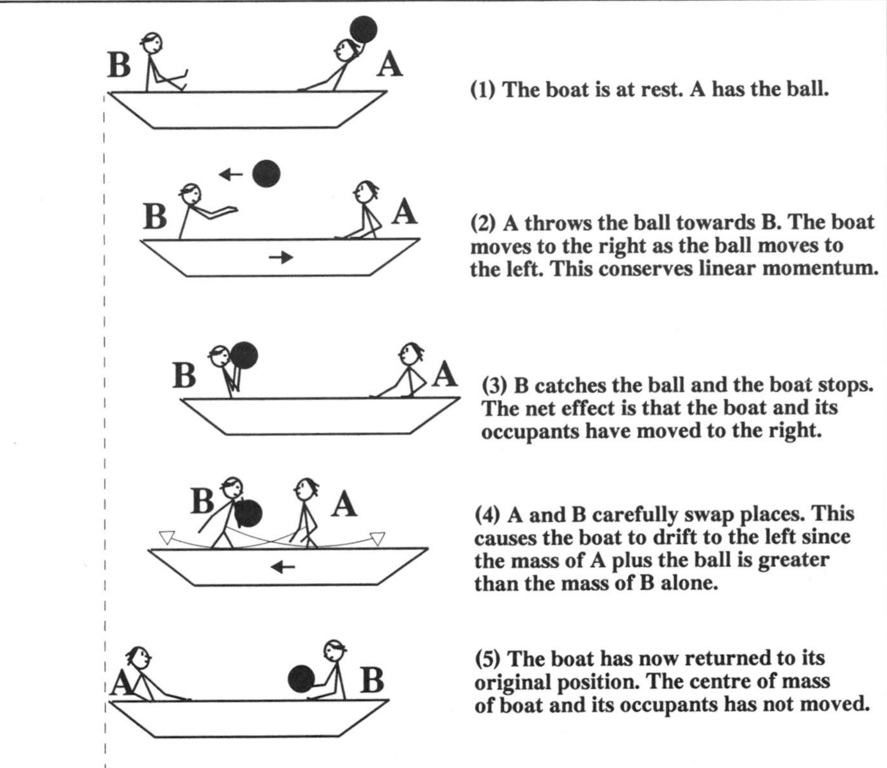

fig 2.35 If A and B throw a massive ball back and forth, there is no net displacement of the boat or its contents. The backwards and forwards drift of the boat keeps the centre of mass of the system in the same place at all times.

The problem posed by Einstein's thought experiment can be resolved if we assume that the radiation transmitted along the cylindrical spacecraft carries mass as well as energy. We can use the condition that the centre of mass of the system cannot move to derive an expression for the mass associated with this amount of radiant energy. The craft will still be displaced when the radiation is in flight but, when A and B are swapped over, B (which has just absorbed the energy) will be

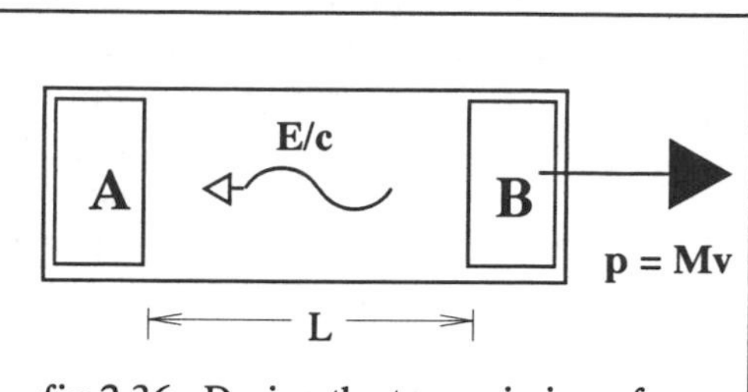

fig 2.36 During the transmission of radiation the linear momentum of the craft and its contents to the right must be equal and opposite to that of the radiation moving to the left.

more massive. The craft will move back to its original position just as the boat does in the Newtonian example above.

2.13.4 A Simple Derivation Of The Mass-Energy Relation.

Consider the transfer of energy E from A to B as in Einstein's thought experiment described above. As A emits the energy it experiences a reaction force due to the radiation pressure which gives the craft and its contents a momentum p to the right. By conservation of linear momentum this is equal but opposite to the linear momentum of the transmitted radiation. Hence we can calculate the velocity of recoil of the craft (if we know its total mass) and, since we know how far the radiation has to travel, we can determine the craft's overall displacement, x, before it is stopped by the impulse applied to B as it absorbs the radiation. The mass transferred by the radiation must then be just the amount necessary to maintain the centre of mass of the system (vessel plus energy) in the same position.

Linear momentum of radiant energy	$p = \frac{E}{c}$	
Velocity of recoil (M is total mass of craft plus A and B)	$v = \frac{p}{M} = \frac{E}{Mc}$	
Time of energy transfer A to B (assuming $v << c$)	$t = \frac{L}{c}$	
Displacement of vessel	$x = vt = \frac{LE}{Mc^2}$	
To maintain the CM in the same position we need (where m is the mass transferred)	$Mx = Lm$	
Substituting for x gives:	$E = mc^2$	(2.11)

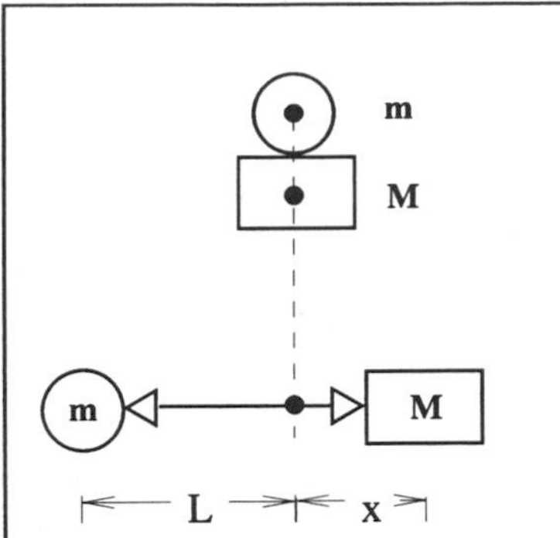

fig 2.37 The circle represents the radiation and the rectangle represents the craft and contents. To keep the centre of mass in the same place xM = Lm. The fact that the radiation does not originate at the centre of the craft does not affect the argument or calculation.

N.B. This derivation assumes that L is much greater than x so that the slight reduction in t and x due to the displacement of the craft during the energy transfer can be neglected. However, a detailed derivation including this produces exactly the same result.

2.13.5 Comments.

- Since the speed of light is very great this formula implies that there is an incredibly large amount of energy associated with the mass of ordinary matter. If it were possible to completely transform 1 kg of matter into pure energy this would be sufficient to fuel a large power station for several years! Unfortunately we are not able to do this for bulk matter. The closest we come on a commercial scale is in nuclear fission where the mass of fission products is typically about 0.1% lower than the mass of the same particles in the original nucleus.

- Note that, although this argument has been based on transfer of energy as electromagnetic radiation, it made no assumptions about the form in which this energy was stored in bodies A and B. It follows that all energy transfers involve mass transfers. If this were not the case it would be possible to transfer mass as light, absorb it and convert it to a different form that has no mass associated and hence violate the conservation laws. (It would also allow us to propel a spacecraft without recourse to any external forces.)

- However, the large value of c^2 means that most energy changes do not produce a noticeable change of mass. For example, normal chemical changes, like combustion, result in a mass change of the order of 0.00001%. In both the examples mentioned, that is fission and combustion, it is significant that the particles change their mass as a result of moving to lower states of potential energy rather than because some particles have been annihilated.

Example: Calculate the 'energy released' in the following induced fission reaction.

$$^{235}_{92}U + ^{1}_{0}n \rightarrow ^{144}_{56}Ba + ^{90}_{36}Kr + 2\,^{1}_{0}n$$

U-235 atomic mass = 235.044 u
Ba-144 atomic mass = 143.923 u
Kr-90 atomic mass =89.920 u
mass of neutron = 1.009 u

Although atomic rather than nuclear masses are given the number of electrons balances on each side of the reaction equation so their masses make no difference to the calculation (the electron binding energies do change slightly, but by a negligible amount compared to the nuclear binding energies). The mass defect for the reaction is:

$$(235.044 + 1.009)\ \text{u} - (143.923 + 89.920 + 2 \times 1.009)\ \text{u} = 0.192\ \text{u}$$

This is the loss in rest mass of the particles taking part in the reaction and equals the kinetic energy that is shared between them after the reaction (ignoring any additional KE brought in by the incident neutron).

Energy equivalent of 1 u is 931.5 MeV so the total kinetic energy generated by the nuclear fission reaction is about 179 MeV.

2.13.6 Mass, Energy And Velocity.

We derived an equation for the variation of relativistic mass with velocity (equation 2.8). If we combine this with the mass-energy relation above we have an expression for the variation of energy with velocity for a massive body. This is:

$$E = mc^2 = \gamma m_o c^2 = \gamma E_o \qquad (2.12)$$

This will obviously behave in much the same way as the mass equation, tending to infinity as v approaches c. Once again we can interpret this as setting c as the limiting velocity for any massive body since it would take more than the total energy in the universe to accelerate one atom up to the speed of light! We can expand the energy equation using the binomial theorem:

$$E = m_o c^2 \left(1 - \frac{v^2}{c^2}\right)^{-½}$$

$$E = m_o c^2 \left(1 + \frac{v^2}{2c^2} - \frac{3v^4}{8c^4} \cdots\right)$$

For velocities $v << c$ this will be a sequence of rapidly diminishing terms that form a convergent series. Only the first term is independent of velocity v. This will remain constant whatever the velocity of the body and so will not take part in physical processes involving the body (unless that body is annihilated). Einstein identified this as the 'rest energy' of the body.

Subsequent terms, which all depend on v, must represent the energy the body gains because of its motion, its kinetic energy. When we look at the second term in the series we can see it is exactly the expression for kinetic energy in Newtonian mechanics.

$$E = m_o c^2 + \frac{1}{2} m_o v^2 \cdots$$

The relativistic 'correction' consists of all the following terms, but these will only make a significant difference if $(v/c)^2$ is not negligible since each successive term is multiplied by $(v/c)^4$, $(v/c)^6$ etc.... The classical kinetic energy is therefore simply a first order approximation to the correct relativistic expression. It works extremely well in most cases because we rarely encounter bodies travelling at velocities comparable to c.

Example: What is the ratio of total mass to rest mass for a 500 GeV proton (i.e. a proton that has been accelerated through a potential difference of 500 GV)? The rest mass of a proton is 1.67×10^{-27} kg.

500 GeV is the kinetic energy of the proton, T. The total energy is the sum of kinetic energy and rest energy.

$$E_o = m_o c^2 = 1.67 \times 10^{-27}\ \text{kg} \times \left(3.0 \times 10^8\ \text{m s}^{-1}\right)^2 = 1.5 \times 10^{-10}\ \text{J} = 0.94\ \text{GeV}$$

$$TE = T + E_o = 500.94\ \text{GeV}$$

$$\frac{TE}{E_o} = \frac{500.94\ \text{GeV}}{0.94\ \text{GeV}} = 533$$

This ratio is γ and from this we can calculate β, about 0.999998, and hence the velocity of the proton, about $0.999998c$.The kinetic energy in this case is so large that the rest energy makes a negligible contribution to the total energy.

Example: What is the velocity of a particle whose total mass is double its rest mass?

$$m = \gamma m_o = 2m_o \quad \text{so} \quad \gamma = \frac{1}{\sqrt{1-\beta^2}} = 2$$

$$1 - \beta^2 = \frac{1}{\gamma^2} \qquad \beta^2 = 1 - \frac{1}{\gamma^2} = \frac{3}{4}$$

$$\beta = 0.87 \qquad \underline{v = 0.87c}$$

2.14 MASS ENERGY AND MOMENTUM.

2.14.1 Classical And Relativistic Momentum.

The most important property of linear momentum in classical mechanics is its conservation in all closed systems. This derives from Newton's laws of motion since all forces arise from interactions and the forces acting on a pair of interacting particles are always equal and opposite. In relativistic mechanics we have seen that energy has mass and so the total mass of a moving body is always greater than its rest mass. This led us to assume that relativistic momentum would have the form mv where m is the relativistic mass. We then proceeded by assuming that relativistic mass and momentum would be conserved in all inertial reference frames. Imposing these two conservation laws produced expressions for the dependence of mass and momentum on velocity:

$$m = \gamma m_o$$

$$\underline{p} = \gamma m_o \underline{v}$$

Linear momentum is a vector quantity and so each component is conserved as well as the magnitude of the momentum vector. This will be true in all inertial reference frames although the actual values of components and magnitudes (which are covariant quantities) will vary from one frame to another.

2.14.2 Mass Energy And Momentum.

We can derive an interesting and very useful relationship between mass energy and momentum as follows (a more satisfying interpretation of this relationship will be given in the next chapter).

Consider the relativistic mass	$m = \gamma m_o$	
square it	$m^2 = m_o^2/(1 - v^2/c^2)$	
rearrange	$m^2c^2 - m^2v^2 = m_o^2c^2$	
use $E = mc^2$ and $E_o = m_oc^2$	$E^2/c^2 - p^2 = E_o^2/c$	
multiply by c^2	$E^2 - p^2c^2 = E_o^2$	(2.13)

The right hand side of this identity is the square of the rest energy of a particle. This is a value determined in the rest frame of the particle and is a constant for that particle. This means that the left hand side of the identity that is constructed from the particle's total energy and the magnitude of its momentum must be an invariant. That is, it will have the same value in all inertial reference frames. This will be true even though the values of E and p are covariant (that is their values will be frame dependent). Invariant quantities are extremely useful in relativity and identifying them can often simplify the solution of problems.

2.14.3 Kinetic Energy.

In classical physics the kinetic energy is quite a significant quantity and has a simple formula. The Newtonian formula is just a first order approximation to the relativistic kinetic energy. In relativity the kinetic energy is given by the difference between the total and rest energy of the particle:

$$T = E - E_o$$

$$T = (\gamma - 1)E_o \tag{2.14}$$

The classical expression must be abandoned when relativistic corrections are significant. In a particle accelerator kinetic energy equals the work done on the particle by the accelerator. For example, an electron accelerated through 25 GV will emerge with kinetic energy equal to 25 GeV and total energy equal to this plus the rest energy of the electron (an extra and almost insignificant, 0.5 MeV).

2.14.4 Units For Mass, Energy and Momentum.

The electron-volt is a convenient non-SI unit widely used in atomic and nuclear physics. Energy, mass and momentum are all closely related, so it is convenient to have a consistent set of units related to the electron volt in which to measure them.

- Energy is measured in MeV or GeV and mass is related to energy by $m = E/c^2$ so convenient mass units are MeV/c^2 or GeV/c^2.
- Momentum is mass times velocity so it is measured in MeV/c or GeV/c.

Example: What is the rest mass, total mass, energy and momentum of an electron accelerated through 1.5 MV? (Rest mass of an electron is 9.1×10^{-31} kg.)

Rest energy $E_o = m_o c^2 = 0.51$ MeV

$$m_o = \frac{E_o}{c^2} = \underline{0.51 \text{ MeV}/c^2}$$

Kinetic energy $T = 1.5$ MeV

Total energy $E = T + E_o = \underline{2.0 \text{ MeV}}$

Total mass $m = \dfrac{E}{c^2} = \underline{2.0 \text{ MeV}/c^2}$

$E^2 = E_o^2 + p^2c^2$ so momentum $p = \dfrac{\left(E^2 - E_o^2\right)^{1/2}}{c} = \underline{1.9 \text{ MeV}/c}$

When the kinetic energy is so large that the rest energy can be neglected the numerical values of total energy, mass and momentum are all equal in this set of units.

2.14.5 Conservation Laws.

In relativity mass, total energy and momentum are all conserved in any closed system of interacting or non-interacting particles. Moreover the conservation laws apply in *all* inertial reference frames although the actual values of these quantities vary from one frame to another. The relativistic expressions are:

- Mass $m = \gamma m_o$ (2.8)
- Momentum $p = \gamma m_o v$ (2.15)
- Energy $E = mc^2 = \gamma m_o c^2 = \gamma E_o$ (2.11,2.12)

The relation between mass and energy is such that these do not really represent separate conservation laws. Conservation of energy guarantees conservation of mass and vice versa. However, this does lead to some confusion when people discuss the 'conversion of mass to energy' or 'energy to mass'. Since energy has mass the energy released in say a nuclear reaction has exactly the same mass as the mass defect of the reaction, all that has happened is that the energy and mass have transformed.

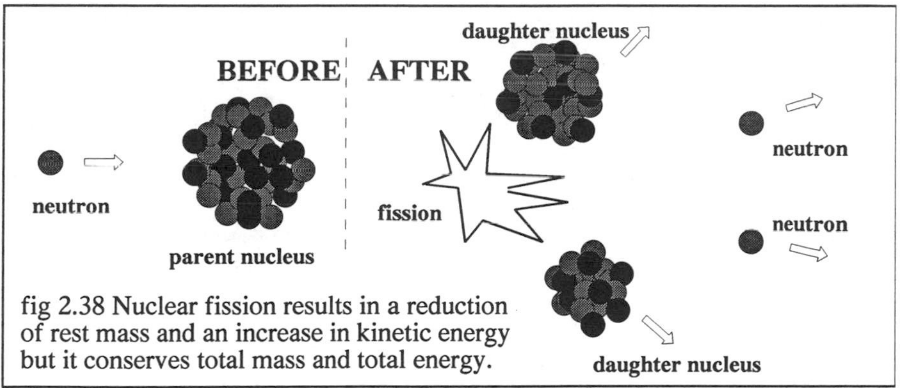

fig 2.38 Nuclear fission results in a reduction of rest mass and an increase in kinetic energy but it conserves total mass and total energy.

This point can be illustrated by two examples. The first is spontaneous nuclear fission (figure 2.38) in which a heavy nucleus splits to form two lighter daughter nuclei releasing some neutrons and a great deal of energy. If we consider the protons and neutrons that take part in the reaction we find there are exactly the same number of each before and after the fission. However, the total *rest* mass of the original nucleus is found to be greater than the sum of the *rest* masses of the products. But the products are not formed at rest, they are flying apart with great speed and therefore possess a great deal of kinetic energy! The mass of this extra kinetic energy exactly balances the mass defect (that is the difference in rest masses of reactants and products) of the reaction. The mass before and after fission is the same, so is the energy. This *must be* the case if the conservation laws for mass and energy are to be obeyed. When we say that mass has been converted to energy what we mean is that the rest mass of a collection of particles has been reduced to pay for the extra mass in their kinetic energy or in the creation of new particles (for example photons).

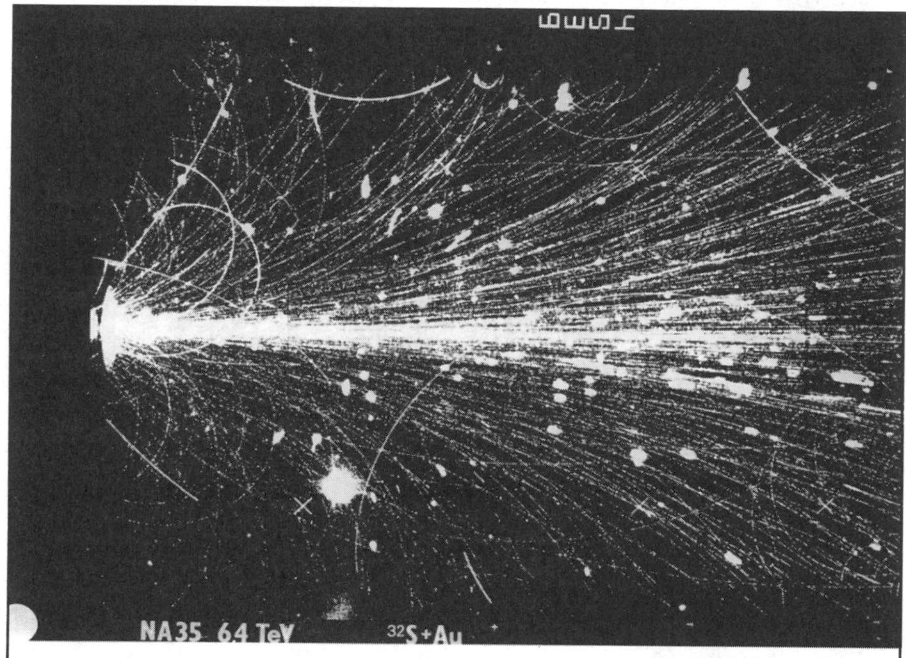

fig 2.39 Matter from energy. A nucleus of sulphur has been accelerated in the Super proton Synchrotron at CERN and collided with a stationary nucleus of gold. A complex shower of secondary particles has been created from the kinetic energy of the sulphur nucleus. Collisions like this recreate conditions that existed in the early universe just after the Big Bang. Photo used courtesy of CERN.

A second example involves particle annihilation (see figure 2.40). An electron and positron collide and annihilate one another to produce a pair of gamma ray photons. In this case the rest mass of the colliding particles completely disappears (along with any kinetic energy they may have had) since photons have zero rest mass. However, photons carry energy and the mass associated with this is exactly equal to that of the original pair of colliding particles. It is clear that the conversion of mass into energy usually refers to a conversion of rest mass into some other form. (The photon mass is given by $m = E/c^2 = hf/c^2$ where h is the Planck constant).

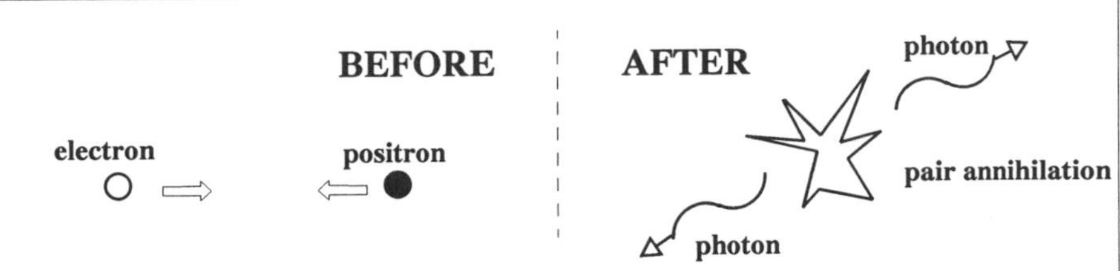

fig 2.40 Pair annihilation also converts rest energy to 'kinetic energy' whilst conserving total energy and total mass.

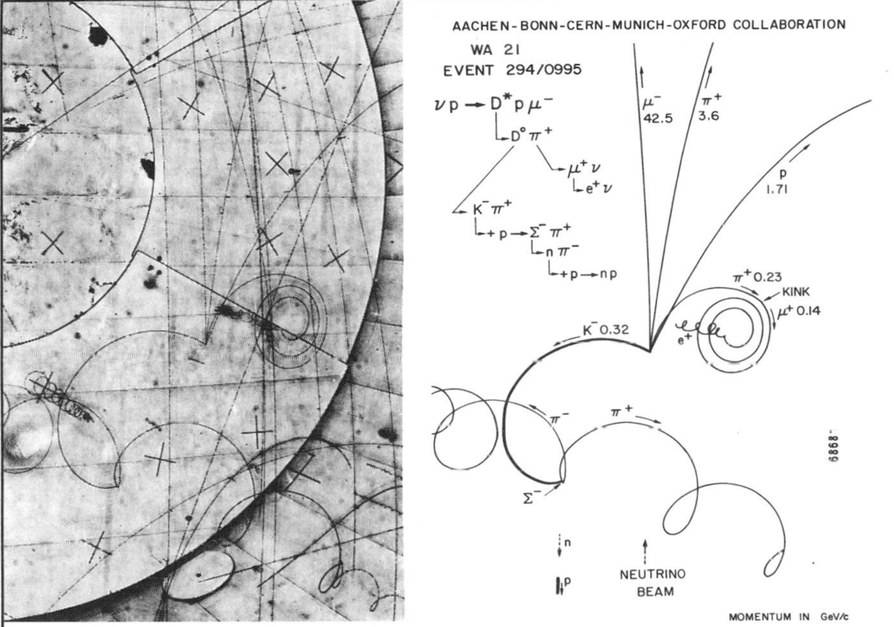

fig 2.41 The photograph at left shows tracks from the Big European Bubble Chamber (BEBC) at CERN following the collision of a neutrino (entering bottom centre) with one of the hydrogen nuclei (a proton). The tracks are interpreted in the line drawing on the right. There is a magnetic field of about 3.5 T directed into the page as you look at it and the diameter of the BEBC was about 3.7 m. The values shown beside the tracks are momenta in GeV/*c*. Compare the curvature of the two π^+ tracks at 1 and 2-o-clock, the radius of curvature is given by *p/Bq* and can be used to calculate momentum. The high energy μ^- goes almost straight. Notice also that the neutral and weakly interacting neutrino leaves no track. Nor does the particle emitted from the π^+ as it decays to a μ^+ where the track 'kinks'. A lot of detective work must be done to interpret these tracks. Try to understand what has happened at the end of the K^- track close to the point at which a $\pi^+ \pi^-$ pair has been formed. How are these two events linked to the displacement of a proton at the bottom left of the picture? Bubble chambers were very important during the 1970s and early 1980s but have been superseded by complex electronic detectors.

Photo used courtesy of CERN.

2.15 MASS ENERGY AND PARTICLE PHYSICS.

2.15.1 What Are Accelerators For?

It is a mistake to think that the prime purpose of a particle accelerator is to make particles travel at ever higher speeds. What accelerators really do is provide conditions for the creation of matter from energy. Compare the velocities achieved by protons at CERN in 1959 using the then revolutionary 25 MeV proton synchrotron with those at Fermilab's 1000 MeV Tevatron.

We calculate the proton speeds using relativistic energy equations.

(The particle has two contributions to its total energy E, rest energy E_o and energy W imparted by the accelerator.)

$$E = E_o + W$$

since $E = \gamma E_o$ we can substitute for γ to find v in terms of total energy.

$$\gamma E_o = E$$

$$\frac{1}{\left(1-\frac{v^2}{c^2}\right)} = \left(\frac{E}{E_o}\right)^2$$

with a little algebra:

$$v = c\sqrt{1-\left(\frac{E_o}{E}\right)^2} \qquad 2.16$$

or:

$$v = c\sqrt{1-\left(\frac{E_o}{E_o+W}\right)^2} \qquad 2.17$$

The proton rest mass is equivalent to about 940 MeV or 1.5×10^{-10} J

At CERN (1959) $W = 25$ GeV $= 4\times10^{-9}$ J giving $v = 0.9993c$

Fermilab (1990) $W = 1000$ GeV $= 1.6\times10^{-7}$ J giving $v = 0.9999995c$

A 40-fold increase in accelerator energy has resulted in a mere 0.7% increase in speed! It would be true to say that in all modern accelerators the speed of the accelerated particles is so close to the speed of light that it is effectively equal to it. A sketch graph of v as a function of accelerator energy E shows (unsurprisingly) that velocity is asymptotic to $v = c$ and that it would require an infinite accelerator energy to reach this asymptote.

However, for a 40-fold increase in particle energy we do achieve a 40-fold increase in mass (since the two are equivalent). The 25 GeV device produces the equivalent of about 27 times the proton rest energy whereas the 1000 MeV device increases the proton mass more than 1000 times. Particle accelerators are used to create new and exotic particles in the laboratory. Some of the energy of the particle collisions is transferred to the rest masses of new particles while the rest goes to their kinetic energy. In 1977 Lederman *et al.* at Fermilab discovered a massive upsilon meson in the debris of a proton collision experiment. The upsilon has a rest mass around 10 times that of the original protons. (The upsilon particle is a meson, consisting of a bound quark/anti-quark pair, and was the first example of a particle containing the 'bottom quark'.)

It might appear that the Tevatron produces much more energy than would be needed to create the upsilon. After all, the available energy amounts to about 100 times that required for the particle itself. But there are two other factors to take into account -

- conservation laws dictate that other particles will also be produced in the collision
- since the incident proton has a great deal of momentum, which is also a conserved quantity, the product particles will not be at rest. Some of the incident energy necessarily ends up as kinetic energy of the products.

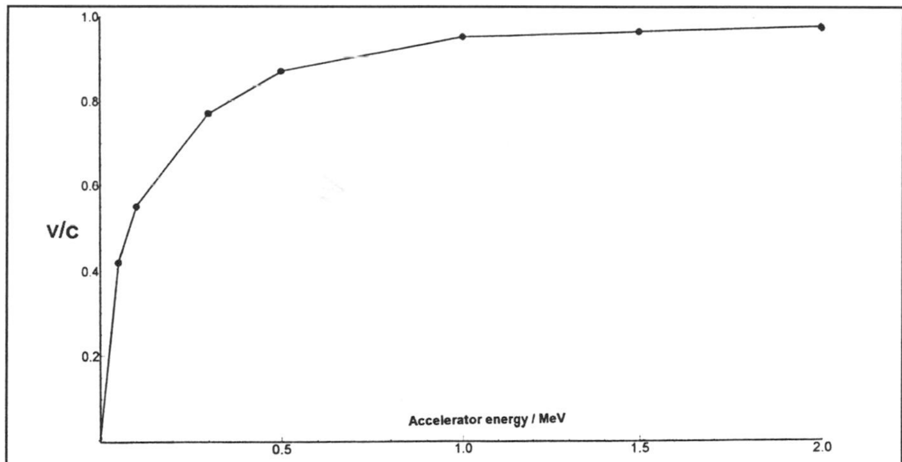

fig 2.42 Variation of $\beta = v/c$ with accelerator energy for an electron. The rest energy of an electron is about 0.51 MeV. The energy plotted on the x-axis here is the kinetic energy, that is eV, where V is the accelerating voltage.

(If our accelerator is used to collide particles of equal mass travelling at equal speeds in opposite directions, for example protons and antiprotons, then the momentum in the laboratory frame is zero and we can obtain the maximum transfer of energy to mass. This is much more efficient for producing particles than firing a particle beam into a stationary target.)

2.15.2 CERN LEP and ALEPH

Like all particle accelerators, the large electron-positron collider (LEP) at CERN is designed to create new particles. Up to mid 1996 the collision energy was 'tuned' to the Z^o resonance close to 100 GeV, an energy at which these particles are readily created. This was important for checking details of the Standard Model in particle physics and in establishing that here are only three generations of quarks and leptons. However, during 1996 LEP was upgraded to nearly 200 GeV in order to produce $W^+ W^-$ pairs. Short pulses of electrons and positrons make about 10 000 circuits of the 27 km ring every second and are brought together inside large detectors by operating electrostatic switches which deflect the beams into a collision course. Energy is supplied to the beam in radio frequency cavities tuned to supply a 'kick' to the particles in the direction they are travelling as they pass through the cavity. This is important since the particles continually lose energy by emitting synchrotron radiation (because they are accelerating in circular motion). They are made to follow the near circular path by a large number of superconducting magnets placed around the ring which create a field perpendicular to the particle motion and exert a 'motor effect' force on the moving charges. The annihilation takes place inside large layered detectors at four points around the ring, ALEPH, OPAL, DELPHI and L3.

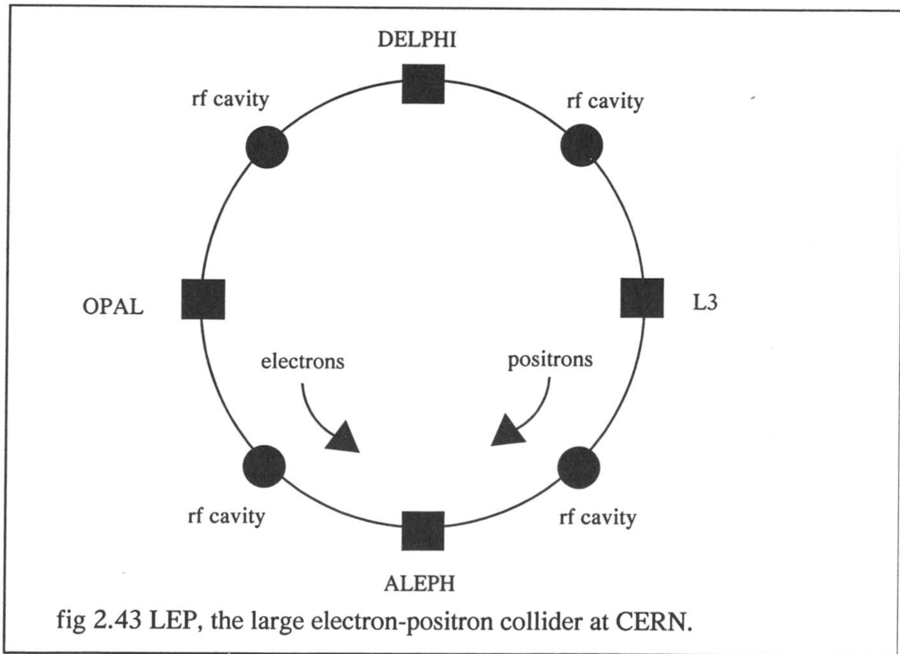

fig 2.43 LEP, the large electron-positron collider at CERN.

The collision and annihilation of high energy electrons and positrons in the LEP collider at CERN transfers most of their total energy (which is mainly kinetic) to the energy of new particles. In order to interpret what has happened in each collision event highly complex and sophisticated layered electronic detectors are used. Figure 2.44 shows a cut-away section through the ALEPH detector. The two beams are kept apart from each other by electrostatic separators and brought together by switching these off. Annihilation takes place in the evacuated beam tube and the particles created leave characteristic signals inside the various layers of the detector. These signals are monitored by computers which are programmed to record any significant events. A photograph of ALEPH and some characteristic results are shown in plate 2.2. The tracks that can be seen in the pictures are reconstructed by computer.

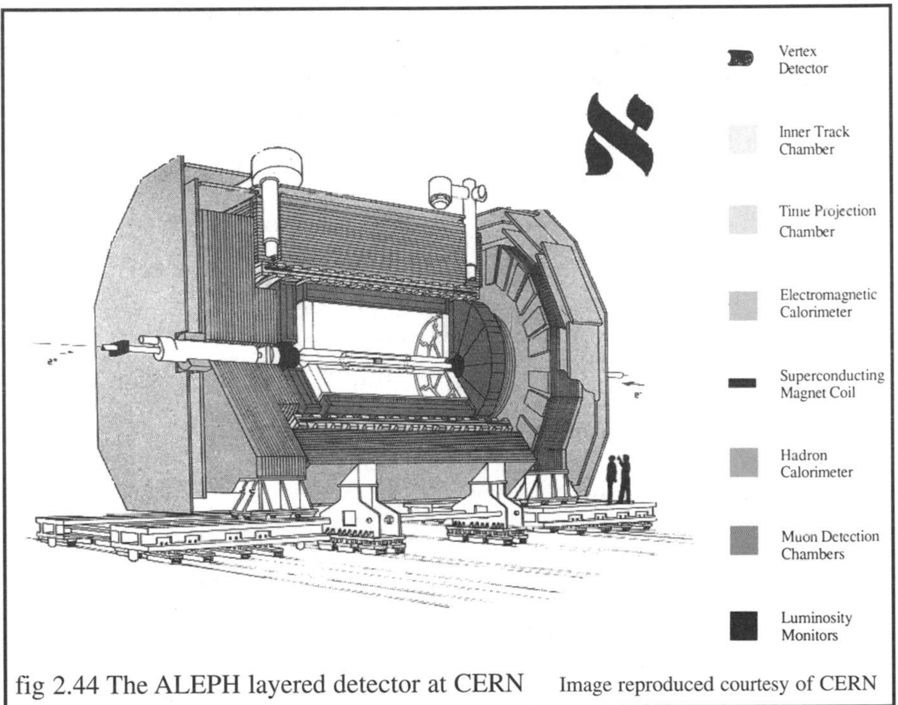

fig 2.44 The ALEPH layered detector at CERN Image reproduced courtesy of CERN

The main layers in ALEPH are:

- **Vertex detector** - this is a small semiconductor detector that surrounds the beam tube close to the collision point. It is used to detect very short-lived particles that decay before they reach the outer layers and must resolve particle tracks that are only a fraction of a millimetre long. It is called a vertex detector because the creation and decay of a short-lived particle are marked by a sudden change in direction and branching of recorded tracks. The separation of vertices in the laboratory may be considerably extended because of time dilation if the particle concerned is moving at close to the speed of light. The inner track chamber also monitors particle tracks close to the collision.

- **Time projection chamber** - this is a huge cylindrical detector through which the ions created by charged particles drift toward electrodes at either end of the chamber. On arrival they create tiny electric pulses. The position of ionization can then be calculated from their time of arrival and the position of the electrode they hit, and the track of the ionizing particle can be reconstructed.
- **Superconducting magnets** - create an axial magnetic field, so the curvature of particle tracks perpendicular to the field, which can be measured using the TPC gives information about the momentum of the charged particles created.
- **Electromagnetic calorimeter** - this traps and measures the energy carried by electrons and photons.
- **Hadron calorimeter** - this stops particles like neutrons, protons, pions and kaons and often incorporates the iron that intensifies the magnetic field.
- **Muon calorimeter** - The only particles that are likely to penetrate all the other layers are muons or neutrinos. The final layer detects the charge on the muons (which are like heavy electrons). The neutrinos that do escape can only be inferred from the difference between the total energy detected and the known energy of the collision.

The various ALEPH results illustrated in plate 2.4 are summarized below. In each case the view is along the axis of the beam tube parallel to the magnetic field (of about 1.5 T). The large black space in every picture is the TPC, the colour of the other layers varies slightly. The bars on the outer calorimeters are measures of the energy detected there, but these need careful interpretation and are not all to the same scale. Events (a) (b) and (c) all took place before the accelerator upgrade in 1996. Event (d) was one of the first W^+ W^- events (11 July 1996) following the increase in accelerator energy.

(a) $$e^+ + e^- \rightarrow Z^o \rightarrow e^+ + e^-$$

This is called a 'Bhabha' event. The electron and positron have annihilated to form a Z^o. The Z^o lifetime is only 10^{-25} s and it has decayed to another e^+ e^- pair. These have such high energy and momentum that their paths appear straight in the TPC. The tracks end at the electromagnetic calorimeters, confirming that no hadrons or muons have been produced.

(b) There is no rule that says the energy released in a collision event must go into a particular group of particles. As long as the conservation laws are obeyed a wide range of outcomes are possible. Here we have the annihilation of an electron and positron to form a quark-antiquark pair:

$$e^+ + e^- \rightarrow q + \overline{q}$$

However, quarks cannot exist as separate particles and their energy 'condenses' into two 'jets' of particles moving off in opposite directions. There are significant responses from all the calorimeters indicating a variety of particle types and energies. Notice the isolated 'V' at about '2-o-clock'. This is 'pair conversion' - a gamma ray photon from the collision has interacted with a molecule in the low-pressure argon-methane gas in the TPC to create an electron-positron pair. These then move apart on oppositely curved paths in the magnetic field. This conversion cannot happen in isolation since it is impossible then to conserve momentum. The original event has created a large number of lower energy particles which is evident from the variety of significantly curved tracks inside the TPC.

(c) This is a clear-cut '3-jet' event. Once again the electron and positron have created a quark-antiquark pair, but this time one of the pair has radiated a gluon (a carrier of the colour force between quarks) and jets have formed around the quark, the antiquark and the gluon. In this picture the more energetic particles are shown in yellow.

(d) This is an event at 161 GeV, just over the W^+ W^- threshold. The main event is the creation of a W^+ W^- pair as the electron and positron collide, but since both W^- particles have lifetimes of about 10^{-25} s they too decay. In this case both have decayed to quark-antiquark pairs and each quark or antiquark has formed a separate jet. The jets are distinguished by color in the picture. It is worth remembering that the rest mass of a single W^+ or W^- is about 160 times that of an electron or positron. Pictures like these, that show cascading particles emerging from a simple collision of high energy but low rest mass incident particles, are a wonderful confirmation of Einstein's mass-energy relation.

2.16 THE TWIN PARADOX, A FIRST LOOK.

2.16.1 Introduction.

If relativity leads to a *genuine* paradox or contradiction in its predictions for observable physical phenomena then we should have to abandon or modify the theory. Time dilation is very strange, but it is not paradoxical because its predictions are completely consistent within any single reference frame. However, we could ask what would happen if clocks which are in relative motion could be brought to rest and then compared. Observers moving with each clock will expect that the other clock runs slow, but when the comparison is made there must be a definite result: either one lags the other or they both read the same time. This is a case in which it might seem that relativity leads to contradictory expectations depending on which of the two observers we choose to believe. If both are equally justified in applying their own arguments to the situation then we have no way to decide between them.

This is the idea behind the so-called 'Twin Paradox' of special relativity, which goes something like this. A(ngela) and B(renda) are twins. On their twenty-first birthday B boards a high speed rocket for a round trip to a distant star. Apart from relatively short periods of acceleration and deceleration at Earth and the star, B maintains a constant velocity of v (which is close to c) relative to her twin A who remains on Earth. To make the argument quantitative we shall assume that the star is twenty light years from Earth and that the velocity v is so high that it produces a time dilation factor of 10 (that is $\gamma = 10$).

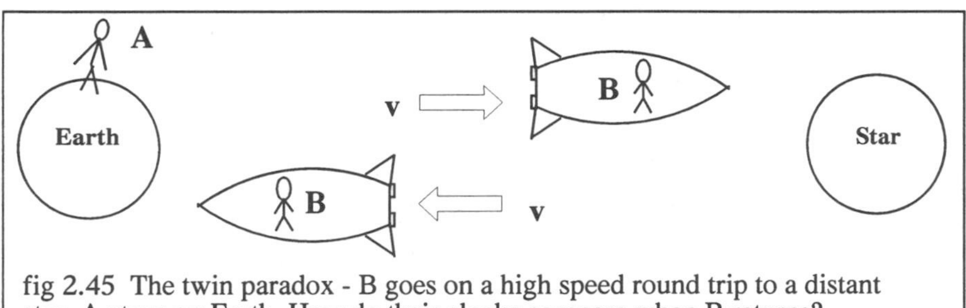

fig 2.45 The twin paradox - B goes on a high speed round trip to a distant star. A stays on Earth. How do their clocks compare when B returns?

2.16.2 'Symmetric' Points Of View.

From A's point of view the round trip is 40 light years and so will take B a little over 40 years to complete (since v is very close to c). A expects B to be 61 years old on her return to Earth when the twins reunite. However, A sees B's time run slowly during the journey (in fact at one tenth of the rate that time passes for A). A therefore calculates that B will have aged by only 4 years during the entire journey. The biological ages of the twins at their reunion should be, according to A, 61 years (for A) and 25 years for B. If the journey had been much longer B might well have returned to find her sister long dead. This peculiar conclusion (which is not in itself a contradiction) is reached by applying the consequences of relativity theory from A's point of view. It is strange but it is not a paradox. The apparent paradox comes about when we try to analyse events from B's point of view. By 'symmetry' we could presumably argue that B remains at rest and the Earth, A, and the star complete a round trip in the opposite direction.

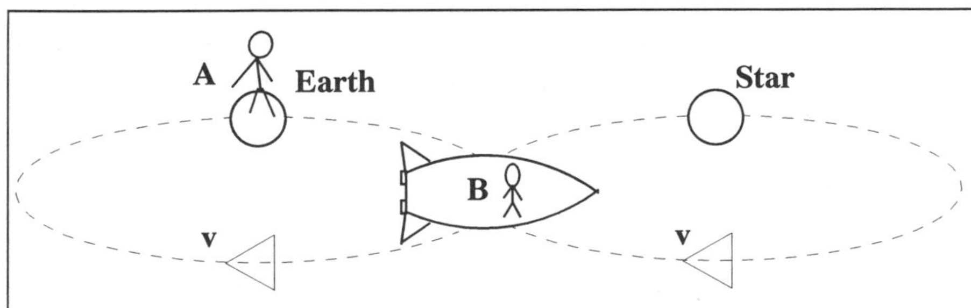

fig 2.46 From B's point of view the Earth moves away as the star approaches and then motions of Earth and star reverse to bring A and B together again.

From B's point of view A's clocks and time run slow and, at the same relative velocity, A will spend a little over 40 of B's years travelling but will (because of time dilation) only age about 4 years. B will therefore expect the age difference to be the opposite to A's prediction at the reunion, that is B will be 61 and A only 25. This is an apparent paradox because it will be obvious, when they reunite, whether their respective ages are 61 and 25 or 25 and 61 respectively. Even if this was not obvious clocks travelling with A and B could be compared instead. So, is relativity wrong? How can this argument be resolved?

2.16.3 An Essential Asymmetry.

Observations of the half lives of moving particles seem to suggest that time will slow for twin B, the twin who leaves the Earth and whose motion has to be arrested in order to make the comparison. If this is actually the case and relativity is to survive there must be some basic error in the way we have applied the theory to the twin paradox. Our basic assumption was that it is equally valid to analyse the motion from B's reference frame as from A's. We referred to the 'symmetry' of the two motions, but are they *really* symmetric?

If we compare what A and B experience during the round trip we see that there is a clear difference. While A stays at home on Earth B goes through three periods of rapid acceleration or deceleration, at the start, middle and end of the journey. During these periods B is in a *non-inertial* reference frame and the laws of special relativity do not apply. Furthermore, B moves in no fewer than four inertial reference frames during the journey: at first she is at rest with respect to Earth, then at a constant velocity moving away from it, then at a constant velocity returning towards it, and finally once again at rest on Earth. Whilst it is true that we can apply an argument from symmetry between A and B when B is in *any one* of these frames, there is no single inertial reference frame at rest with respect to B from which we can analyse the *entire* journey. Special relativity does not tell us how to deal with the transition from one inertial frame to another.

A's argument is a valid application of relativistic principles since A does occupy a single inertial reference frame throughout the entire process. Therefore A will have aged more than B and their ages at the reunion will be 61 and 25 respectively. Somehow the short periods of acceleration experienced by B must be responsible for the time difference which develops between the two twins.

It is possible to produce a version of the twin paradox in which the two travellers age exactly the same amount. Imagine the triplets, Albert, Bill and Charlie. On their twenty-first birthday Albert and Bill set off in identical high speed rockets to visit a pair of stars which are both twenty light years from Earth in opposite directions. They too travel so fast that the time dilation factor is 10 and reunite on Earth with Charlie at the end of their journeys. For A and B the motions are now completely symmetric and so they have aged exactly the same amount (a little over 4 years). Charlie, of course, is in the same situation as the stay-at-home twin in the original example and has aged more than 40 years.

2.16.4 Mr Tompkins And The Brakeman.

In George Gamow's popular and entertaining book 'Mr Tompkins in Wonderland' a bemused bank clerk with an interest in modern science find himself in a world where relativistic effects are hugely exaggerated since the velocity of light there is comparable to the velocity of ordinary human movements. He is amazed one day to see a man, obviously in his forties, alight from a train to be met by a little grey haired old lady who addresses him as 'Dear Grandfather'. When Mr Tompkins asks the man about this strange situation the man's answer shows how counter-intuitive these relativistic effects are:

'.... the thing is really quite simple. My business requires me to travel quite a lot, and, as I spend most of my life in the train, I naturally grow old much more slowly than my relatives living in the city. I am so glad that I came back in time to see my dear old granddaughter still alive!'

This doesn't really help Mr Tompkins so, in a last attempt to make sense of it all, he turns to:

'....a solitary man in railway uniform sitting in the buffet.
"Will you be so kind, sir," he began, "will you be good enough to tell me who is responsible for the fact that the passengers in the train grow old so much more slowly than the people staying at one place?"
"I am responsible for it," said the man, very simply.
"Oh!" exclaimed Mr Tompkins. "So you have solved the problem of the Philosopher's Stone of the ancient alchemists. You should be quite a famous man in the medical world. Do you occupy the chair of medicine here?"
"No," answered the man, being quite taken aback by this, "I am just a brakeman on this railway."
"Brakeman! You mean a brakeman....," exclaimed Mr Tompkins, losing all the ground under him. "You mean you - just put the brakes on when the train comes to the station?"
"Yes, that's what I do: and every time the train gets slowed down, the passengers gain in their age relative to other people. Of course," he added modestly, "the engine driver who accelerates the train does his part in the job."
"But what has it got to do with staying young?" asked Mr Tompkins in great surprise.
"Well, I don't know exactly," said the brakeman, "but it is so. When I asked a university professor travelling in my train once, how it comes about, he started a very long and incomprehensible speech about it, and finally said that it is something similar to 'gravitational redshift' - I think he called it - on the sun. Have you heard anything about such things as redshifts?"
"No-o," said Mr Tompkins, a little doubtfully; and the brakeman went away shaking his head."

2.16.5 A Walk In The Park.

Many of the strange effects with time in relativity seem more acceptable if they can be related to a spatial analogy. Imagine walking from one tree to another in a park. You could choose to walk in a straight line path which takes you directly from one to the other or you might take a triangular path via a third tree. Let's say you go directly and a friend goes via the third tree. When you reunite you will both have the same displacement from your starting position, but your friend has obviously walked farther. The corners he turned as he left you, at the middle tree and when he rejoins you and you walk on together, are in a sense 'responsible' for the different distances you cover. In a similar way, as twin B accelerates and decelerates, she 'turns a corner in space and time' so that when the twins reunite she has experienced a different distance and time from A. (Notice here that the twin 'turning the corners' experiences *less* time than the twin who does not - a significant clue to the geometric properties of space and time!).

The next chapter will deal with a geometric interpretation of 'spacetime' and we will return to the twin paradox there. The gravitational redshift is discussed in chapter 4.

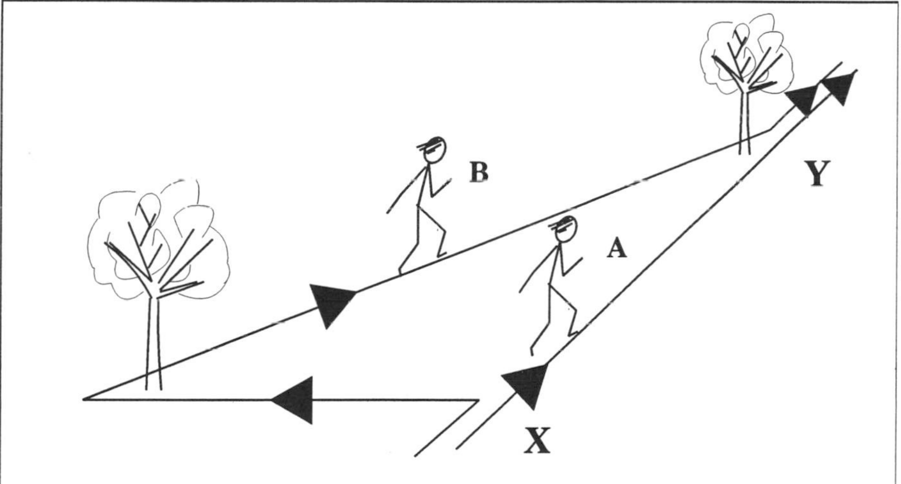

fig 2.47 A and B separate at X and reunite at Y. They achieve the same displacement but cover different distances on different paths through space.

2.17 THE LORENTZ TRANSFORMATION.

2.17.1 Introduction.

Many texts on relativity begin with the Lorentz transformations and proceed to derive the consequences of relativity from them. We have chosen a different route, emphasizing the effects of relative motion on measurements in different reference frames. However, the Lorentz transformation equations are essential to the further mathematical development of the theory so we shall end this chapter by deriving them and discussing how they can be used.

We have seen that observers in relative motion disagree about the duration and separation of events. The Lorentz transformation equations convert the co-ordinates of an event in one reference frame into the co-ordinates of the same event relative to another reference frame in uniform motion. Here the term 'event' is taken to mean a set of co-ordinates specifying both position (x,y,z) and time (t). For simplicity we shall assume that the origins of both reference frames coincide at $t = 0$ (that is clocks at the origin of both frames read zero as they pass one another). We shall also assume that spatial axes have been chosen so that the relative motion occurs only along the x-axis (and hence along the x'-axis). There will therefore be no motion along y, y', z, or z' axes and so $y = y'$ and $z = z'$

2.17.2 The Galilean Transformation.

To derive the Lorentz transformation we first consider the simpler Galilean transformation of classical physics. In this case time proceeds at the same rate in both reference frames and so $t = t'$.

Let an event occur at (x,y,z,t) in frame A. If frame B passes in the positive x direction at velocity v then the co-ordinates of this event in B will be reduced by the distance B travels in the x direction in time t, that is vt. The transformation is therefore:

$$
\begin{aligned}
x' &= x - vt \qquad & x &= x' + vt' \\
y' &= y & y &= y' \\
z' &= z & z &= z' \\
t' &= t & t &= t'
\end{aligned}
\tag{2.18}
$$

2.17.3 The Lorentz Transformation, Derivation 1.

In this first derivation we shall use some of the results we derived previously to calculate how the x and t co-ordinates of an event in frame A transform to frame B that moves past at velocity v in the positive x direction.

At $t = t' = 0$ the origins of A and B coincide. Some time t later in A a flash of light is emitted from a point distance x from the origin along the x-axis. According to A, B will now have moved a distance vt in that direction so that the event occurs at a distance $x - vt$ from B. This *will not* be the x' coordinate measured in B's frame of reference (remember the effect of length contraction). To obtain the transformation we must consider the position of B relative to A at the moment (in A) that the flash is emitted.

ACCORDING TO A: Time of flash is t
Position of B along x-axis is $x = vt$
Distance of B from flash when flash occurs is $x - vt$
This distance is length contracted, since B is in motion so distance of flash from B's origin will be $\gamma (x - vt)$

giving $$x' = \gamma (x - vt) \qquad (1)$$

ACCORDING TO B: Time of flash is t'
Position of A along x'-axis is $x' = -vt'$
Distance of A from flash when flash occurs is $x' + vt'$
B sees this distance length contracted so distance of flash from A's origin will be $\gamma (x' + vt')$

giving $$x = \gamma (x' + vt') \qquad (2)$$

This gives transformations for x and x'. We now have to derive the transformations for t and t'. These can be derived directly from (1) and (2).

From (2): $$t' = \frac{x}{\gamma v v} - \frac{x'}{v}$$

now substitute from equation (1) for x' to obtain t' in terms of A's co-ordinates x and t alone.

This gives: $$t' = \gamma\left(t - \frac{xv}{c^2}\right)$$

A similar manipulation gives: $$t = \gamma\left(t' + \frac{x'v}{c^2}\right)$$

The complete Lorentz transformation is therefore:

$$\begin{aligned} x' &= \gamma (x - vt) & x &= \gamma (x' + vt') \\ y' &= y & y &= y' \\ z' &= z & z &= z' \\ t' &= \gamma\left(t - \frac{xv}{c^2}\right) & t &= \gamma\left(t' + \frac{x'v}{c^2}\right) \end{aligned} \qquad (2.19)$$

2.17.4 The Lorentz Transformation, Derivation 2.

If we are to consider the Lorentz transformation as a starting point for many relativistic effects we ought to be able to derive it on the basis of just a few assumptions from the principle of relativity itself. This is in fact the case and in this section we shall do that. The approach we shall take is to look for a simple linear transformation of A's co-ordinates into B's. By 'linear' we mean that x' will depend on a sum of terms involving x and t but not powers of these or more complicated functions of one or both (x will depend in a symmetric way on x' and t'). We can also be guided in the knowledge that, at low velocities ($v<<c$) the Lorentz transformation must reduce to the Galilean transformation.

For these reasons we shall assume the transformation has the form:

$$x' = ax - bt \qquad (3)$$
$$x = ax' + bt' \qquad (4)$$

and proceed to determine the unknown coefficients a and b.

If $x = 0$ in equation (4) then the relation between x' and t' must be the co-ordinates of the origin of A as seen in B. This gives us:

$$\frac{dx'}{dt'} = -\frac{b}{a}$$

This is clearly a velocity and must be the velocity of A's origin in B's frame as it moves in the negative x' direction. If we put $x' = 0$ in equation (1) this gives us the trajectory of B's origin in A and leads to:

$$\frac{dx}{dt} = +\frac{b}{a}$$

therefore
$$\frac{b}{a} = v \qquad (5)$$

Now consider a light pulse that leaves the origin of both reference frames at the instant ($t = t' = 0$) when they coincide. Light is used since we know this will have the same velocity, c, relative to both observers. The x and x' co-ordinates of the light pulse will be:

In A: $x = ct$

In B: $x' = ct'$

By substituting for x' in equation (4) and rearranging we obtain:

$$a^2 - \frac{b^2}{c^2} = 1 \qquad (6)$$

Equations (5) and (6) can be used to derive expressions for A and B:

$$a = \frac{1}{\sqrt{1 - \frac{v^2}{c^2}}} \qquad \text{or} \qquad \underline{a = \gamma}$$

$$\underline{b = \gamma v}$$

These can now be substituted back into equations (3) and (4) to obtain the Lorentz transformations for x and x' previously derived. From these we can also derive the transformations for t and t' as in the last section.

2.17.5 Comments.

- Starting from these equations we can derive time dilation, the relativity of simultaneity, length contraction, velocity addition, etc....
- The co-ordinates (x, y, z, t) and (x', y', z', t') are alternative representations of the location of an event in 'spacetime'. The co-ordinates are described as 'covariant'. This means that the co-ordinates differ from one reference frame to another but that the law connecting the co-ordinates is the same and is independent of the event considered.
- The transformation itself can be regarded as equivalent to a rotation of a four dimensional vector in spacetime, an approach that will be developed in some detail in the next chapter.
- The equations above describe the transformation of spacetime co-ordinates but the same transformation applies for other physical quantities like momentum (whose three co-ordinates transform like the x, y and z co-ordinates) and energy (which transforms like time t).
- It is noteworthy that the Lorentz transformation equations were introduced by Lorentz in 1904, a year before Einstein independently discovered them. Lorentz was attempting to modify classical electromagnetism in a way that could both explain the Michelson-Morley experiment and retain the notion of an absolute reference frame in the ether.

Example: Rocket B passes rocket A at velocity $v = 0.7c$ in the direction of A's x-axis which is parallel to B's x'-axis. They synchronize clocks at their origins to read $t = t' = 0$ as they pass. A later observes two events which occur at the same place in his reference frame but at times $t_1 = 100$ s and $t_2 = 120$ s respectively. The co-ordinates of the two events are:

$$x_1 = x_2 = 5 \times 10^{10}\ \text{m} \qquad y_1 = y_2 = 10^{10}\ \text{m} \qquad z_1 = z_2 = 0\ \text{m}$$

(a) What are the co-ordinates of these events in the frame of rocket B?
(b) What is the time and distance between the events in each frame?

(a) Use the Lorentz transformation (primes refer to B's frame).

$$\gamma = \frac{1}{\sqrt{1-\beta^2}} = 1.40$$

$$x_1' = \gamma\left(x_1 - vt_1\right) = 4.1\times10^{10}\ \text{m} \qquad x_2' = \gamma\left(x_2 - vt_2\right) = 3.5\times10^{10}\ \text{m}$$

$$y_1' = y_1 = 10^{10}\ \text{m} \qquad y_2' = y_2 = 10^{10}\ \text{m}$$

$$z_1' = z_1 = 0 \qquad z_2' = z_2 = 0$$

$$t_1' = \gamma\left(t_1 - \frac{vx_1}{c^2}\right) = \underline{-23\ \text{s}} \qquad t_2' = \gamma\left(t_2 - \frac{vx_2}{c^2}\right) = \underline{4.7\ \text{s}}$$

Notice that the time order of some events is changed by relative motion. In B's frame event 1 occurs *before* the moment when A and B pass whereas in A's frame it comes later. Notice also that event 2 occurs at a *different place* from event 1 in B's frame whereas both occur at the same place in A's frame.

(b)

$$\Delta t = t_2 - t_1 = 20\ \text{s} \qquad \Delta t' = t_2' - t_1' = 28\ \text{s}$$

$$\Delta x = x_2 - x_1 = 0\ \text{m} \qquad \Delta x' = x_2' - x_1' = -6\times10^9\ \text{m}$$

(Negative sign is because second event occurs closer to B's origin than the first.)

Example: Two events occur on the x-axis in frame S at co-ordinates (x_1, t_1) and (x_2, t_2) respectively. Frame S' moves at velocity v along the positive x-axis such that $x = x' = 0$ at $t = t' = 0$ and the x'-axis is parallel to the x-axis. Calculate a value of v for which the two events will be simultaneous in S' and state the conditions under which this could be possible.

To be simultaneous in S', t_1' and t_2' for the two events in S' must be the same.

$$t_1' = \gamma\left(t_1 - \frac{vx_1}{c^2}\right) \qquad t_2' = \gamma\left(t_2 - \frac{vx_2}{c^2}\right)$$

$$t_2' - t_1' = \gamma\left(t_2 - \frac{vx_2}{c^2} - t_1 + \frac{vx_1}{c^2}\right) = 0$$

$$t_2 - t_1 = \frac{v\left(x_1 - x_2\right)}{c^2}$$

$$v = \frac{c^2\left(t_2 - t_1\right)}{\left(x_2 - x_1\right)}$$

However, v must be less than or equal to c so:

$$\frac{c^2\left(t_2 - t_1\right)}{\left(x_2 - x_1\right)} \geq c \qquad \underline{\left(x_2 - x_1\right) \leq c\left(t_2 - t_1\right)}$$

That is, this will be possible so long as the spatial separation exceeds the distance light could travel in the time between the two events.

2.18 USING THE LORENTZ TRANSFORMATION.

2.18.1 Time Dilation.

As usual A and B are inertial reference frames moving with a relative velocity v. Identical clocks are placed at the origins of A and B. How do the clock rates compare when viewed from either reference frame? To find out we use the Lorentz transformation to convert time co-ordinates (that is clock readings) from one frame into time co-ordinates for the other frame. We will assume that $x = x' = 0$ when $t = t' = 0$ and that all motion is along the positive x- and x'-axes.

ACCORDING TO A: $$t = \gamma\left(t' + \frac{vx'}{c^2}\right)$$

But B's clock is at $x' = 0$ (the origin) so: $$t' = \gamma v$$

The reading on A's clock is greater than on B's. B's clock, which is moving with respect to A, runs slow. This is the time dilation formula we derived previously.

ACCORDING TO B: $$t' = \gamma\left(t - \frac{vx}{c^2}\right)$$

but A's clock is at $x = 0$ so $$t' = \gamma v$$

B's clock always shows a greater time than A's. A's clock runs slow.

The Lorentz transformation leads to the same symmetrical results we derived previously.

2.18.2 Length Contraction.

Now consider a measuring rod at rest in A's frame lying along the x-axis so that its ends have co-ordinates:

$$x_1 = 0 \qquad (l_o \text{ is the rest length of the rod})$$
$$x_2 = l_o$$

B measures this rod as she passes A. To do this she has to mark the positions of the rod's ends in her own frame at the same instant (of her time). Let us assume for the sake of simplicity that she decides to make this measurement at $t' = 0$, that is the instant she passes A's origin. To find the rod's measured length we need to transform the co-ordinates from A's frame to B's and then find their difference.

$$x'_1 = 0 \qquad \text{(origins coincide at } t = t' = 0\text{)}$$
$$x'_2 = \gamma\left(x_2 - vt\right) \qquad (1)$$

Now it is tempting to suppose that, since $t = t' = 0$ at $x = x' = 0$, we simply substitute $t' = 0$ to obtain x'_2. This is incorrect. It is wrong because A and B do not agree on what constitute simultaneous events at different points in space. The time at which B measures the far end of the rod is different from the time at which A measures it. They disagree about the moment at which the far end of the rod should be measured. The value of t that must be used in the equation above is the value at which $t' = 0$, which we can obtain from the Lorentz transformation of time:

$$t' = \gamma\left(t - \frac{vx_2}{c^2}\right) = 0 \qquad \text{giving} \qquad t = \frac{vx_2}{c^2}$$

substituting in (1) $x'_2 = \gamma\left(x_2 - \dfrac{v^2 x_2}{c^2}\right) = \gamma x\left(1 - \dfrac{v^2}{c^2}\right) = \dfrac{\gamma x_2}{\gamma^2} = \dfrac{x_2}{\gamma}$

therefore
$$l' = x'_2 - x'_1 = \frac{l_o}{\gamma} \qquad (2.5)$$

This is the familiar length contraction formula. B obtains a value less than that in the rod's rest frame - moving rods contract. This derivation emphasizes the need to take care when doing relativistic calculations. It is all too easy to slip back, albeit unconsciously, into the habit of using an absolute time and assuming that events that are simultaneous in one frame will be simultaneous in all others. In this case the 'synchronization error' at the end of the rod (which manifests itself as a disagreement over when to measure that end's position) is vx/c^2. According to A, B will note the position of the rod at the origin and then wait a time vx/c^2 (on A's clock) before measuring the position of the other end of the rod. During this time B moves along the rod and so A is not surprised that B's measurement indicates a contraction. Of course for B the two measurements are simultaneous.

2.18.3 Velocity Addition.

Consider a projectile moving at velocity u relative to B's reference frame while B moves past A at velocity v. What is the speed w of the projectile measured relative to A? Once again we assume all motion is along the parallel x and x' axes and that the origins coincide at $t = t' = 0$. The x' coordinate (in B's reference frame) of the projectile is then $x' = ut'$ so that $u = x'/t'$. We must use the Lorentz transformation to find how the x coordinate varies with t (in A's reference frame), in other words to find the velocity $w = x/t$ relative to A. (This is mathematically far simpler than the approach adopted in section 2.11.4, although that method emphasizes the way different relativistic effects contribute to the measurement of relative velocities.)

$$w = \frac{x}{t} \quad (1)$$

$$u = \frac{x'}{t'} \quad (2)$$

Now use the Lorentz transformation to express x and t in terms of x' and t'

$$x = \gamma\left(x' + vt'\right) \quad (3)$$

$$t = \gamma\left(t' + \frac{vx'}{c^2}\right) \quad (4)$$

Substitute (3) and (4) into (1) and divide top and bottom by t' to obtain a formula for w:

$$w = \frac{\gamma\left(x' + vt'\right)}{\gamma\left(t' + \frac{vx'}{c^2}\right)} = \frac{\left(x'/t' + v\right)}{\left(1 + vx'/t'\right)}$$

Now use (2):

$$w = \frac{(u + v)}{\left(1 + \frac{uv}{c^2}\right)} \quad (2.7)$$

This is the velocity addition formula.

2.18.4 Transformation Of Velocities.

The derivation of velocity addition given in the previous section is simply a special case of velocity transformation when the relative velocity is parallel to the direction of motion of the moving frame. The general transformation of velocities is a little more awkward to derive and is one of many examples in relativity where algebraic care is rewarded. In particular it is important to distinguish co-ordinates belonging to each of the two reference frames. The transformation is derived in the following way:

- Write down the Lorentz transformation for co-ordinates x, y, z and x', y', z'.
- Differentiate each set with respect to t or t' respectively..

The trick is to obtain dx/dt from the product $(dx/dt') \times (dt'/dt)$ rather than by direct differentiation with respect to t. This is because the transformation equations give x in terms of x' and t'. A similar approach is used for the y- and z- components of velocity in the 'rest' frame and for the x'-, y'-, and z'-components of velocity in the moving frame. Ultimately we are looking for a set of equations that give the velocity components in one frame (for example u_x, u_y, and u_z) solely in terms of velocity components in the other frame (u'_x, u'_y, u'_z). The transformation is derived below.

In the moving frame: $u'_x = \frac{dx'}{dt'} \quad u'_y = \frac{dy'}{dt'} \quad u'_z = \frac{dz'}{dt'}$

In the rest frame: $x = g(x' + vt') \quad y' = y \quad z' = z$

differentiating with respect to t': $$\frac{dx}{dt'} = g\left(\frac{dx'}{dt'} + v\right) = g(u'_x + v) \qquad (1)$$

$$\frac{dy}{dt'} = \frac{dy'}{dt'} = u'_y \qquad (2)$$

$$\frac{dz}{dt'} = \frac{dz'}{dt'} = u'_z \qquad (3)$$

from the Lorentz transformation: $t = g\left(t' + \frac{vx}{c^2}\right)$

$$\frac{dt}{dt'} = g\left(1 + \frac{vu_x}{c^2}\right) \qquad (4)$$

From (1) to (4) $$u_x = \frac{dx}{dt} = \frac{dx}{dt'} \cdot \frac{dt'}{dt} = \frac{(u'_x + v)}{\left(1 + \frac{vu'_x}{c^2}\right)}$$

$$u_y = \frac{dy}{dt} = \frac{dy}{dt'} \cdot \frac{dt'}{dt} = \frac{u'_y}{g\left(1 + \frac{vu'_x}{c^2}\right)}$$

$$u_z = \frac{dz}{dt} = \frac{dz}{dt'} \cdot \frac{dt'}{dt} = \frac{u'_z}{g\left(1 + \frac{vu'_x}{c^2}\right)} \qquad (2.20)$$

similarly, if we are transforming the other way:

$$u'_x = \frac{u_x - v}{\left(1 - \frac{vu_x}{c^2}\right)} \qquad u'_y = \frac{u_y}{\gamma\left(1 - \frac{vu_x}{c^2}\right)} \qquad u'_z = \frac{u_z}{\left(1 - \frac{vu_x}{c^2}\right)} \qquad (2.21)$$

2.18.5 Comments.

- Notice that the expression for u_x is simply the relativistic velocity addition of v and u'_x.
- Note also that, since the velocity transformation is different for different components, the direction of a velocity relative to the reference axes will be different in different frames too. This is the relativistic explanation of stellar aberration.

- It is clear from the velocity addition formula that light propagating in the moving frame at velocity c will have the same velocity c relative to the rest frame. To see this substitute $u'_x = c$ in the equation for u_x above:

$$u_x = \frac{c+c}{\left(1+\frac{c^2}{c^2}\right)} = \frac{2c}{2} = c$$

  This is consistent with the assumption of relativity that light has a constant velocity independent of the motion of the source.
- It must also follow that the magnitude of a velocity vector representing light in any reference frame is: $u_x^2 + u_y^2 + u_z^2 = u'^2_x + u'^2_y + u'^2_z = c^2$
- For a light beam the quantity $u_x^2 + u_y^2 + u_z^2 - c^2 = 0$ is an invariant. In the next chapter we shall see that this represents the magnitude of a spacetime velocity vector.
- The approach used to derive the transformation of velocities can also be used to transform accelerations. The fact that accelerations parallel and perpendicular to the direction of motion have different transformation rules explains why acceleration itself, which is such a central quantity in Newtonian mechanics, is far less significant in relativistic mechanics.

Example The stellar aberration α for a distant star lying along a line perpendicular to the ecliptic is approximately v/c. This was derived previously on the (false) assumption of an all-pervading ether. Show that relativity leads to the same result.

We must transform the velocity vector for the light emitted by the star to the Earth's moving reference frame. Take the direction of the Earth's motion as along an x-axis parallel to the x'-axis of a co-ordinate system at rest relative to the star. The y- and y'-axes are then parallel to the Earth-star separation.

velocity components relative to the star: $u'_x = 0 \qquad u'_y = c$

velocity components relative to Earth:

$$u_x = \frac{u'_x + v}{\left(1+\frac{vu'_x}{c^2}\right)} = v$$

$$u_y = \frac{u'_y}{g\left(1+\frac{vu'_x}{c^2}\right)} = \frac{c}{g}$$

$\tan\alpha = \frac{u_x}{u_y} = \frac{\gamma vv}{c}$ but $v<<c$ and $\gamma<<1$ so $\underline{\alpha \approx \frac{v}{c}}$

2.19 SUMMARY OF IDEAS AND EQUATIONS IN CHAPTER 2.

- ***Galilean relativity:*** The laws of mechanics are the same in all inertial reference frames.
- ***Principle of relativity:*** The laws of physics are the same in all inertial reference frames.
- ***Speed of light:*** Has the same value for all inertial observers.
- ***Covariant quantities:*** Quantities such as co-ordinates that depend on reference frame but transform according to some standard rule - in this case the Lorentz transformation equations.
- ***Invariant quantities:*** Quantities which have the same value in all inertial reference frames, for example the speed of light or the rest mass of a particle.
- ***Proper length etc...*** The proper length is the length of an object measured in its own rest frame. The proper time is the time between two events registered by the same clock fixed at some point in a particular inertial reference frame. Rest mass is the mass of an object measured in its own rest frame.

Time dilation: $t' = \gamma t$ (prime indicates 'moving' frame)

Length contraction: $l' = l/\gamma$ (prime indicates 'moving' frame)

Velocity addition: $w = \dfrac{u+v}{(1 + uv/c^2)}$ (u is velocity relative to 'moving' frame)

Relativistic mass: $m = \gamma m_o$ (m_o is rest mass)

Radiation pressure: $R = I/c$ (radiation intensity I is absorbed)
$R = 2I/c$ (radiation is reflected)

Doppler effect $f = f'\dfrac{(1 \mu b)^{1/2}}{(1 \pm b)^{1/2}}$ (top signs if source moves away)

Photon momentum: $p = E/c$ (E is the photon energy)
Mass-energy relation: $E = mc^2 \quad E_o = m_o c^2$

Relativistic momentum: $p = \gamma m_o v$

Energy-momentum invariant: $E^2 - p^2c^2 = E_o^{\,2}$ or $m^2c^4 - p^2c^2 = m_o^{\,2}c^4$

Kinetic energy: $KE = E - E_o = (\gamma - 1)E_o$

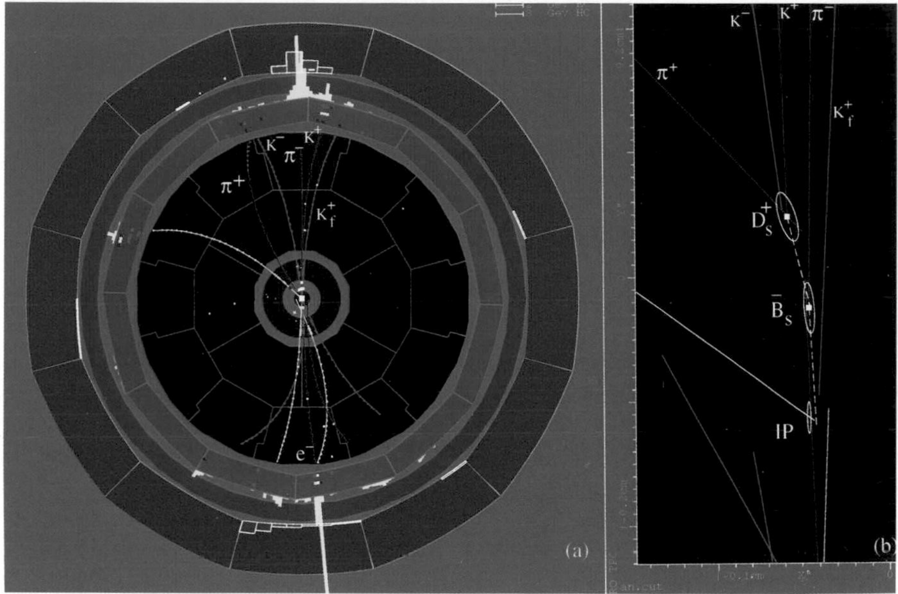

plate 2.1 Top left is a reconstruction of particle paths following an electron-positron collision in the ALEPH detector at CERN. The vertex detector (bottom) resolves paths of very short-lived particles close to the interaction point (IP). It shows (top right) that the two jets contained B-mesons that decayed rapidly to D-mesons which then decayed to longer lived particles that left tracks in the main part of the detector. The lengths of the B and D meson tracks are about 1 mm even though light could only travel about 0.3 mm in their proper lifetime. This is possible because they are travelling so fast that their lifetime in the laboratory frame is increased by time dilation. Results and photo used courtesy of CERN.

plate 2.2 The photograph at right shows part of the 27 km LEP tunnel which runs underground between Geneva airport and the Jura mountains. Above is the ALEPH cavern, 140 m beneath Echenevex in France. The LEP beam tube passes through the centre of ALEPH which is used to determine, as accurately as possible, the nature, direction and energy of each of the particles created in an electron-positron collision at the centre of the detector. On average about 40 particles are created per collision.

Photos used courtesy of CERN.

plate 2.3 The ALEPH detector. This is an end view of the detector with its end caps removed. The small polished metal circle at the centre is the end of the beam tube. The various layers of detectors can be clearly seen (see section 2.15.2 for more information). The complete detector is about 12 m by 12 m by 12 m and has a mass of about 3000 tonnes. The outputs are monitored on 500 000 separate electronic channels. It cost about 70 million Swiss francs to build. The man standing second left is Jack Steinberger who shared the 1988 Nobel Prize for Physics for his work in neutrino physics.

Photos used courtesy of CERN.

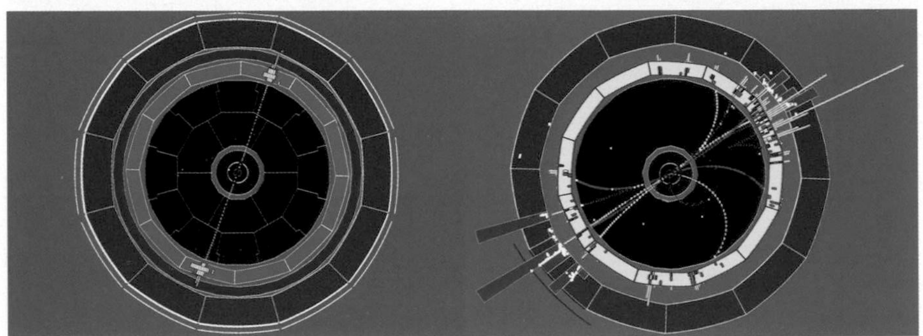

plate 2.4 False-colour, computer generated images showing particle tracks through the ALEPH layered detector at CERN (see section 2.15.2 for more information): (a) (Top left) an electron-positron annihilation creating a Z° particle (unseen) which decays to another electron-positron pair.

(b) (Top right) here the Z° decays to a quark and antiquark pair which immediately form two jets of particles

(c) (Bottom left) similar to (b) but this time one of the quarks radiates a gluon which forms a third jet.

(d) (Bottom right) a higher energy event in which the electron and positron annihilate to form a W^+ W^- pair. Each of the Ws decays to a quark-antiquark pair and each of the quarks generates a jet. This is a 4-jet event.

Results reproduced courtesy of CERN.

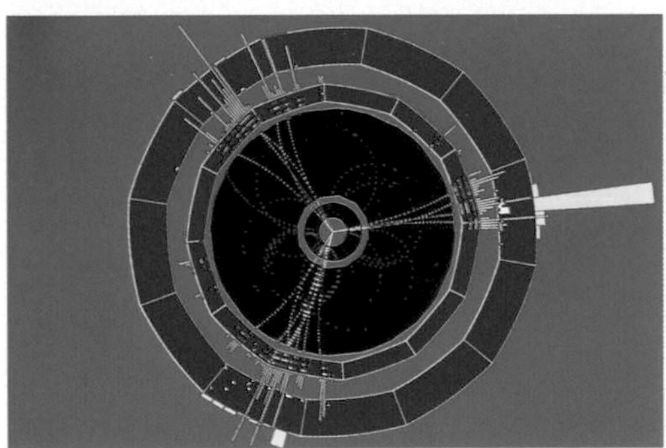

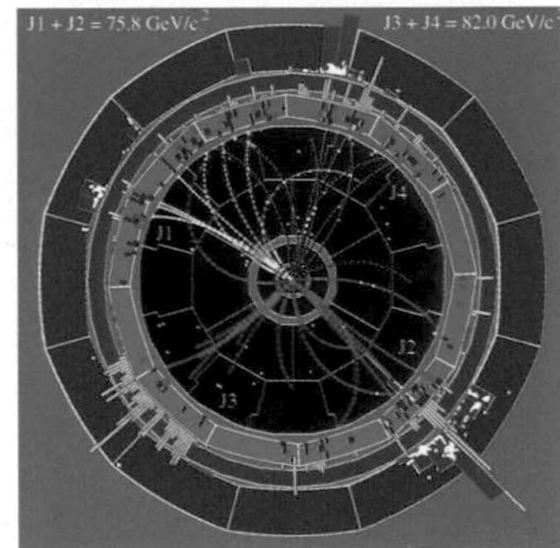

plate 4.1 Supernova 1987A appeared suddenly on 23 february 1987 at all wavelengths from gamma to radio. It was heralded, 20 hours earlier, by a burst of energetic neutrinos several million million of which passed through your body. Light takes a little longer to cover the 170 000 light years to us because the refractive index of space is just more than one. The explosion lit the gas rings surrounding the star making them radiate. The time delay before this was observed allowed astronomers to calculate the size of the rings and use them to make an accurate measurement of the star's distance using parallax and then recalibrate the cosmic distance scale.

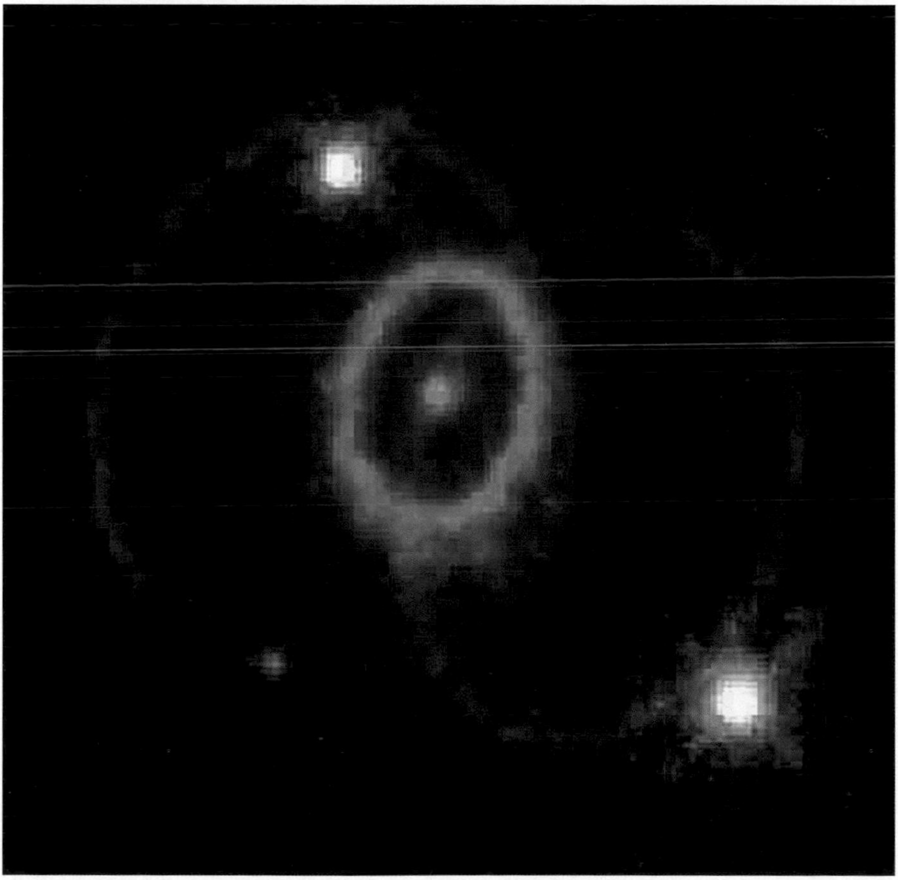

Top photograph copyright © The Royal Observatory Edinburgh/UK Schmidt telescope, credit: D.F.Malin. Bottom photograph courtesy of NASA/STSI.

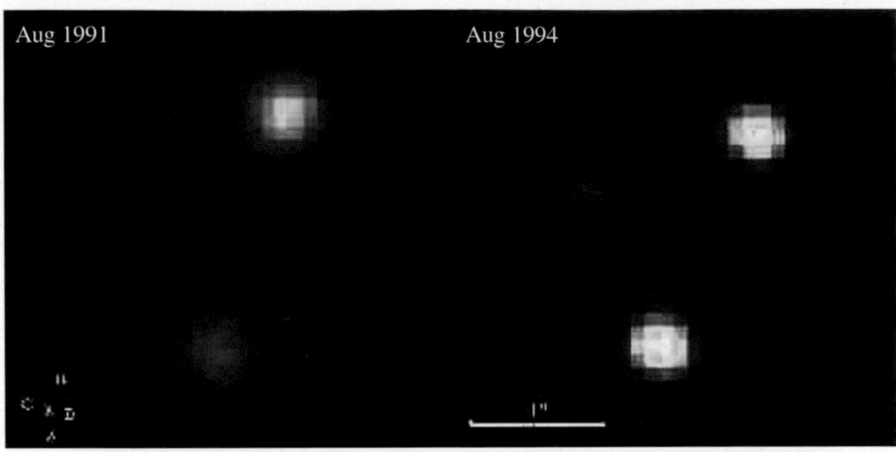

plate 4.2 Gravitational lenses. The top picture is a gravitational microlens forming four separate images of a distant quasar (an Einstein Cross). The variation in relative image intensity is caused by the motion of stars in the foreground galaxy. The two images were taken about 3 years apart. The lower picture shows some cluster galaxies (yellow) acting as gravitational lenses for very distant galaxies and forming multiple images of them (all the blue 'smudges').

Top photograph credit: G. Lewis & M. Irwin, William Herschell Telescope.

Bottom photograph credit: J. Tyson AT&T Bell Laboratories and NASA.

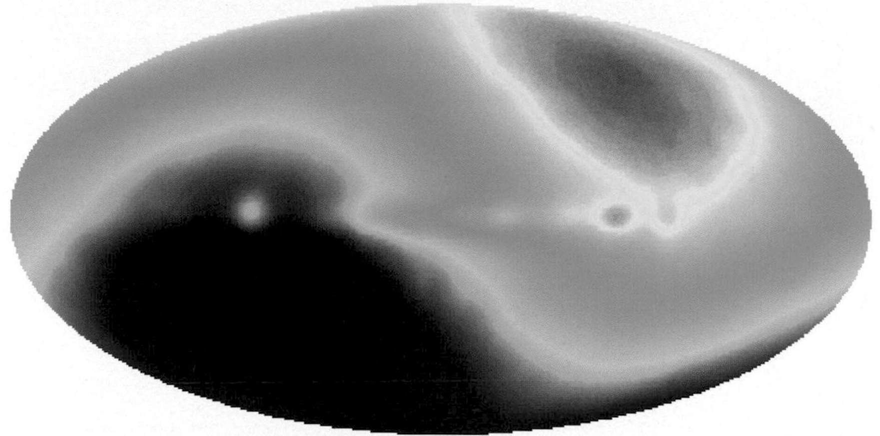

Plate 4.3a This COBE image shows the intensity distribution in the background radiation reaching Earth from the full sky. The yellow-red colours indicate lower than average intensity (red shifts) and the blue-purple colours indicate higher than average intensity (blue shifts). This pattern indicates that the Earth, carried along in our Local Group of galaxies, is moving at about 600 km s^{-1} relative to the background radiation. The effect on intensity is due to the Doppler effect, but the reason for the motion is unexplained.

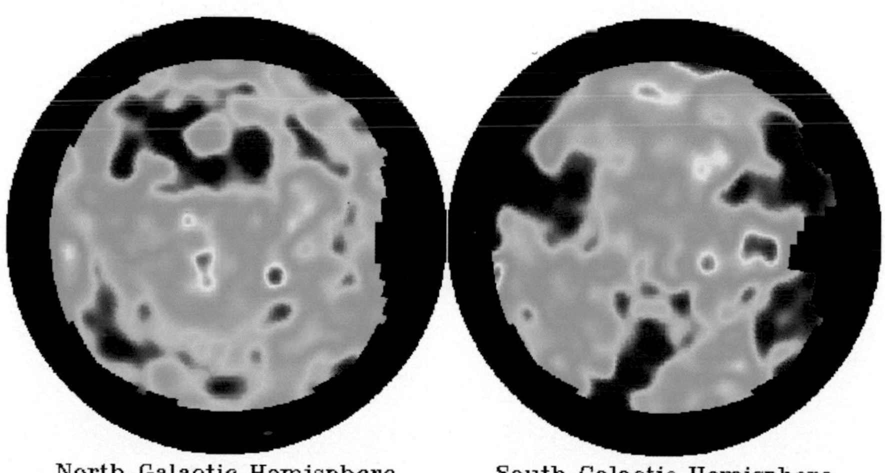

plate 4.3b The oldest structures known. These sky surveys have have been corrected to remove the effects of local motion but still show small variations in intensity of the cosmic background radiation. These 'wrinkles' are thought to correspond to the density variations in the early universe that eventually led to galaxy formation, and are probably the oldest structures humanity will ever see, having formed a million years after the Big Bang.

Images used courtesy of NASA Goddard Flight Centre and COBE Science Working Group.

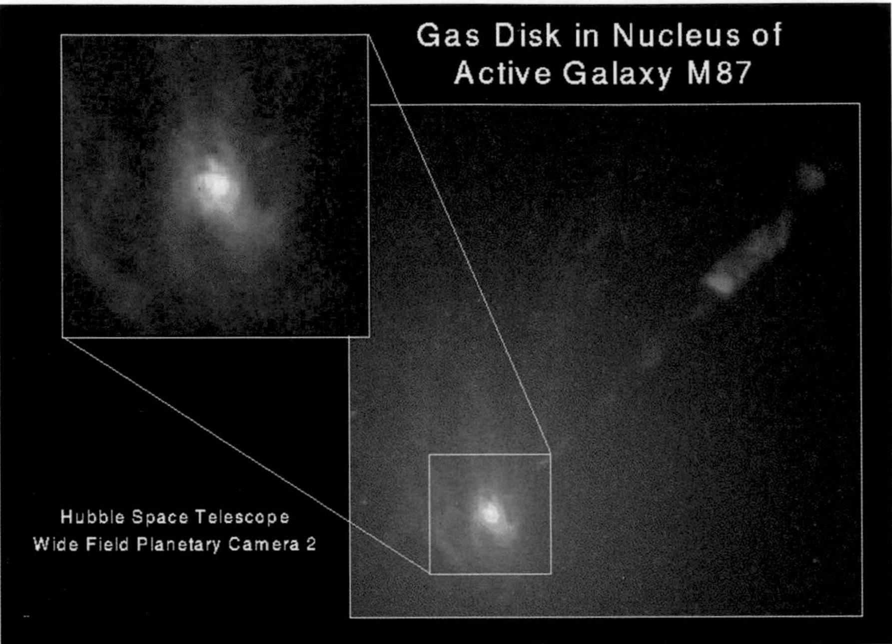

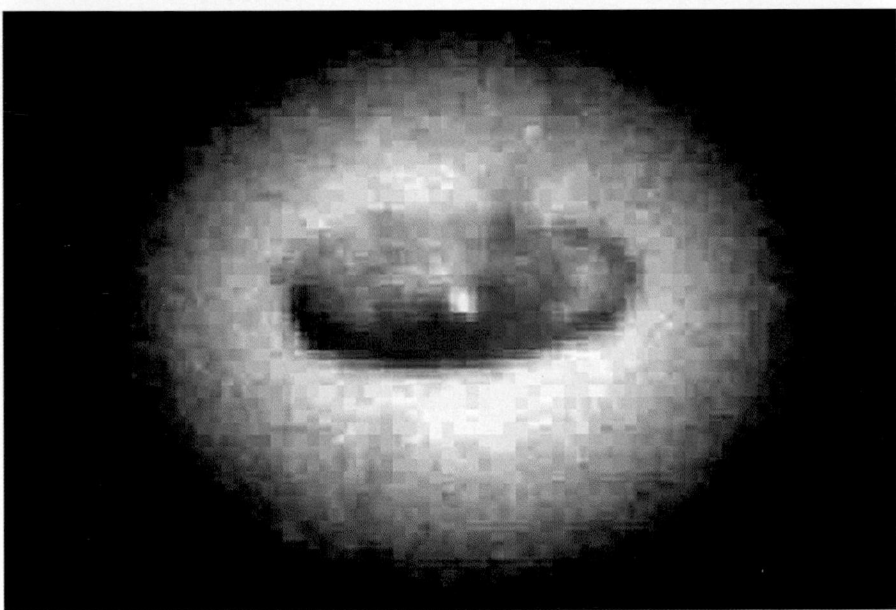

plate 4.4 Two candidates for Black Holes. The gas disc surrounding M87 (top) must be bound by a tiny extremely massive central object. The black hole in NGC4261 (bottom) contains about 1.2 billion times the mass of the Sun and appears to have collided with another galaxy.
Photograph credits: H.Ford & R.Harms, NASA, HST.

Lorentz transformation:

$$x' = \gamma\,(x - vt) \qquad x = \gamma\,(x' + vt')$$
$$y' = y \qquad y = y'$$
$$z' = z \qquad z = z'$$
$$t' = \gamma\,(t - xv/c^2) \qquad t = \gamma\,(t' + x'v/c^2)$$

Velocity transformations:

$$u_x = \frac{(u'_x + v)}{\left(1 + \frac{vu'_x}{c^2}\right)} \quad u_y = \frac{u'_y}{g\left(1 + \frac{vu'_x}{c^2}\right)} \quad u_z = \frac{u'_z}{g\left(1 + \frac{vu'_x}{c^2}\right)}$$

$$u'_x = \frac{u_x - v}{\left(1 - \frac{vu_x}{c^2}\right)} \quad u'_y = \frac{u_y}{g\left(1 - \frac{vu_x}{c^2}\right)} \quad u'_z = \frac{u_z}{g\left(1 - \frac{vu_x}{c^2}\right)}$$

2.20 PROBLEMS

2.1. Sound is a mechanical wave in the air. Light on the other hand requires no material medium.
a. How does the speed of sound relative to an observer in the atmosphere depend on the velocity of sound relative to the air, the velocity of the sound source relative to the air and the velocity of the observer relative to the atmosphere? Compare your answer with the answer given to 2.1a above.
b. How does the speed of light relative to an observer in space depend upon the velocity of light in a vacuum, the velocity of the light source, and the velocity of the observer? (Take care over the interpretation of the last two velocities.)
c. How does the difference between light and sound affect our approach to calculating the Doppler shift in each case?

2.2. Expectations based on the ether were tested in many experiments. One of these involved looking for a shift in the focus of an astronomical telescope pointed at a particular star as the Earth moved around its orbit.
a. To understand why this was expected sketch a ray diagram for a thin lens forming a real image of a distant object (give it a focal length of about 5 cm). Since it takes time for light to travel from the lens to the image the lens will have moved as the image forms. Assuming the angles involved are small and the lens is moving at a velocity $c/10$ through the ether use your diagram to determine the approximate change in focal length of the lens as a result of its motion towards or away from the light source.
b. How would the focus of a terrestrial telescope be expected to vary if the ether hypothesis is correct? This effect has been looked for but has never been observed.

2.3. In a Michelson-Morley type experiment the apparatus is set up with one arm of the interferometer parallel to the expected ether wind and the other perpendicular to it. Both arms are of length l and the expected ether wind speed is v.

a. Explain why the experimenters expected interference fringes to shift when they rotated their apparatus.

b. A student considers the set-up and argues that light travelling along the perpendicular arms will be delayed because it is forced to take a diagonal path through the ether that is longer than $2l$. On the other hand rays travelling on the parallel arm are delayed in one direction and sped up in the other direction. Since these effects cancel the time taken on this path will be just $2l/c$, the same as if the apparatus was at rest in the ether. He concludes that the perpendicular path takes longer. Do you agree?

c. Calculate the times t_1 and t_2 for light to travel perpendicular and parallel to the ether wind expressing each in terms of l_1, l_2, v and c.

d. The result of the experiment was that there is no delay and so empirically the ratio of the times calculated above is unity. What does this imply about the ratio of lengths l_1 to l_2? Compare this result with that derived relativistically for length contraction.

e. Can you think of any physical reason to support the Lorentz-Fitzgerald contraction of moving objects?

2.4. Maxwell commented that since all terrestrial measurements of the velocity of light involved timing light on a return trip, the effect of the Earth's motion through the ether would only show up as a second order correction and so would probably not be detectable.

a. What did he mean by this?

b. Derive an expression for the expected experimental value of the measured speed of light in a laboratory experiment carried out by timing a beam over a return path of length l (each way) in a laboratory moving at speed v through the ether. The direction of the light path is parallel to the motion through the ether.

c. The Earth's orbital speed is about 3.0×10^4 m s^{-1}. What fraction of c is this? What time delay would the Earth's motion through the ether be expected to produce in an experiment to measure the speed of light over a return path 10 m in each direction? How far does light move in this time?

2.5. One of the (ultimately unsuccessful) attempts to save the ether theory and explain the Michelson-Morley result was the Fresnel drag coefficient. According to this theory the earth would interact with the ether and partially drag it along as it moved in its orbit. Ether layers close to the Earth's surface would be almost at rest with respect to the Earth and so no fringe shift would be expected in the M-M experiment. To test Fresnel's idea experiments were carried out to see if other substances could drag the ether with them. In one experiment the velocity of light in flowing and stationary water was measured. The result confirmed that the speed of light through the flowing water (as measured in the laboratory) differed from the speed in stationary water. If the ether theory is wrong, how can special relativity account for this effect?

2.6. Stellar aberration is consistent with the theory that the Earth moves relative to the ether. Imagine a terrestrial telescope at the North pole focused on a star whose light enters the telescope in a direction perpendicular to the plane of the Earth's orbit. Assume the star is far enough away for parallax effects to be neglected. The velocity of the Earth in its orbit is v. During the year the telescope has to be adjusted to keep the star at the centre of field of its view.

a. Derive an expression for the angle through which the telescope must be tilted between December and June assuming the ether hypothesis.

b. Use the value of the Earth's orbital speed from question 2.4(c) to calculate the aberration angle α caused by the Earth's motion (that is half the angle derived above).

c. How good is the approximation $\alpha = v/c$ radians for this observation from Earth?

d. If our telescope had been focused on a star whose light did not reach us at right angles to our orbital plane, how would the aberration vary throughout the year? Give a qualitative answer first and then derive expressions for the maximum variation of both components of the aberration.

e. In practice it is easier to measure the variation of North-South aberration. Why is this?

2.7. Special Relativity must also be able to account for stellar aberration. The argument runs as follows. Light reaches the solar system from the fixed stars. The Earth moves at speed v relative to this reference frame because of its orbital motion. The direction of starlight in the Earth's reference frame is found by transforming from the reference frame of the fixed stars.

a. Carry out this transformation for the simple case in which starlight arrives perpendicular to the plane of the Earth's orbit. (Hint: choose axes in the frame of the fixed stars so that the z-axis is perpendicular to the orbital plane. In this frame the starlight has velocity components $u_x = 0$, $u_y = 0$, $u_z = -c$. Now transform to the Earth's frame when the earth is moving at velocity v along the x-axis. The Earth frame will have axes x', y', z' parallel to x, y, and z. Hence find the angle between the starlight velocity in the Earth's frame and the y'-axis.)

b. How does the result compare with that obtained on the basis of the ether theory?

c. Show that the velocity of the starlight is still c in the Earth's reference frame.

2.8. A rocket emits a laser pulse toward a distant mirror. The rocket is approaching the mirror at velocity v. In the questions below 'the velocity relative to the mirror' means measured in a reference frame at rest with respect to the mirror.

a. What is the velocity of the light pulse relative to the rocket?

b. What is the velocity of the light pulse relative to the mirror?

c. What is the velocity of the reflected light pulse relative to the mirror?

d. What is the velocity of the reflected light pulse relative to the rocket?

e. How are these answers changed if the rocket is 'at rest' and the mirror approaches it at speed v?

2.9. A bright light flashes at the origin of a reference frame that is moving past us at speed v parallel to our x-axis. The flash is emitted at the moment our origin coincides with that of the moving reference frame.
a. The wavefronts in the moving reference frame form spherical surfaces. What shape are they in our reference frame?
b. Consider a photon travelling in the (x', y', z') direction in the rest frame of the source. What is its distance from the origin of this frame at time t'?
c. Write down this distance in terms of x', y', and z', using Pyhtagoras's theorem.
d. Use the Lorentz transformation to find the photon co-ordinates in the x, y, z, t frame.
e. Find the distance of the photon from the origin in the x, y, z, t frame.. Confirm that the speed of the photon in this frame is c.

2.10. In the text a light clock perpendicular to the direction of motion was used to derive the time dilation equation. A particular rocket carries two identical light clocks, one lying parallel to the rocket motion and one perpendicular to it. The rocket passes a 'stationary' observer at velocity v.
a. How do the times on the two light clocks compare for an observer travelling in the rocket?
b. The 'moving' clocks will both appear to the 'stationary' observer to run slowly. Derive the time dilation factor for the clock lying parallel to the motion and confirm that it is the same as we derived previously.

2.11. A lambda particle has a lifetime of 2×10^{-10} s in its own rest frame. A lambda is created in a cosmic ray impact and has a velocity relative to the laboratory of $0.98c$. It leaves a track in a cloud chamber.
a. Roughly how long would the track be according to classical physics?
b. Roughly how long would the track be according to relativity?
c. How fast would an observer moving with the lambda see the cloud chamber approaching?
d. Assuming this moving observer survives the decay of the lambda and continues moving through the laboratory at $0.98c$, what length track does he observe in the cloud chamber?

2.12. A simple model of the hydrogen atom has the electron moving in a circular orbit around the proton bound by mutual electrostatic attraction.
a. Taking the radius of the atom to be of order 10^{-10} m calculate the electron's orbital velocity (equate electrostatic attraction to centripetal force).
b. Hence calculate γ. Comment on how important relativistic effects are likely to be in calculating the properties of the hydrogen atom.

2.13. A space touring company in the future takes passengers on a round trip. The speed relative to the Earth during most of the journey is a constant $0.75c$ and 5 years pass on Earth during the trip.

a. How much time has elapsed for the passengers during the trip?

b. How far do the tourists travel relative to the Earth?

c. How far do the tourists think they have travelled?

d. At one point in the journey the rocket passes a spherical star of radius R. What shape does it appear to the travellers?

e. One Earth year after the rocket departs a radio signal is sent to it. The signal is received and a reply sent back immediately. Calculate the time at which the signal reaches the rocket according to observers on Earth and on the rocket.

f. Calculate the time at which the reply is received on Earth.

g. How far from Earth is the rocket at the time the reply is received on Earth? (There are two answers to this question, one according to an Earthbound observer and one according to an observer on the rocket.)

2.14. Imagine the speed of light is a mere 3 m s^{-1}. A lightbulb hangs at the centre of a square room whose walls are 12 m long. It is switched on at $t = 0$.

a. At what time does light reach opposite walls?

b. Is it true that opposite walls are illuminated simultaneously?

c. An observer moving through the room passes its centre at the instant the light is switched on. She is walking directly between one pair of opposite walls at a speed of 1 m s^{-1}. Would she agree (after correcting for time of flight of light) that the walls are illuminated simultaneously?

d. What does she actually see?

e. The moving observer and the person who switched on the light begin to argue about simultaneity. They agree to carry out an experiment in which light sensitive detectors are placed on opposite walls of the room and wired up to a central alarm which only goes off if both sensors detect light at the same time. Will the alarm go off? Will this resolve the argument?

2.15. Radiation from a moving source suffers a Doppler shift in frequency and wavelength. If the source approaches us the frequency rises and wavelength falls. If the source recedes from us the opposite occurs. It is as if the approaching source compresses the waves and the receding source stretches them. The fact that the velocity of light is independent of the source velocity makes the relativistic expression for the Doppler shift different from that obtained for, say, sound in the atmosphere and time dilation also reduces the effective emission frequency relative to the Earth. A rocket moving away from us emits radar pulses at frequency f_o.

a. What is the effective emission frequency of the source?

b. By considering the time and position of emission of successive pulses in our rest frame derive an expression for the frequency at which the pulses are received.

c. How do the frequency and wavelength of the received radio waves compare with the frequency and wavelength of the emitted waves?
d. Show that the fractional change in wavelength is approximately v/c if $v<<c$.

2.16. A jet aircraft is flying at 500 m s^{-1} above the surface of the Earth.
a. By what factor does its length shrink when viewed from the Earth?
b. How much time would an atomic clock 'lose' per hour of flight in the aircraft relative to a similar clock on the Earth's surface. (Ignore complications such as the plane's curved path and the important effect of the Earth's gravitational field.)

2.17. An electron deflection tube has an accelerating voltage of 2500 V. The electron beam is then deflected into a semi-circular path by a magnetic field of strength 0.1 T before entering a detector.
a. What is the final speed of the accelerated electrons? ($m_e = 9.1 \times 10^{-31}$ kg)
b. By what factor does their mass increase?
c. What is the rest energy, the total energy and the kinetic energy of the accelerated electrons?
d. How sensitively must the detector position be measured if we are to determine the increase in mass due to electron kinetic energies?

2.18. 1kg of water at 10°C is heated to 90°C. ($c_w = 4200$ J kg^{-1} K^{-1})
a. What is its mass increase?
b. How is energy conserved in this heating process?

2.19. When we derived $E=mc^2$ from the 'Einstein Box' thought experiment we neglected the effect of the box movement on the time of flight of the light. Show that, if we analyse this situation more rigorously, we obtain the same expression for the mass-energy relation.

2.20. A Thorium-228 nucleus at rest decays by alpha emission to Radium-224. The rest masses of thorium, radium and helium atoms respectively are: 228.02873 u; 224.02020 u and 4.00260 u. Other values you will need are: $c = 2.9979 \times 10^8$ m s^{-1}, $e = 1.602 \times 10^{-19}$ C, u = 1.6606×10^{-27} kg.
a. Show that the energy equivalent of 1 u is about 931.5 MeV.
b. Explain how conservation laws of mass, energy and momentum apply to this decay.
c. What is the mass defect for this reaction?
d. What is the energy released?
e. Calculate the velocity of the emitted alpha particle (You can do this using Newtonian mechanics: why is this valid?).

2.21. A particle of rest mass m_o travelling at $c/2$ in the positive x direction collides with a particle of rest mass $2m_o$ that is at rest. The two coalesce and move off together.

a. What is the total energy and mass of the incident particle?
b. What is the kinetic energy of the incident particle?
c. What is the momentum of the incident particle?
d. What is the total energy and mass of the composite body?
e. What is the velocity of the composite body?
f. What is the rest mass of the composite body?

2.22. It is impossible for a single photon to be completely absorbed by an isolated electron.
a. Show why this is.
b. Prove that it is also impossible for an electron and positron to mutually annihilate to produce a single photon.

2.23. Compare and comment on the difference between the rest mass of the composite particle created when:
a. Two particles of rest mass m_o approach each other from opposite directions with equal kinetic energies T.
b. One of the particles has kinetic energy $2T$ and collides with the other which is initially at rest.

2.24. What is the minimum kinetic energy that must be given to a proton if it is to collide with a second stationary proton to create a pion? (That is, after the collision there are two protons and one pion.) The rest energies of the proton and the pion are 940 MeV and 140 MeV respectively. (Hint: consider the collision in the centre of mass frame and then transfer to the frame in which one proton is at rest, or use the equation derived in 2.23b.)

2.25. A train of length 200 m passes through a station at velocity 40 m s^{-1}. What must the speed of light be if two clocks, one at either end of the train, have a synchronization error (when seen from the platform) of 1 second?

2.26. A heavy nucleus moving at velocity $0.75c$ in the x-direction through the laboratory undergoes a beta decay and the emitted electron leaves the nucleus at a relative velocity (in the frame of the nucleus) of $0.6c$. What is the velocity of the electron relative to the laboratory? (Neglect the recoil of the nucleus.)

2.27. A square frame in a moving laboratory lies in the x', y' plane. One corner is at the origin and the corner diagonally opposite is at $x' = a, y' = a$. At $t = t' = 0$ the origin of this laboratory frame coincides with the origin of our own laboratory. The velocity of the laboratory relative to us is v along the positive x-axis.
a. What are the co-ordinates of the far corner in our own laboratory at $t = 0$?
b. What is the area of the frame in our reference frame?

2.28. A rocket maintains a constant acceleration a'_x relative to its own instantaneous inertial reference frames. The rocket left us at $t = 0$ and has a velocity v relative to us.

a. Find the acceleration of the rocket as a function of its velocity relative to us.

b. Sketch a graph of a_x versus v. Comment on the significance of this relation.

c. How does velocity change with time?

2.29. The Moon is about 400 000 km from Earth.

a. If we could travel at 99% of the speed of light how far would we have to travel to reach the Moon?

b. How long would it take us?

c. How much time would pass on Earth as we complete our journey?

2.30. The extract that follows is from Roger Penrose's book, 'The Emperor's New Mind': '*Even with quite slow relative velocities, significant differences in time-ordering will occur for events at great distances. Imagine two people walking slowly past each other in the street. The events on the Andromeda Galaxy...judged by the two people to be simultaneous with the moment they pass one another, could amount to a difference of several days ... For one of the people, the space fleet launched with the intent to wipe out life on the planet Earth is already on its way; while for the other, the very decision whether or not to launch the fleet has not yet been made*!'

a. Explain what is meant by 'simultaneous' for each person.

b. Explain why separate events that are simultaneous for one of the observers cannot be simultaneous for the other.

c. Let us say that, at the moment of passing, the space fleet has already been launched for person A but no decision has yet been made for person B. To what extent, if any, does this imply that B's future has already been decided, even at the moment he passes A? Is this an argument for determinism?

d. Would it ever be possible, even in principle, to publish a Universal Atlas listing the positions of every object in the Universe at one instant of time?

2.31. We have argued that the increase of relativistic mass with velocity implies that no material particle can travel at the speed of light. Photons are often treated as particles. They transmit energy and energy has mass. They also travel at the speed of light. How is this consistent with the ideas of relativity?

2.32. This question is a variation on a well known 'paradox'. A rocket passes at high speed along the axis of a large hollow cylindrical space station. The rest length of the rocket is l_o, that is identical to the rest length of the space station axis. An observer on the space station argues that the rocket will Lorentz contract and therefore fit entirely inside the space station. The rocket commander disagrees, saying that the approaching space station will contract and so there will be a moment when both ends of the rocket protrude beyond the ends of the space station. Take $\gamma = 0.5$ and $l_0 = 100$ m.

a. Do you agree with one observer, both observers or neither observer?
b. How can the argument be resolved?
c. The two observers fail to agree so the person on the space station suggests they settle it by experiment. He sets up an alarm system. It consists of a light beam and detector at each end of the space station. If the rocket cuts both beams simultaneously (as it would if it was longer than the space station) then the alarm will sound. Otherwise it remains off. Will the alarm ring?
d. If the alarm does ring does this mean that the observer on the space station is correct and the rocket observer is wrong?
e. Is the experiment fair?
f. Is there a similar experiment that could be set up by the observer on the rocket? If so, what result would it produce?

3

SPACE-TIME

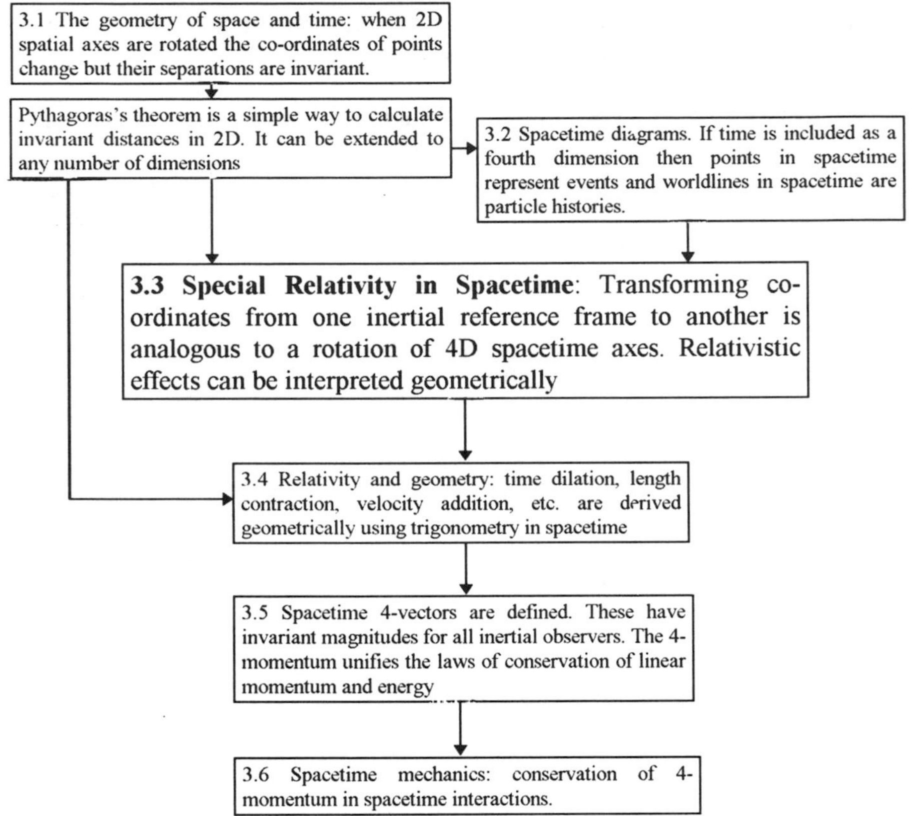
3.1 The geometry of space and time: when 2D spatial axes are rotated the co-ordinates of points change but their separations are invariant.
Pythagoras's theorem is a simple way to calculate invariant distances in 2D. It can be extended to any number of dimensions
3.2 Spacetime diagrams. If time is included as a fourth dimension then points in spacetime represent events and worldlines in spacetime are particle histories.
3.3 Special Relativity in Spacetime: Transforming co-ordinates from one inertial reference frame to another is analogous to a rotation of 4D spacetime axes. Relativistic effects can be interpreted geometrically
3.4 Relativity and geometry: time dilation, length contraction, velocity addition, etc. are derived geometrically using trigonometry in spacetime
3.5 Spacetime 4-vectors are defined. These have invariant magnitudes for all inertial observers. The 4-momentum unifies the laws of conservation of linear momentum and energy
3.6 Spacetime mechanics: conservation of 4-momentum in spacetime interactions.

3.1 THE GEOMETRY OF SPACE AND TIME.

3.1.1 Introduction.

At Cologne, on 21 September 1908, Hermann Minkowski delivered an address to the 80th assembly of German Natural scientists and Physicians. He began with the following words:

fig 3.1 Hermann Minkowski (1864-1909)
'I should like to show how it might be possible, setting out from the accepted mechanics of the present day, along a purely mathematical line of thought, to arrive at changed ideas of space and time.'

'The views of space and time which I wish to lay before you have sprung from the soil of experimental physics, and herein lies their strength. They are radical. Henceforth space by itself, and time by itself, are doomed to fade away into mere shadows, and only a kind of union of the two will preserve an independent reality.'

We call this union of space and time, 'space-time'. Minkowski had realized that the Lorentz transformations of special relativity were very similar, mathematically, to the familiar transformations of Cartesian co-ordinates when the reference axes are rotated. He discovered that all the results of special relativity can be reproduced geometrically if we extend our usual three-dimensional view of space to a four-dimensional view with an extra dimension related to time. This startling insight provided a novel and convincing interpretation of relativity and paved the way for many later theories that also attempt to describe the Universe geometrically, the first of these being general relativity that interprets gravity as a distortion of space-time geometry.

3.1.2 Cartesian Co-ordinates.

In Cartesian geometry we locate the position of a point in space by giving co-ordinates measured along mutually perpendicular axes issuing from some common origin. In a 1-dimensional (1D) world this would be the same as stating the distance from a fixed point. In 2D it is like giving a map reference and in 3D it is like adding an altitude to the map reference so that position is determined by three independent numbers. There is no reason this approach cannot be extended to any number of dimensions and in an n-dimensional space a point is located by giving n independent co-ordinates. The approach can also be generalized so that positions are referred to curvilinear axes (as is the case for positions on the curved surface of the Earth - latitude and longitude) but in this chapter we shall be concerned only with simple perpendicular axes.

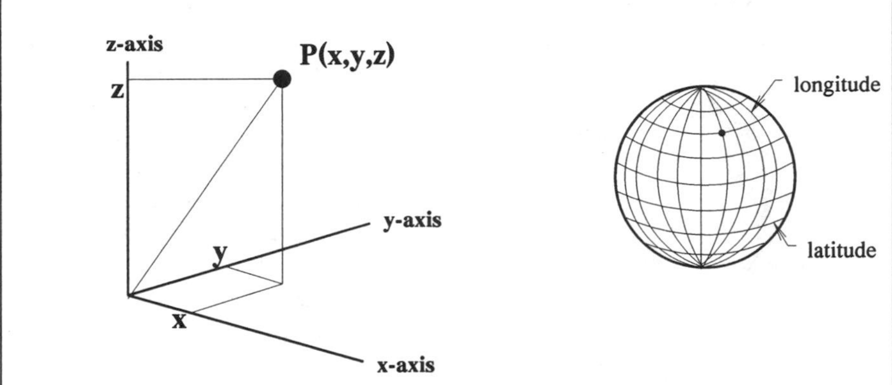

fig 3.2 Cartesian co-ordinates give positions with respect to a set of mutually perpendicular (orthogonal) axes. Positions on curved surfaces are given in terms of 'curvilinear co-ordinates' relative to axes defined within the surface itself (e.g. latitude and longitude on Earth).

3.1.3 Co-ordinates And Distances.

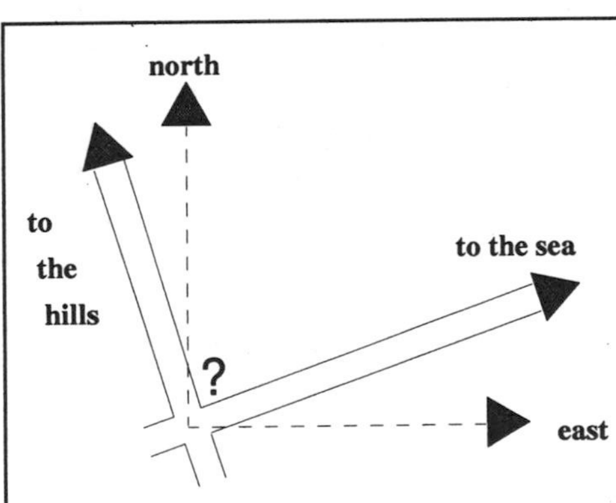

fig 3.3 The road grid and compass directions define two sets of Cartesian axes at an angle to one another.

Fortunately many of the most interesting and significant ideas in 4D space-time can be illustrated using a 2D model. This also makes it much easier to draw meaningful diagrams! The example that follows seems fairly trivial but it emphasizes the fact that the reference frame we use to view the world is chosen arbitrarily and significant quantities in the world itself will be independent of our choice.

A stranger arrives in a country whose roads are set out on a rectangular grid system. In the grid one set of parallel roads heads toward the sea whilst those at right angles to them head toward the hills. The stranger arrives at a cross-roads and asks the locals for directions to a particular town. He gets two different responses. The first is that the town is 3 km east and 4 km north of the junction. The second is that he must walk 4 km towards the sea and 3 km towards the hills. Since the roads do not align with compass directions the traveller is confused and spends some time debating what to do. Eventually he sets off towards the sea and is relieved having followed the second set of directions to arrive safely at the town he sought. He needn't have worried, both sets of directions were accurate. The map below shows how this can be.

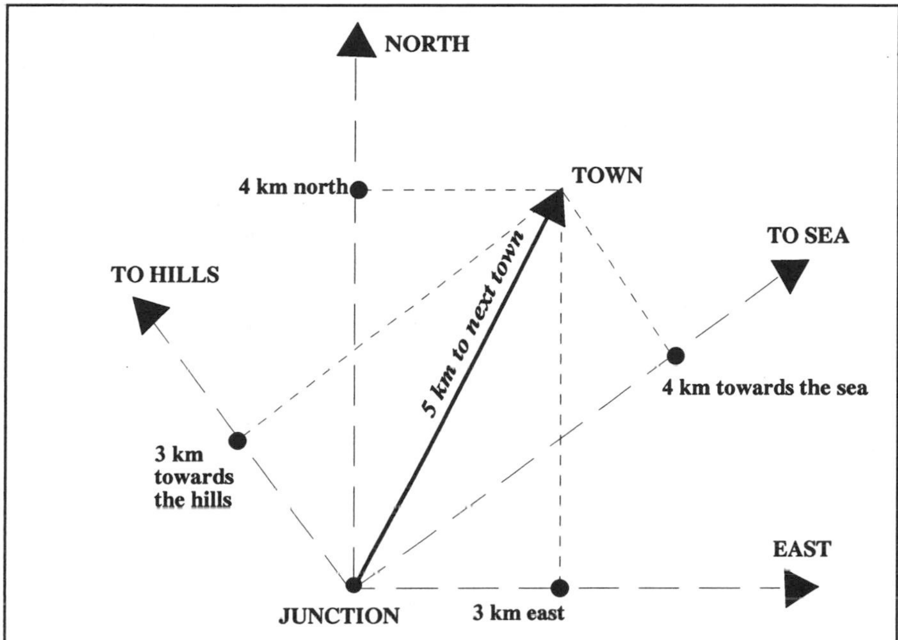

fig 3.4 The co-ordinates of a point depend on the axes along which they are measured. However, the distance between the points is independent of the choice of axes.

What has confused the traveller is being given the co-ordinates of the same point referred to two different sets of Cartesian axes. His starting position and the position of the town are not in any way affected by the choice of axes, but co-ordinates are. Let us refer to the two sets of reference axes as compass and geographic axes. We can then represent the position of the town from the original junction by a displacement vector in each system:

Compass vector: $\underline{C} = (3, 4)$
Geographic vector $\underline{G} = (4, 3)$

The axes in each reference frame are perpendicular so we can use Pythagoras's theorem to calculate the distance of the town from the junction. Let this distance be s_c in the compass frame and s_g in the geographic frame:

$$s_c^{\,2} = 3^2 + 4^2 = 25 \qquad so \qquad s_c = 5 \text{ km}$$
$$s_g^{\,2} = 4^2 + 3^2 = 25 \qquad so \qquad s_g = 5 \text{ km}$$

No one is going to be surprised by this result. The straight-line distance of the town from the junction is obviously a property of the country being visited and not of the axes to which we refer our measurements. (The symmetry of the numbers used here also makes this obvious.) We say that distance is an invariant quantity with respect to rotation of the co-ordinate axes. On the other hand the co-ordinates themselves are covariant quantities - that is they vary according to some fixed rule.

3.1.4 Rotation Of 2D Axes.

The previous example had rather conveniently chosen numbers (two 3-4-5 triangles) and so it will be worthwhile to show that the conclusions drawn in that special case also apply more generally. First of all we will need to derive a set of equations that transform the co-ordinates from one set of Cartesian axes to another set that have been rotated about the origin. The diagram below shows a point P that has co-ordinates (x,y) relative to the first set of Cartesian axes. Its new co-ordinates with respect to the rotated axes will be (x',y'). The relation between (x',y') and (x, y) can be determined trigonometrically from the diagram.

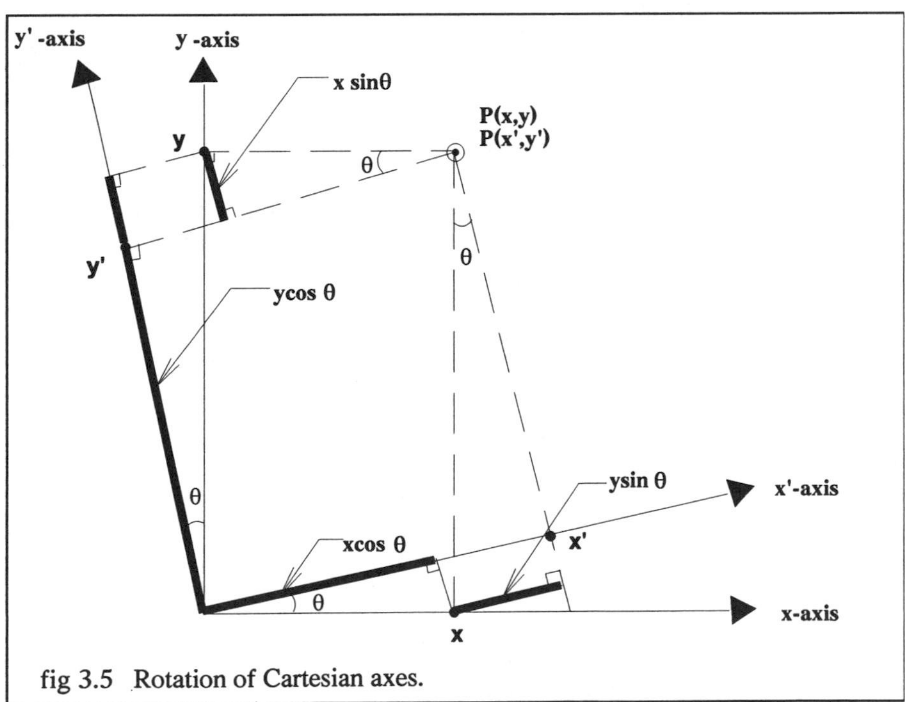

fig 3.5 Rotation of Cartesian axes.

The transformation from (x,y) to (x',y') is found by projecting the x and y co-ordinates onto the new x' and y' axes as shown above. The result is:

$$\begin{aligned} y' &= -x\sin\theta + y\cos\theta \\ x' &= x\cos\theta + y\sin\theta \end{aligned} \tag{3.1}$$

The inverse transformation is derived similarly by projecting the x' and y' co-ordinates onto the old x and y axes. This gives:

$$\begin{aligned} x &= x'\cos\theta - y'\sin\theta \\ y &= x'\sin\theta + y'\cos\theta \end{aligned} \tag{3.2}$$

(If the rotation is clockwise rather than anti-clockwise the sign of θ changes. This changes the sign of terms involving sines but not those with cosines.)

We can now calculate an expression for the length of the displacement vector OP in each reference frame:

(i) in xy frame: $OP^2 = x^2 + y^2$

(ii) in $x'y'$ frame $OP'^2 = x'^2 + y'^2$

We can substitute for x' and y' from the transformation equations (3.1):

$$OP'^2 = x^2\cos^2\theta + 2xy\cos\theta\sin\theta + y^2\sin^2\theta + x^2\sin^2\theta - 2xy\cos\theta\sin\theta + y^2\cos^2\theta$$

$$= x^2(\cos^2\theta + \sin^2\theta) + y^2(\cos^2\theta + \sin^2\theta)$$

$$= x^2 + y^2 = OP^2$$

The result is $x^2 + y^2 = x'^2 + y'^2$ for all rotation angles.

This Pythagorean expression has the same magnitude in all rotated reference frames even though the individual co-ordinate values change. The general 2D result (in Euclidean plane geometry) is that the co-ordinates of a pair of points P (x_1,y_1) and Q (x_2,y_2) (neither necessarily the origin) obey the transformation equations (3.1) and (3.2) when the axes are rotated. The expression below, (in which $\Delta x = x_2 - x_1$ etc...) involving co-ordinate differences is then an invariant with respect to rotation:

$$\Delta x^2 + \Delta y^2 = \Delta x'^2 + \Delta y'^2 \tag{3.3}$$

The interpretation of this is that the distance between two points is an invariant with respect to rotation of the reference axes.

3.1.5 Pythagoras In Many Dimensions.

A little thought will convince you that the result (3.3) we derived for 2-dimensional space can be generalized into the familiar 3-dimensional space of our experience. The separation s of two points $P(x_1, y_1, z_1)$ and $Q(x_2, y_2, z_2)$ in 3D is given by:

$$s^2 = \Delta x^2 + \Delta y^2 + \Delta z^2$$

and s itself must be independent of our choice of axes so expressions of the same form as the right hand side of the equation above will be *invariants* when the axes are rotated (or translated). If the prime in the equations below indicates another arbitrarily chosen set of Cartesian axes then:

$$\Delta x'^2 + \Delta y'^2 + \Delta z'^2 = \Delta x^2 + \Delta y^2 + \Delta z^2 \tag{3.4}$$

We don't have to stop at three-dimensional space. We may not be able to visualize higher dimensional spaces but we can certainly write down equations for them. If they behave like the Cartesian spaces we have already discussed they will have invariant quantities of the form:

$$s_n^2 = \sum_{i=1}^{i=n} \Delta x_i^2 \tag{3.5}$$

Minkowski recognized that the invariant quantities of special relativity did have this form and that, if a suitable set of 4D space-time co-ordinates are chosen, the transformation itself becomes a rotation in space-time. In the next section we shall develop the description of events in space-time and then derive and re-interpret the results of special relativity. The beauty of this approach is that laws of physics are seen to correspond to compelling geometric relationships.

3.2 SPACE-TIME DIAGRAMS.

3.2.1 A Simplification.

The geometry of familiar '3-space' is static. Time does not enter into the picture. Points in 3-space are defined for all eternity and the relations between them are the theorems of Euclidean geometry (we shall meet some examples of non-Euclidean geometries when we discuss General Relativity later). However, *events* take place in space *and* time and require a fourth component to define them. For example, it would be pointless to arrange to meet your friend on a certain corner unless you also agreed *when* you should be there. The location alone is not enough, the *event* of the meeting involves something that happens at a particular place *and* at a particular time. Here is another example: a particular apple may have struck Newton on the head as he sat beneath a tree in his home village of Woolsthorpe at 10:15 am on 23 May 1666. This event could be located by a set of four numbers - the map reference of Newton's tree, the altitude of his head and the time at which the apple struck. Where 3-space deals with points 4-space is concerned with events, and an event is specified by a set of four co-ordinate values: x, y, z, and t relative to some original event which occurred at (0, 0, 0, 0).

Just as it is possible to draw a map or chart of a 2D or 3D world, so it is possible to map out events in 4-space. Such 'maps' are called 'space-time diagrams', and are extremely useful in relativity. Unfortunately a complete space-time diagram would itself be 4-dimensional and so impossible to represent accurately on a 2-dimensional page. Fortunately for us events of interest in space-time (particularly in many simple examples) often occur in a 2-dimensional slice of the 4-dimensional space-time continuum. This is analogous to the way a 2-dimensional map is perfectly adequate for terrestrial navigation despite the fact that we inhabit 3-dimensional space. Our journeys, by being constrained to remain on the surface of the Earth, consist of 2-dimensional movements only (that is they can be defined by the use of only two co-ordinates).

When we consider two observers in uniform relative motion we are only interested in the way their separation changes along the line joining them and two of the spatial directions can be neglected. This allows us to reduce the space-time diagram to a 2-dimensional form that contains all of the essential physics and can be drawn accurately on the page (although with some strange scale distortions as we shall see later).

In many of the examples we have considered, two inertial observers A and B move at a constant velocity relative to one another. Their relative motion is along a line. We can define our axes so that this line is the x-axis of our co-ordinate system. All the essential physics can be represented on a *space-time* diagram that shows just this one space dimension and time. For example, the diagram below shows the trajectory of B's laboratory relative to A's reference frame. Each point on the trajectory represents an event in space and time and the line joining these is called a 'worldline'. The space-time diagram shows B's worldline in A's reference frame.

3.2.2 Worldlines And Mechanics.

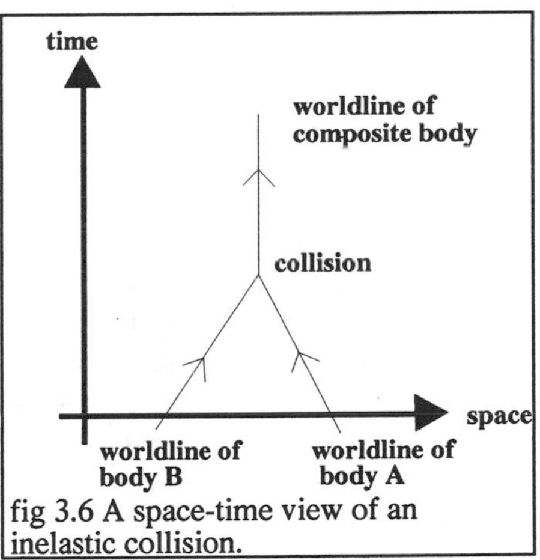

fig 3.6 A space-time view of an inelastic collision.

If A and B are together when A's clock reads zero then, as time goes on, B will be progressively farther from A. The diagram is effectively a time-displacement chart of B's motion as seen by A and the gradient of this line is related to B's velocity in A's reference frame. Since it is conventional to plot the time axis vertically the velocity is the reciprocal of the gradient (we shall soon see that the 'time' axis is not simply time in seconds, but this does not affect our arguments in this section). The faster B moves the larger the angle his worldline makes with A's time axis. A's own worldline runs up his own time axis, it is a vertical straight line on the diagram since, for A, time goes on but his position is unchanging.

The worldline of a particle is a representation of its dynamic history. In the absence of applied forces this will be a straight line through space-time. If a resultant force does act the worldline will be deflected or curved. Newton's first law of motion will therefore take a form like: 'The worldline of a free particle is a straight line through space-time.'

In a similar way the second law will determine the curvature of the worldline when a resultant force acts on a particular mass. The idea that free particles move on straight worldlines is central to general relativity where space-time becomes distorted by matter and the 'straightest line' is a geodesic (like a great circle path on the surface of a sphere).

Figure 3.6 shows an inelastic collision between two similar bodies. The collision is observed from the reference frame of the centre of mass of the system. During the collision itself the worldlines of A and B curve in opposite directions before joining to form the worldline of the composite body. This is the result of an interaction and Newton's third law can be interpreted as saying that individual worldlines cannot curve by themselves, there will always be other worldlines elsewhere in the universe which curve in the opposite way. This is true whether the interaction involves an apparent 'contact' or is mediated at a significant distance by some kind of field.

3.2.3 The Light Cone.

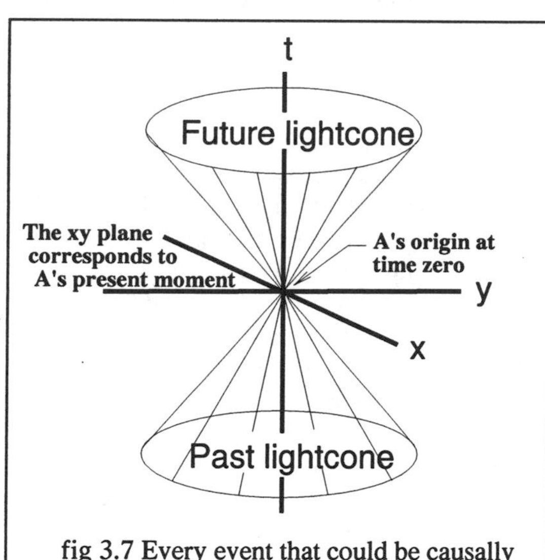

fig 3.7 Every event that could be causally connected with A's present moment must lie inside A's past or future lightcones.

The velocity of light represents a limiting velocity in the universe. If A emits a light pulse at time $t = 0$ then the wavefront of this pulse will travel away from A at velocity c in all directions. On our 2D space-time diagram this will be represented by two diagonal worldlines spreading out from the origin of co-ordinates in both the $+x$ and $-x$ directions. The worldlines of all material bodies passing A at $t = 0$ must lie between these diagonals and A's time axis (since smaller velocities must be inclined to A's time axis at a smaller angle than the angle made by the light worldline). This limitation is extremely significant. It effectively divides the world (that is the totality of space-time events) into distinct regions with respect to each individual space-time event.

The 3D 'surface' defined by the worldlines of light rays leaving the origin forms a 'lightcone' in space-time. This is represented using just two space directions in fig 3.7 above. Viewed from the origin of A's co-ordinates the lightcone divides space-time into three distinct regions.

- **The past.** This is the region of space-time between the negative time axis and the lightcone. All events inside the past lightcone could possibly have affected

A by the time $t = 0$. That is, some influence travelling at or below the velocity of light could have travelled from an event in A's past lightcone and reached A on or before $t = 0$. Events at the origin can be causally connected with events that occur in the past lightcone.

- **The future.** This is the region of space-time between the positive time axis and the lightcone. All events inside the future lightcone could be affected by events at the origin. Any event in the future lightcone can be causally connected with events occurring at (x, y, z) at or after $t = 0$.
- **Absolute elsewhere.** This is the rest of space-time. In classical physics this would be half past and half future. Its significance in relativity is that it contains all the events that cannot be causally connected with an event at the origin (or with A at $t = 0$). The reason for this is quite simple. If causal influences can travel at or below the velocity of light then an event in absolute elsewhere at $t < 0$ will produce effects which are too far from A's origin to reach it before $t = 0$. The condition for this to be the case is very easy to calculate. If the distance of an event from A is d and it occurs at time t then it will be inside the lightcones only if d is less than the magnitude of ct. If d is equal to ct the event occurs on the lightcone and can only be connected with A by influences traveling *at* the speed of light. If d is greater than ct then the event is not causally connected with A.

It is worth pointing out that this limit on faster than light communication between events seems to be violated by quantum theory in which non-local interactions apparently occur (for example during the 'collapse of the wavefunction'). However this does not seem to result in a reversal of causal connections because the effects concerned involve only the correlation between distant events and not their independent outcomes. This is a subtle point but has great importance in the interpretation of quantum theory.

3.3 SPECIAL RELATIVITY AND GEOMETRY.

3.3.1 Measuring Time In Metres.

Before we can discuss space and time on an equal footing we must confront the problem of units. In the S.I. system it is conventional to measure spatial displacements in metres and times in seconds. However, the dimensions of space and time are naturally linked by a constant of nature, the speed of light. This is an invariant quantity that has the same value in all inertial reference frames. It is possible to measure time in metres simply by using the distance light travels in the time to be measured. Whereas the second was an arbitrarily defined unit the new unit is more useful because it removes the artificial distinction between measurements in space and time. One second will now be 3×10^8 m in the new unit. We can convert times t in seconds to times in metres using the expression ct. This may not seem to offer any great advantage over the old system of units but it makes life much more straightforward in 4D space-time. There is one extra complication which we shall soon meet - to make the geometry consistent with special relativity the 'time' axis will become 'imaginary'. This does not affect the magnitude of measurements along it but it has a very important effect on the way we calculate 'distances' along the worldline.

Defining the metre. In October 1983 the General Conference on Weights and Measures met in Paris to redefine the metre. Previously one metre had been a fixed number of wavelengths of an emission line from a particular isotope of krypton. Before that it was the length of a standard platinum bar. The new standard is a derived value based on the distance light travels in a certain time. The reason for the change is that physicists can measure time intervals much more accurately than they can measure distances.

- *The speed of light is defined to be exactly 299 792 458 ms^{-1}.*
- *The second is defined to be 9 192 631 770 periods of the radiation emitted during the transition between two hyperfine levels of the groundstate of the cesium-133 atom.*
- *The new metre is therefore 1/(299 792 458) of the distance light moves in one second.*

One peculiar consequence of this system of definitions is that any future refinement in our ability to measure c will not change the speed of light (which is a defined number), but it will change the length of the meter!

3.3.2 The Space-time Of A Moving Observer.

In the new system light will travel one unit of space in one unit of time. This means the surface of the lightcone will make an angle of 45° with the space and time axes. If moving observers pass with a relative velocity v then they can each represent the worldline of the other with a straight line inclined at some angle less than 45° to their own 'time' axis.

Both observers are in inertial reference frames so time and space will seem normal to them. The surprising effects of special relativity: time dilation, length contraction, simultaneity etc. come about when one observer compares his measurements using apparatus moving with him against measurements made by the other observer using their own apparatus. To understand these effects geometrically it will be important to switch our point of view from one reference frame to another and to understand how these observations are related to each other.

An observer's worldline corresponds with the 'time' axis of his own reference frame (since an observer's self-experience is of staying put but aging). If we adopt A's point of view the effect of B's motion is to rotate B's 'time' axis in A's space-time. This is similar to the spatial effect of rotation discussed previously in the story of the traveller. Just as the directions North and East each resolved into components parallel to the road grid so two events separated only in time for B *will be separated in both space and time for A*. This is a radical departure from the way we would conventionally interpret a graph of displacement and time.

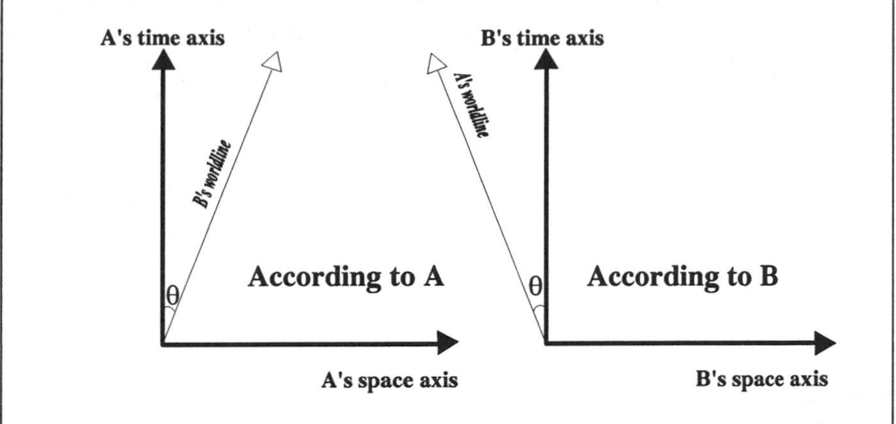

fig 3.8 B's motion relative to A is shown by B's 'worldline' through A's spacetime. A is also represented by a worldline in B's spacetime.

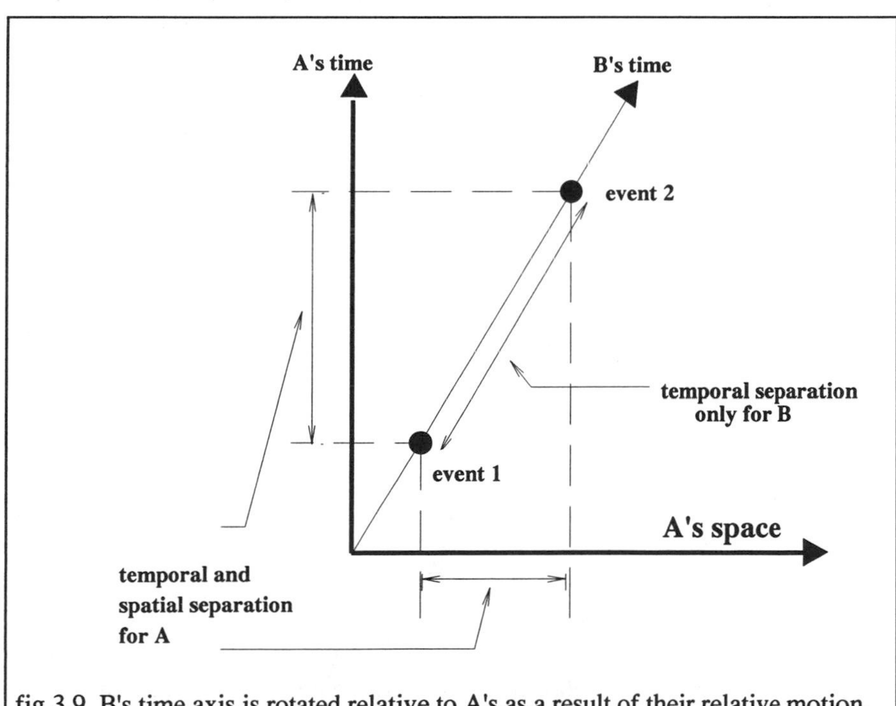

fig 3.9 B's time axis is rotated relative to A's as a result of their relative motion. The faster B goes the greater the angle of rotation.

What is new in the relativistic geometry is equality of treatment of space and time. In the classical world we could draw a very similar diagram but the trajectory of B in A's reference frame would not be identified with B's time axis. In classical physics there is but one absolute and unalterable time axis which is (on these diagrams) always vertical and carrying a common scale for all observers. The effect of relative motion in space-time is to cause the transformation of space into time and time into space so that the observers in relative motion cannot agree on the co-ordinate separations of events (even though they will be able to establish some invariant properties for the separation of those events).

- *Motion in space-time is equivalent to a rotation of space-time axes and results in what may have been a purely spatial or temporal separation for one observer being resolved into spatial and temporal components for another*

3.3.3 Rotation Of Space-time Axes.

So far, by drawing B's worldline in A's space-time or vice versa we have shown how B's 'time' axis is rotated by motion. We have not shown how to represent the space axis for the same moving observer. This too will be rotated, but how? Since both observers are in inertial reference frames they will encounter the same laws of physics and measure the same velocity of light. We can use the motion of light relative to each observer to construct B's rotated space axis in A's reference frame.

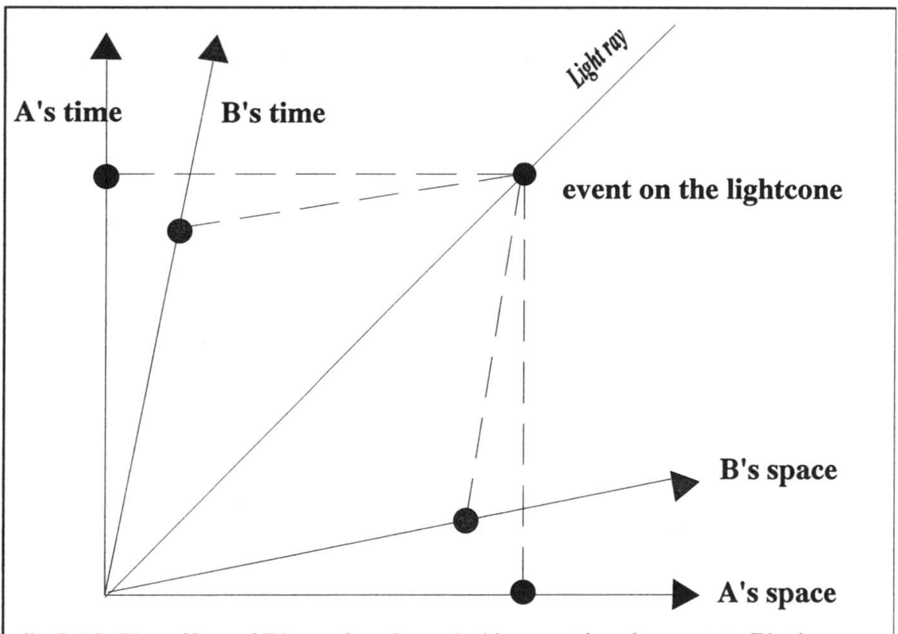

fig 3.10 The effect of B's motion through A's spacetime is to rotate B's time and space axes with respect to A's.

Let A and B synchronize their clocks at the instant their origins cross. At this moment a flash of light is emitted from the origin and travels out into space. Both observers will see spherical waves travelling at the speed of light and in each reference frame an event lying on this lightcone must be symmetric with respect to the space and time axes (since the light rays move one unit of space in one unit of 'time'). This is shown in fig 3.10. Notice that the angle through which B's space and time axes rotate is the same but the direction of the rotation is in an opposite sense.

If a particular event is projected onto the two sets of reference axes its co-ordinates in each frame will differ. It also turns out that the scale along the axes is different so that, for example, the time elapsed along an inclined worldline is less than the time elapsed along a vertical worldline of the same (apparent) length. We shall get to grips with this more quantitatively in the next section. For the moment we shall account qualitatively for the relativistic effects derived in chapter 2.

3.3.4 Simultaneity.

Consider two events X and Y that are simultaneous for A. They will occur at the same time co-ordinate in A's reference frame but may have different spatial co-ordinates. This places them on a horizontal line in A's space-time diagram.

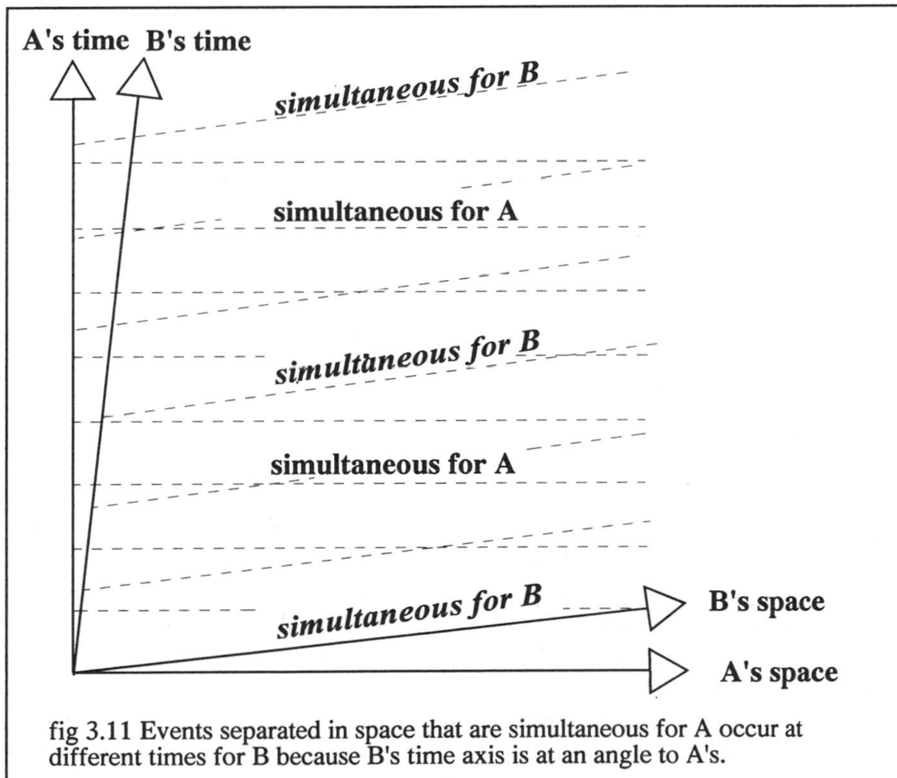

fig 3.11 Events separated in space that are simultaneous for A occur at different times for B because B's time axis is at an angle to A's.

However, simultaneous events for B lie on lines parallel to B's space axis. Unless the two events occur at the same place B will see them occur at *different* times. The rotation of B's axes, which is how B's motion is represented in space-time, accounts for the relativity of simultaneity. In the example above Y will occur before X in B's frame even though both are simultaneous in A's frame. If B's motion had been reversed the temporal order of the events would also have been reversed. Moreover, this is clearly a reciprocal effect, if events are simultaneous in B's frame and occur at different places then they will not be simultaneous in A's frame. It is also clear that events which occur at the same place for A occur at different places for B and vice versa.

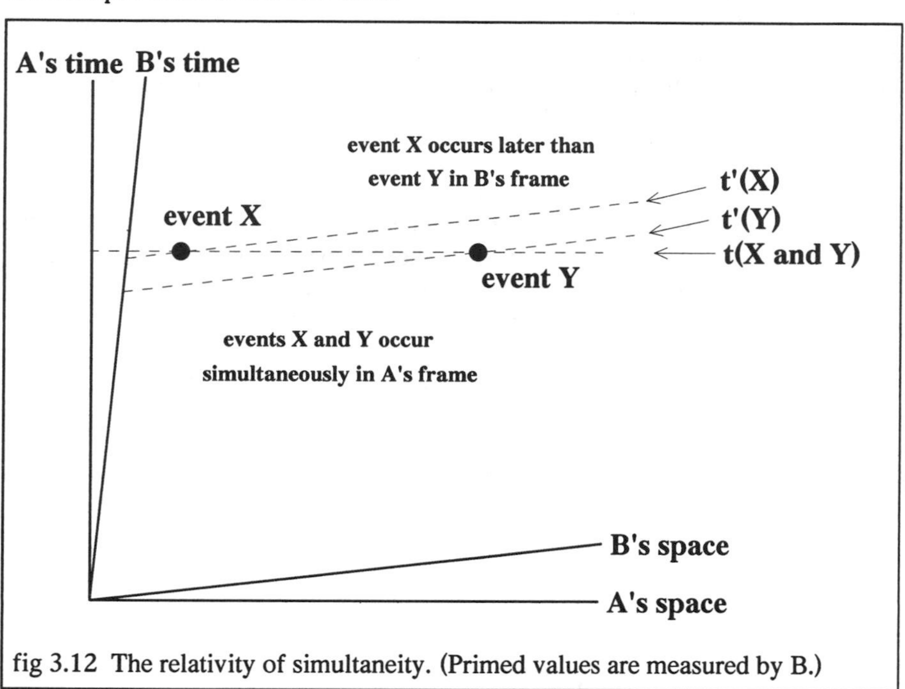

fig 3.12 The relativity of simultaneity. (Primed values are measured by B.)

3.3.5 Length Contraction.

Consider a rod at rest relative to B. The worldlines of its ends are shown on A's space-time diagram in fig 3.13. They define a band through A's space-time. When A measures the length of the moving rod he does so by measuring the *simultaneous separation of the rod's ends in his own reference frame*. On the other hand the proper or rest length of the rod in B's frame is determined by the simultaneous end positions in that frame. Since A and B disagree about simultaneity they will also disagree about the length of the rod. The rod appears shorter relative to A, it is length contracted. A concludes that moving rods contract along the direction of their motion. The lengths of l and l' on the space-time diagram can be used to derive the length contraction formula, but only if the lengths *as they appear* are calibrated. It is then possible to use simple trigonometry to derive l' in terms of l_o. The ratio of the lengths on this diagram will result in length contraction of the moving rod as seen by A, but the effect is completely symmetric. To repeat the calculation from B's point of view we should have to put a rod at rest in A's frame and then have B record the positions of its ends simultaneously in B's frame. In this case worldlines of the rod ends are vertical lines on the space-time diagram.

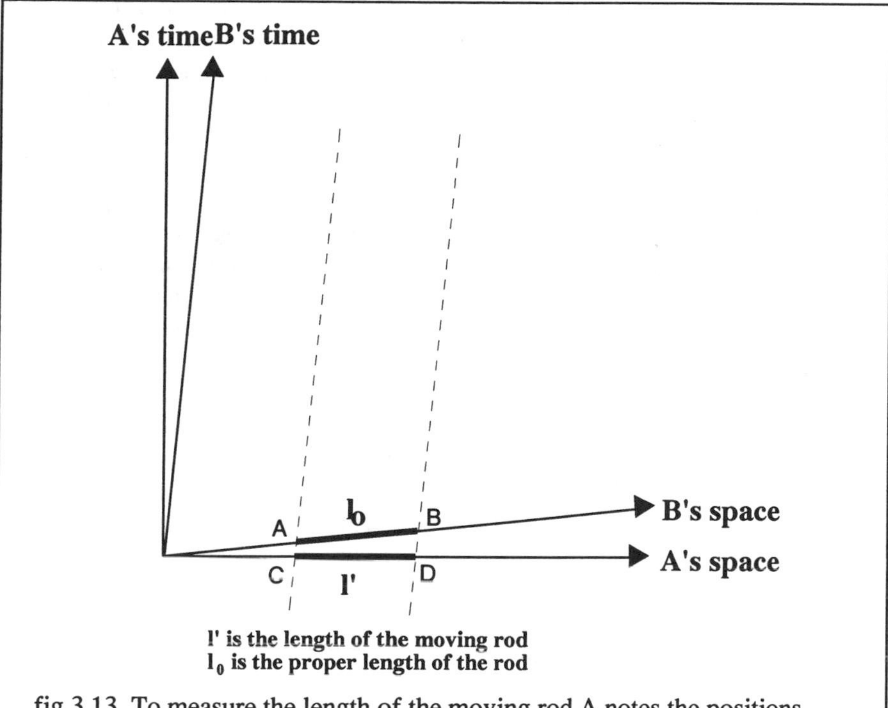

fig 3.13 To measure the length of the moving rod A notes the positions of its ends at t = 0. From the point of view of the moving rod (B's frame) this means the right hand end is recorded before the left hand end, so A's result is a length shorter than that of the rod in its own rest frame. This explains length contraction.

3.3.6 Time Dilation.

A and B separate at constant velocity. According to A, B's inclined worldline up to any particular time in A's frame is longer than his own 'vertical' worldline. In the geometry of space-time this results in a shorter elapsed time. If we had analysed the situation from B's point of view A's worldline would appear longer and less time would have elapsed for A. Of course, simultaneity has an essential role here, A and B do not agree on the pair of moments at which their clocks should be compared. It is this disagreement that allows time dilation to be a reciprocal effect and yet not to cause a rupture in the fabric of space-time!

3.3.7 The Twin Paradox In Space-time.

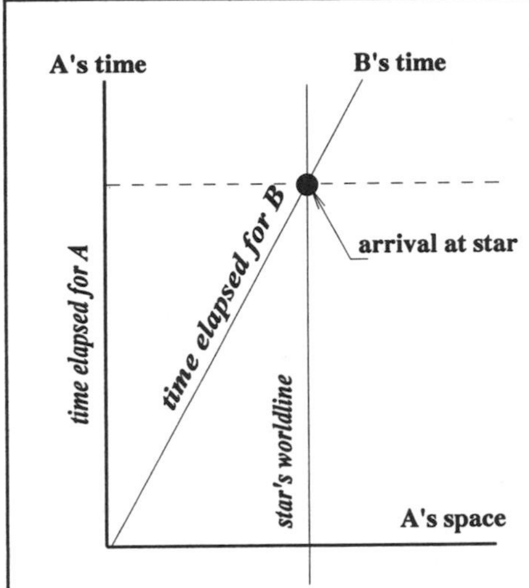

fig 3.14 A and B disagree about the difference in time between two events. In this case the events are B's departure from Earth and B's arrival at a distant star.

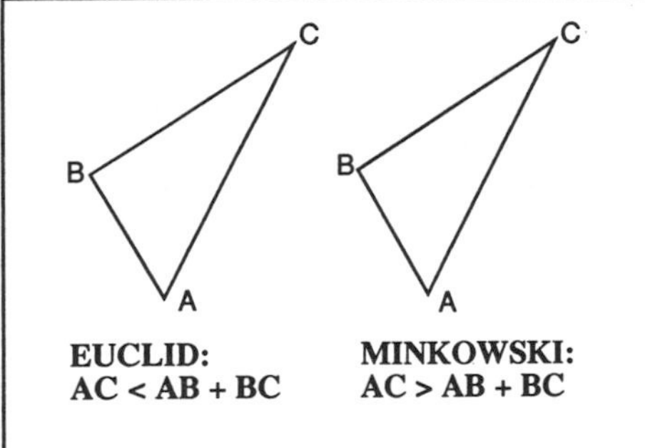

fig 3.15 In Euclidean geometry the hypotenuse of a right angled triangle is the longest side, in Minkowski's spacetime it is the shortest.

In Euclidean geometry we are familiar with the 'triangle inequality' which states that, if there are three non-collinear points A, B, C lying on a common plane then the distance direct from A to C is less than the distance A via B to C. In space-time things are rather different. The route via B produces a longer worldline and therefore a shorter space-time interval than the direct route. Since a traveller's time axis always lies along their worldline this means that less time will pass for an observer moving along the indirect route. (The exact meaning of the term 'interval' will be explained in the next section.)

This feature of space-time geometry can be used to explain the twin paradox as is shown on the space-time diagram in fig 3.16. B's worldline relative to A's reference frame is much longer than A's so less time passes for B than A between separation and reunion. This difference in the 'length' of their worldlines is experienced as a difference in time only since they reunite at a common point. The effect is not reciprocal because there is no single inertial reference frame for B in which we can plot A's worldline. If we were to consider any one part of the motion, for example as B travels away from A, then the separation in space leads to a disagreement over clock synchronization and simultaneity which allows each to think that the other's time runs slow compared to their own This is exactly the situation discussed under Time Dilation above.

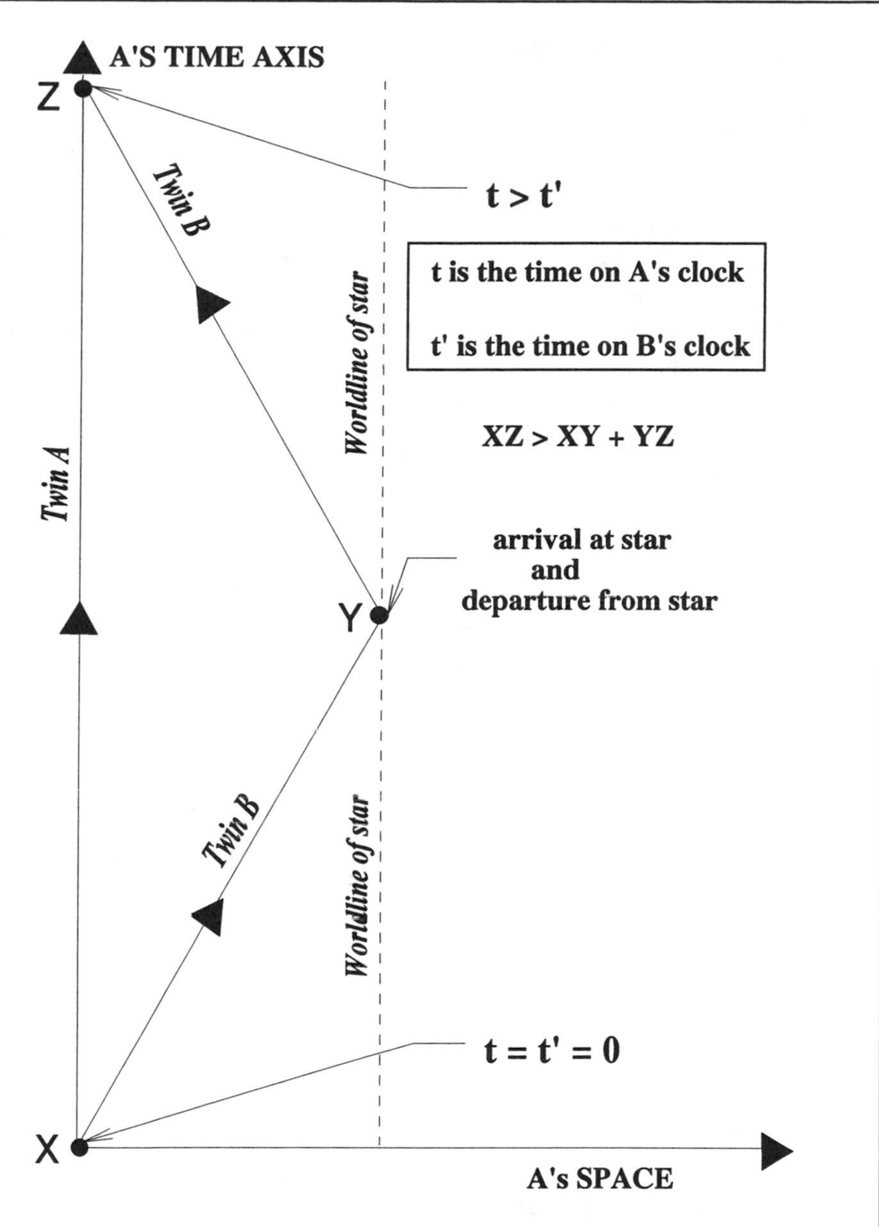

fig 3.16 The twin paradox in spacetime. A and B separate at X when t = t' = 0 and reunite at Z when t > t' because of their different paths through spacetime. The faster the journey the greater the difference in 'lengths' between A and B's worldlines and the greater the difference in the times they record for the journey.

3.4 RELATIVITY AND GEOMETRY.

3.4.1 The Interval.

Imagine a flash of light emitted at the instant B's moving reference frame passes A's origin. A and B must both observe spherical wavefronts travelling out from their own origin at the speed of light. If, a little later, the light scatters from some object then the emission and scattering define two events in space-time. A and B will disagree about the spatial and temporal separations of the two events, but they must agree about the speed of light. This can be expressed mathematically:

For A $\qquad \Delta x^2 + \Delta y^2 + \Delta z^2 = c^2 \Delta t^2$

For B $\qquad \Delta x'^2 + \Delta y'^2 + \Delta z'^2 = c^2 \Delta t'^2$

For the two events considered it is clear that:

$$\Delta x^2 + \Delta y^2 + \Delta z^2 - c^2 \Delta t^2 = \Delta x'^2 + \Delta y'^2 + \Delta z'^2 - c^2 \Delta t'^2 = 0$$

The form of this final expression is similar to that of Pythagoras's theorem for the invariant distance when spatial axes are rotated. This similarity suggests that the quadratic expression, $\Delta s^2 = \Delta x^2 + \Delta y^2 + \Delta z^2 - c^2 \Delta t^2$ might be an invariant between *any two events* in space-time. To test this hypothesis we need to transform the expression from one frame to another using the Lorentz transformation equations. As usual we shall assume the axes have been chosen so that B's motion is in the positive x-direction.

$$\Delta x^2 = \gamma^2 (\Delta x' + v\Delta t')^2 \qquad \Delta y^2 = \Delta y'^2 \qquad \Delta z^2 = \Delta z'^2 \qquad \Delta t^2 = \gamma^2 \left(\Delta t' + \frac{v\Delta x'}{c^2} \right)^2$$

$$\Delta s^2 = \Delta x^2 + \Delta y^2 + \Delta z^2 - c^2 \Delta t^2 \qquad (3.6)$$

$$\Delta s^2 = \gamma^2 \left(\Delta x'^2 + 2\Delta x' v \Delta t' + v^2 \Delta t'^2 \right) + \Delta y'^2 + \Delta z'^2 - c^2 \gamma^2 \left(\Delta t'^2 + \frac{2v\Delta t' \Delta x'}{c^2} + \frac{v^2 \Delta x'^2}{c^2} \right)$$

$$= \gamma^2 \left(1 - \frac{v^2}{c^2} \right) \Delta x'^2 + \Delta y'^2 + \Delta z'^2 - c^2 \gamma^2 \left(1 - \frac{v^2}{c^2} \right) \Delta t'^2$$

$$= \underline{\Delta x'^2 + \Delta y'^2 + \Delta z'^2 - c^2 \Delta t'^2}$$

This invariant quantity is called the *interval* and plays an analogous role in space-time to distance in space.

Example Events P and Q occur simultaneously 100 m apart along the x-axis of reference frame A. Frame B is moving at constant velocity v along the x-axis and its x'-axis is parallel to the x-axis. What is the separation of the two events in B if event Q occurs 1.0 μs before event P in this frame?

This problem could be solved using the Lorentz transformation, but it is easier to use the invariant property of the interval between the two events:

In A: $\Delta s^2 = \Delta x^2 - c^2 \Delta t^2$

In B: $\Delta s'^2 = \Delta x'^2 - c^2 \Delta t'^2$

But $\Delta s^2 = \Delta s'^2$

so $\Delta x^2 - c^2 \Delta t^2 = \Delta x'^2 - c^2 \Delta t'^2$

$$100^2 - 0 = \Delta x'^2 - \left(3 \times 10^8 \text{ m s}^{-1} \times 10^{-6} \text{ s}\right)^2$$

$$\Delta x'^2 = 10^5$$

Separation of events in B = $\Delta x' = 320$ m

Example Events P and Q occur at co-ordinates P (x_1, ict_1) and Q (x_2, ict_2) in frame A. What is the condition that must be satisfied if they are to occur simultaneously in B which is also an inertial reference frame but which moves at constant velocity v in the positive x direction?

Previously we solved this problem using the Lorentz transformation, here it is solved using the invariance of the interval.

$$\Delta s^2 = (x_2 - x_1)^2 - c^2 (t_2 - t_1)^2$$

$$\Delta s'^2 = (x'_2 - x'_1)^2 - c^2 (t'_2 - t'_1)^2 = (x'_2 - x'_1)^2 \quad \text{(simultaneous in this frame)}$$

$$\Delta s^2 = \Delta s'^2$$

$$(x_2 - x_1)^2 - c^2 (t_2 - t_1)^2 = (x'_2 - x'_1)^2$$

The RHS is obviously positive, so the LHS is also positive:

$$(x_2 - x_1)^2 - c^2 (t_2 - t_1)^2 \geq 0$$

$$(x_2 - x_1)^2 \geq c^2 (t_2 - t_1)^2$$

This means their spatial separation in A must exceed the distance light could travel during the time between the two events. In space-time they are separated by a *spacelike* interval. It also implies that they cannot be connected by any causal influence.

Example Two twins separate at $t = t' = 0$. Twin A remains behind while twin B travels at velocity v to a star a distance Δx away along A's x-axis. What is the time in each frame when B arrives at the star?

Two events are defined - departure and arrival. The interval between these two events has the same magnitude in both frames and can be used to relate t to t' for the arrival of B at the star.

$\Delta x^2 - c^2\Delta t^2 = 0 - c^2\Delta t'^2$ ($\Delta x'^2 = 0$ because B's origin travels with b to the star!)

Also $\Delta x = v\Delta t$

$v^2\Delta t^2 - c^2\Delta t^2 = -c^2\Delta t'^2$

$$\Delta t'^2 = \left(\frac{c^2 - v^2}{c^2}\right)\Delta t^2 = \left(1 - \frac{v^2}{c^2}\right)\Delta t^2$$

$$\Delta t' = \frac{\Delta t^2}{\gamma}$$ (the time dilation formula!)

3.4.2 The Fourth Dimension.

If we compare the invariant interval s between two events in space-time with the distance d between points in space we can introduce a new unit along the 'time' axis so that space-time geometry acts like Euclidean geometry in 4 dimensions.

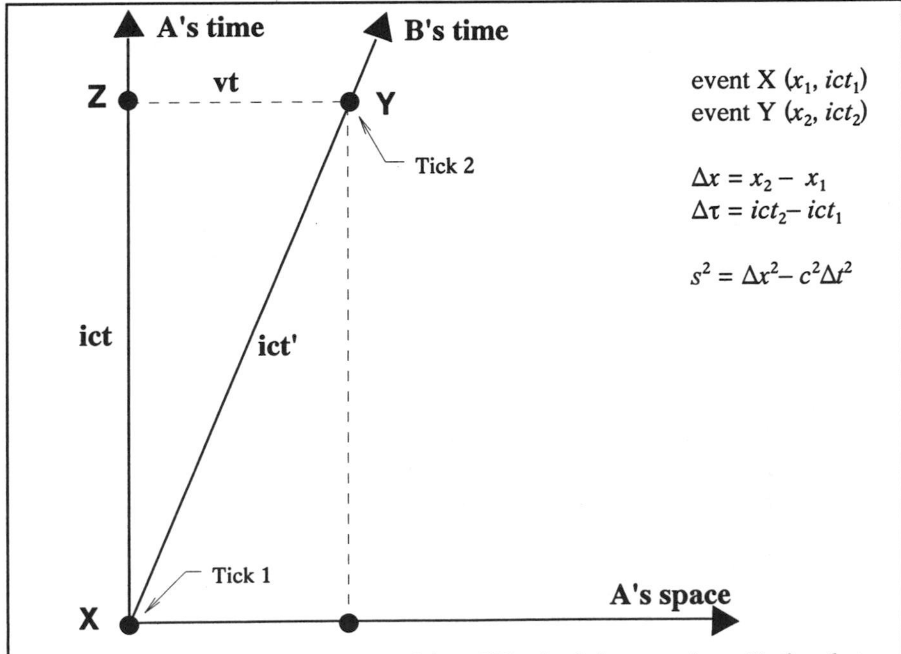

fig 3.17 The interval between two ticks of B's clock in spacetime. Notice that time and space co-ordinates of X and Y are different on each set of axes.

Euclidean 3D distance: $d^2 = \Delta x^2 + \Delta y^2 + \Delta z^2$

$$s^2 = \Delta x^2 + \Delta y^2 + \Delta z^2 - c^2 \Delta t^2$$

Define a new 'time' co-ordinate τ such that $\tau^2 = -c^2t^2$ and the expression for the interval becomes:

$$s^2 = \Delta x^2 + \Delta v^2 + \Delta z^2 + \Delta \tau^2 \tag{3.7}$$

τ itself has the form, $\tau = ict$ (where $i^2 = -1$) and the co-ordinates of an event in space-time are now (x, y, z, τ) or (x, y, z, ict). The advantage of this is that transforming from one inertial reference frame to another in this co-ordinate system is analogous to a rotation in 4D space-time and the invariance of the interval between events can be interpreted as the invariance of the magnitude of a 4-dimensional space-time 'displacement vector' or *4-vector*.

3.4.3 Four Vectors.

We can define vectors in space-time in much the same way as we define them in normal 3-dimensional space. These vectors are called '4-vectors'. In space-time points have four co-ordinates and represent events so a displacement 4-vector will point from one event to another and could be represented by:

$$s_4 = (x, y, z, ict)$$

In this case the vector points from the origin. Otherwise it would have co-ordinate differences rather than co-ordinates as components. Its magnitude is found by using Pythagoras's theorem:

$$s_4^2 = x^2 + y^2 + z^2 - c^2t^2$$

What makes this particularly useful is the fact that it is an invariant quantity and therefore has the same value in all inertial reference frames. Knowing the interval in one reference frame helps in calculating the co-ordinates of events in another.

Consider two events which occur at the origin of B's laboratory moving at velocity v in A's positive x-direction. (As usual we assume that the origins coincide at $t = t' = 0$ and that event 1 occurs at the origin as the origin cross.)

For B:	event 1 (0,0)	event 2 (0,*ict'*)
For A:	event 1 (0,0)	event 2 (x,*ict*)

Interval s between events $\quad s^2 = 0 - c^2t'^2 = x^2 - c^2t^2$

But $\quad x = vt \quad$ so $\quad -c^2t'^2 = v^2t^2 - c^2t^2$

leading to

$$t = \frac{t'}{\sqrt{1 - \frac{v^2}{c^2}}} = \gamma t'$$

This is the time dilation formula! The two events at B's origin could be ticks of a lightclock. The period t' would be the proper time between ticks and t would be the dilated time observed from A's reference frame. The moving clock takes longer to tick than a similar stationary clock. Although the result is familiar the method used to derive is now geometric. If we cast relativity into a space-time form then time dilation is a consequence of the invariance of the interval.

We can further emphasize the role of space-time geometry by deriving the same formula pictorially using Pythagoras's theorem.

Co-ordinates of tick 1:	(0,0)	for A	and	(0,0)	for B
Co-ordinates of tick 2:	(vt,ict)	for A	and	$(0,ict')$	for B

By Pythagoras's theorem in triangle XYZ $\quad (vt)^2 + (ict)^2 = (ict')^2$

Leading to the time dilation formula $\quad t = \dfrac{t'}{\sqrt{1-\dfrac{v^2}{c^2}}}$ $\quad$ as before.

3.4.4 Types Of Interval.

It is clear from the expression for the interval that there will be three distinct classes of interval between space-time events. These are:

- $s^2 = \Delta d^2 - c^2\Delta t^2 < 0$ a 'spacelike' interval
- $s^2 = \Delta d^2 - c^2\Delta t^2 = 0$ a 'null' or 'lightlike' interval
- $s^2 = \Delta d^2 - c^2\Delta t^2 > 0$ a 'timelike' interval.

(In each case $\Delta d^2 = \Delta x^2 + \Delta y^2 + \Delta z^2$, that is the 3D distance.)

- **Spacelike intervals** occur between events with a spatial separation less than the distance light could travel between the times of their occurrence. The speed of light is the fastest causal speed in the Universe so points separated by a spacelike interval *can* be causally connected. In other words, if A is the first event then the effects of event A could have some influence on the event B. B will be inside the future ligthcone of A and A will be in the past lightcone of B. A spacelike interval is represented by a displacement 4-vector which makes an angle of less than 45° with the time axis of a space-time diagram.
- **Null or 'lightlike' intervals** represent events that could just be connected by signals travelling at the speed of light. If A precedes B then a light ray leaving A at the moment A occurs will just arrive at B as B occurs. B lies on the surface of A's future lightcone, A lies on the surface of B's past lightcone.
- **Timelike intervals** separate events that cannot be causally connected (at least not by signals travelling at or below the speed of light). The spatial separation of these events exceeds the distance that light can travel in the time between

them. A will lie outside B's past lightcone and B will lie outside A's future lightcone.

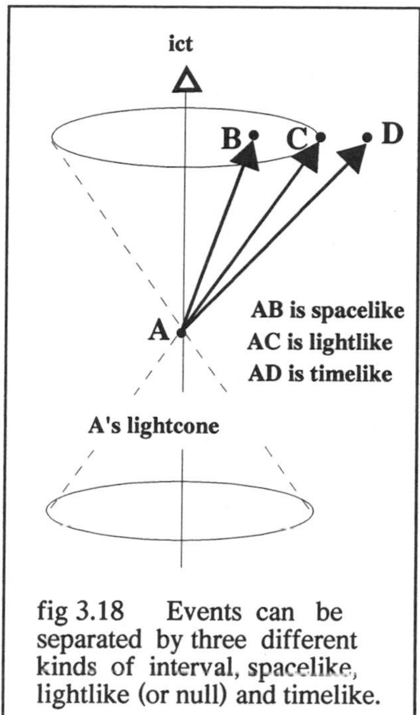

fig 3.18 Events can be separated by three different kinds of interval, spacelike, lightlike (or null) and timelike.

The invariance of the interval ensures that the nature of an interval (spacelike, timelike, or null) will be the same for all observers regardless of their motion. This is reassuring, it means that the order of causally connected events will be the same for all observers. (This is one of the areas where relativity and quantum theory sometimes come into conflict. In quantum theory the effect of observing one part of a system is to collapse the wavefunction for the whole system into a new state. This means that the probability of particular events being measured elsewhere changes instantaneously. There is therefore a correlation between the outcome of a particular measurement in one place and the possible outcomes of events elsewhere, even though the separation of the two measurement events is timelike!)

3.4.5 Space-time Rotation.

The effect of motion is to rotate space-time axes. The higher the velocity of the moving object the greater the angle of inclination of its worldline to that of the laboratory through which it moves. This angle can be used to measure velocity and then we can use simple trigonometric relations to determine how velocities add. Fig 3.19 illustrates this idea. The velocity is directly proportional to the tangent of the angle. In fact, if we rewrite the Lorentz transformation equations in terms of x and ict co-ordinates it is clear that they have the same form as the equations for a 2D rotation of axes. This can be used to define values for the sine, cosine and tangent of the space-time rotation angle ϕ, and this will help us to interpret the effects of special relativity geometrically.

Lorentz transformation to co-ordinates x', ict' in a reference frame B moving at velocity v along the positive x-axis of a reference frame A:

$$x' = \gamma(x - vt) \quad \rightarrow \quad x' = \gamma x - \gamma \frac{v}{ic}(ict) \quad \rightarrow \quad x' = \gamma x + i\beta\gamma(ict) \qquad (1)$$

$$t' = \gamma\left(t - \frac{vx}{c^2}\right) \quad \rightarrow \quad (ict') = \gamma(ict) - \gamma \frac{iv}{c} x \quad \rightarrow \quad (ict') = -i\beta\gamma x + \gamma(ict) \qquad (2)$$

Compare equations (1) and (2) above with the equations for a clockwise rotation of 2D axes:

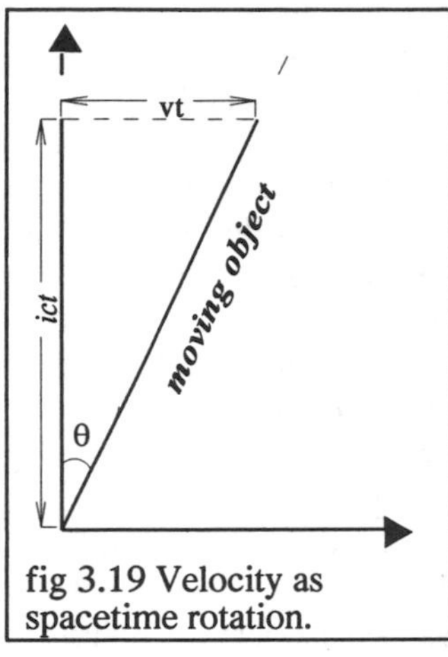

fig 3.19 Velocity as spacetime rotation.

$$x' = x\cos\phi + y\sin\phi$$
$$y' = -x\sin\phi + y\cos\phi$$

The Lorentz transformation would have the same form as a 2D rotation if:

$$\sin\phi = i\beta\gamma$$
$$\cos\phi = \gamma$$
$$\tan\phi = i\beta$$

The final one of these three shows that velocity is indeed directly proportional to the tangent of the angle of rotation. Here we have related 2D spatial rotation to 2D space-time rotation, but space-time is actually 4D. This is not a problem. Whenever relative motion is involved the space-time axes can be chosen so that the velocity involved is parallel to the common x and x' axes, thus reducing the problem effectively to 2D. This is very convenient if we wish to draw space-time diagrams!

The Lorentz transformation is shown below in matrix form (for both 4D and 2D).

$$\begin{pmatrix} x' \\ ict' \end{pmatrix} = \begin{pmatrix} \gamma & i\beta\gamma \\ -i\beta\gamma & \gamma \end{pmatrix} \begin{pmatrix} x \\ ict \end{pmatrix} \qquad \begin{pmatrix} x_1' \\ x_2' \\ x_3' \\ x_4' \end{pmatrix} = \begin{pmatrix} \gamma & 0 & 0 & i\beta\gamma \\ 0 & 1 & 0 & 0 \\ 0 & 0 & 1 & 0 \\ -i\beta\gamma & 0 & 0 & \gamma \end{pmatrix} \begin{pmatrix} x_1 \\ x_2 \\ x_3 \\ x_4 \end{pmatrix} \tag{3.8}$$

In the 4D version x_1 to x_4 are x, y, z, and ict. The y and z co-ordinates are unaffected by relative motion along the common x and x' axes, hence the 1s.

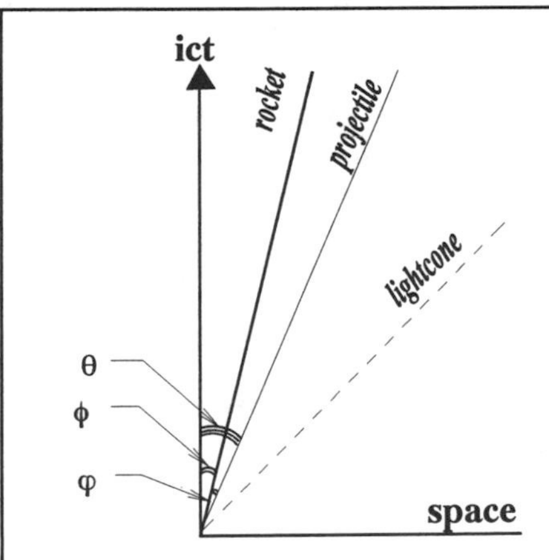

fig 3.20 Velocity addition as successive rotations

Consider a situation in which a rocket moves past us at velocity v and fires a projectile in the same direction at velocity u relative to the rocket (as measured from the rocket). This is shown in the space-time diagram at left. The worldline of the rocket makes an angle ϕ with our own worldline and the projectile makes an angle φ with that of the rocket. The projectile's worldline makes an angle $\theta = (\phi + \varphi)$ with our worldline, so the velocity w of the projectile relative to our reference frame can be found from the tangent of θ. This is derived from the addition formula for tangents in trigonometry.

$$\tan\theta = \tan\ (\phi + \varphi) = \frac{\tan\phi + \tan\varphi}{1 - \tan\phi\ \tan\varphi}$$

But $\tan\theta = i\beta_w \qquad \tan\phi = i\beta_v \qquad \tan\varphi = i\beta_u$

Therefore: $$i\beta_w = \frac{i\beta_u + i\beta_v}{1 - i\beta_u i\beta_v} \quad \rightarrow \quad \beta_w = \frac{\beta_u + \beta_v}{1 + \beta_u\beta_v}$$

giving: $$w = \frac{u + v}{1 + \frac{uv}{c^2}}$$

This is, of course, the familiar velocity addition formula.

3.4.6 Space-time Trigonometry.

We have already seen that relativistic effects like time dilation, length contraction and simultaneity disagreements have a simple geometric interpretation. Trigonometry can also be used to derive quantitative equations for them.

fig 3.21a

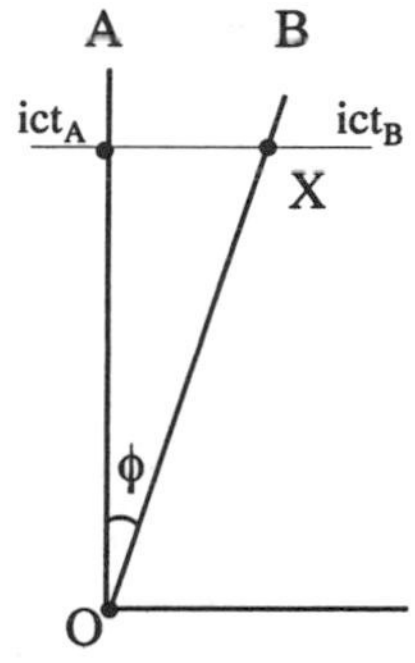

fig 3.21b

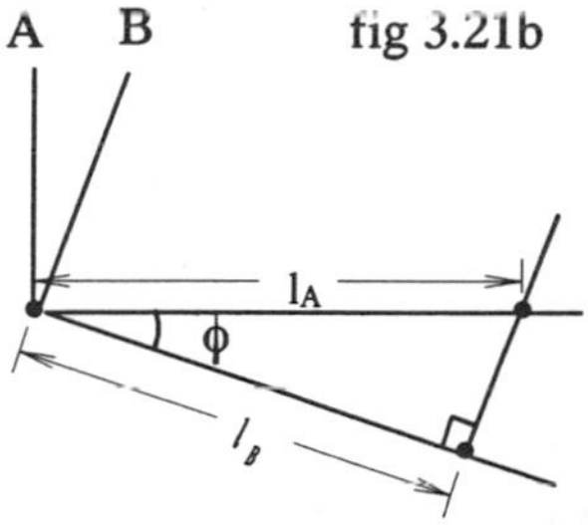

fig 3.21c

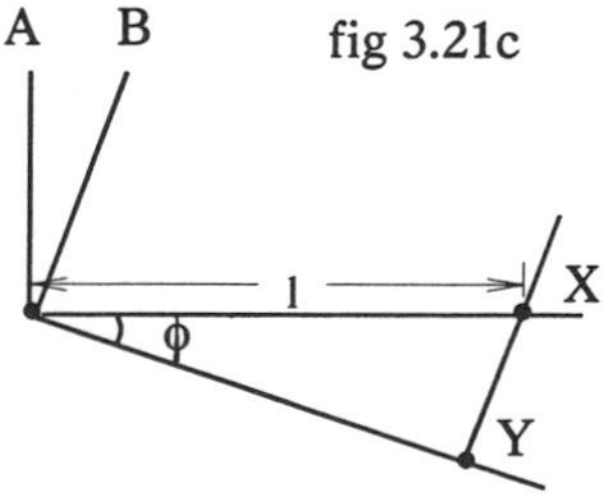

fig 3.21 Spacetime trigonometry

(a) Time dilation.

(b) Length contraction.

(c) Simultaneity.

Time Dilation (fig 3.21a): A and B start their clocks at $t = t' = 0$ and compare their readings when B arrives at X.

$$ict_A = ict_B \cos\phi = ic\gamma t'_B$$

$$t_A = \gamma t_B \quad \text{or} \quad t_B = \frac{t_A}{\gamma}$$

Since $\gamma > 1$ more time passes on A's clock than B's. This is time dilation (as seen by A). Of course, if we drew the diagram from B's point of view the cosine would change sides of the equation giving the symmetrical result for B.

Length contraction (fig. 3.21b): A rod has proper length l_B in B's frame, but moves at velocity v relative to A. The length of the rod in A's frame is then l_A (equal to the simultaneous equation of the rod ends in A's frame).

$$l_B = l_A \cos\phi = \gamma l_A$$

$$l_A = \frac{l_B}{\gamma}$$

This is the familiar effect of the contraction of moving rods (this rod is at rest with respect to B but moving relative to A).

Simultaneity (fig 3.21c): A and B pass at relative velocity v. Event X occurs a distance l along the x axis at $t = 0$ (for A). B determines that this event occurs at a time t' which is not zero. The interval XY is parallel to B's worldline so it gives a measure of the simultaneity 'error' between the two observers for events at X.

$$ic\Delta t_B = l\, sin\,\phi = li\beta\gamma$$

$$\Delta t_B = \frac{lv\gamma}{c^2} \tag{3.9}$$

$$\text{If } v << c \text{ then} \quad \Delta t_B \approx \frac{lv}{c^2} \tag{3.10}$$

Note that the space-time diagrams used in fig 3.21 treat the space axis differently to our previous diagrams. This is part of the analogy with 2D rotations, but can lead to problems interpreting the time order of events. The previous form is more useful for qualitative interpretations of physical phenomena, and when it comes to general relativity a proper tensor method must be used.

Example: If $\tan\phi = iv/c$ what do $\sin\phi$ and $\cos\phi$ represent?

$$\tan\phi = \frac{iv}{c} = i\beta$$

$$\sin\phi = \sqrt{\frac{\tan^2\phi + 1}{\tan^2\phi}} = \frac{-\beta^2}{1-\beta^2} = i\beta\gamma \tag{3.11}$$

$$\cos\phi = \sqrt{\frac{1}{1+\tan^2\phi}} = \sqrt{\frac{1}{1-\beta^2}} = \gamma$$

3.4.7 Calibrating Spacetime.

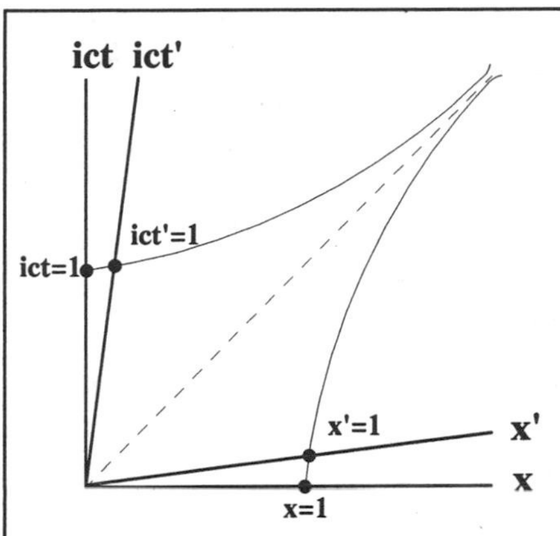

fig 3.22 The hyperbolic curves join all points unit interval from the origin.

When we discussed time dilation and the twin paradox it was clear that the length of the worldline on the spacetime diagram is not always a good indication of the magnitude of the interval represented by that length. It turned out that the longer worldline of the travelling twin represented a smaller interval and less time passed than that of the stay at home twin whose spacetime reference frame we were using. We have also seen that the interval measured along the worldline of a light ray will always be of zero 'length' (since events lying on the surface of a lightcone are separated by a lightlike or null interval).

Putting these observations together suggests that the spacetime scale gets 'stretched' as we rotate the spacetime axes. This can be demonstrated more formally using the invariance of the interval. Consider the locus of events that occur at a unit interval from the origin:

$$x^2 - c^2t^2 = 1$$

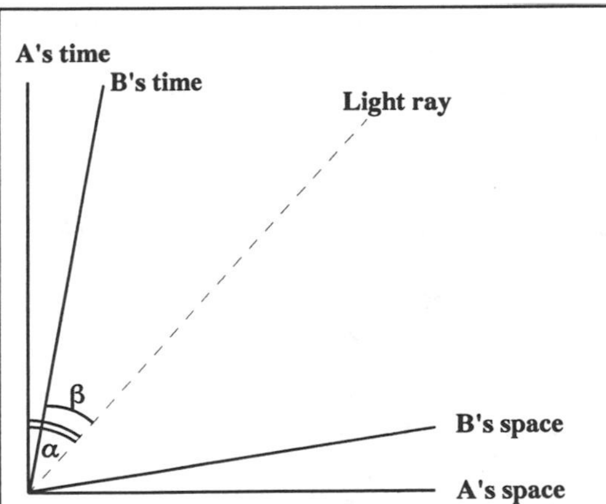

fig 3.23 The velocity of light is a constant for all observers so the angle beta in A's spacetime diagram appears compressed.

If $t = 0$ this intercepts the x-axis at $x = +1$ or -1. If $x = 0$ it intercepts the 'time' axis when $ict = -1$ or $+1$. For large times and distances x and t will approach infinity together and the locus becomes asymptotic to the line $x = ct$, that is the surface of the lightcone originating at the origin of co-ordinates. The equation describes a rectangular hyperbola as shown in fig 3.22.

The significance of this hyperbola is that it will intercept rotated axes at unit intervals too. This follows from the invariance of the interval:

$$x^2 - c^2t^2 = 1 = x'^2 - c^2t'^2$$

An event lying on the hyperbola is one unit interval from the original event in all reference frames.

It is probably worth noting that the distortion of scale also affects angles. This is obvious if we consider the velocity of light which is an invariant for all observers. A light ray moving through a laboratory will have a worldline at 45° to that of an observer at rest in the laboratory. If a second observer moves through the laboratory at velocity v in the same direction as the light ray her worldline will lie between those of the laboratory observer and the light ray. The angle this makes with the lightcone is clearly less than 45° on our diagram, but it must be the same angle in the moving frame since it represents the velocity of light. Just as the scale expands with rotation, angles are 'compressed', see fig 3.23.

3.5 VELOCITY MOMENTUM AND ENERGY.

3.5.1 The Velocity 4-Vector.

The importance of four-vectors stems from their invariant magnitude and the fact that their co-ordinates transform according to the Lorentz transformation equations. We can use this information to 'construct' more 4-vectors. In 3-dimensional classical mechanics velocity is simply the rate of change of displacement; in space-time we can use this as a starting point to define a new '4-velocity'.

Consider an object moving through A's laboratory at velocity v in the positive x-direction. The 4-displacement s will change by:

$$\delta s = (\delta x, \delta y, \delta z, ic\delta t)$$

It is quite straightforward to show that this increment is itself a 4-vector. However, a moment's reflection is sufficient to realize that the quantity $\frac{\delta s}{\delta t}$ is not a 4-vector and so cannot represent a 'space-time velocity'. The problem is with the time increment δt which is measured in A's reference frame. Whereas the numerator has an invariant magnitude the denominator has a magnitude that depends on reference frame. This is clear if we write down the magnitude of this 'vector' in two reference frames that move relative to one another:

$$\frac{\delta s^2}{\delta t^2} = \frac{\delta x^2 + \delta y^2 + \delta z^2 - c^2 \delta t^2}{\delta t^2} \quad \text{and} \quad \frac{ds'^2}{dt'^2} = \frac{dx'^2 + dy'^2 + dz'^2 - c^2 dt'^2}{dt'^2}$$

The numerators of both expressions are equal. The denominators are not. The solution is to find the rate of change of 4-displacement with respect to the *proper time of the moving object* and not the subjective time of each observer. This works because the proper time is itself an invariant and equal to: $\frac{dt}{\gamma} = \frac{dt'}{\gamma'}$ etc.

(If it is not obvious that the proper time is an invariant you should prove that it is using the Lorentz transformation equations.)

We can now define the 4-velocity: $v_4 = (\gamma v_x, \gamma v_y, \gamma v_z, ic\gamma)$

This will have an invariant magnitude: $|v_4|^2 = \gamma^2 v^2 - \gamma^2 c^2 = -c^2$

This is interesting. The magnitude is c and is independent of the actual 3-dimensional velocity. What does this mean? In the frame of reference of the moving body the 4-velocity measures the rate at which this body moves through space-time - that is it is the increase in interval per unit proper time. The unique value of the invariant 4-velocity magnitude could be interpreted as saying that all material objects move through space-time at the same rate, the speed of light. Even though we are 'at rest' in our own reference frame time moves on and our progress through time (along our own worldline) is at the speed of light. When we observe a moving object its time axis is tipped relative to ours so, although we see it make progress at the same overall rate through space-time, it moves partly through space and partly through time and so time appears to run slow for it. In this sense the velocity of light is seen as the universal rate of progress along our worldlines and perhaps the 'speed of time' would be a more appropriate title for it! Pushing this idea a little further, light rays are special in moving through space so fast (at the speed of light) that they have no motion through time at all (that is no time passes in the light reference frame between emission and absorption although time does pass for external observers watching the passage of light).

The components of the velocity 4-vector will transform like the components of the displacement 4-vector. This means we can apply the Lorentz transformation equations to individual components to find their value in any other inertial reference frame (we previously derived these transformations in chapter 4).

Example To derive an expression for 4-velocity the 4-vector displacement was differentiated with respect to the proper time. Show that the proper time $\delta t'$ between two events is equal to $\delta t'/\gamma$ in all inertial reference frames.

In the rest frame of the events both events occur at the same place but at different times so $\delta x=0$.

$$-c^2dt^2 = dx'^2 - c^2dt'^2$$

$$dx' = vdt'$$

$$-c^2dt^2 = vdt - c^2dt'^2$$

$$dt^2 = \frac{d't}{g}$$

3.5.2 The 4-Momentum.

Previously we assumed a relativistic momentum of the form $p = m(v)v$ and used this to look for an expression for the relativistic mass $m(v)$ such that the conservation laws for mass and energy would hold in an interaction. The result was that:

$$m(v) = \gamma m_o \qquad \text{and} \qquad p = \gamma m_o v \qquad (m_o \text{ is rest mass})$$

The expression for momentum is simply rest mass multiplied by γv that we now recognize as the first three components of the 4-velocity. This suggests that the space-time momentum can be constructed as:

$$p_4 = m_o v_4$$

Since m_o is an invariant scalar this is guaranteed to be a 4-vector. It will have components:

$$p_4 = (\gamma m_o v_x, \gamma m_o v_y, \gamma m_o v_z, ic\gamma m_o) \qquad (3.12)$$

We are familiar with the first 3 components but the fourth is new. Its significance becomes clearer if we express it differently:

$$ic\gamma m_o = icm = \frac{imc^2}{c} = \frac{iE}{c}$$

so $$p_4 = \left(\underline{p}, \frac{iE}{c}\right) \qquad \text{where} \qquad \underline{p} = \gamma m_o \underline{v}$$

In space-time the fourth component of the momentum 4-vector is proportional to total energy. This is a startling result. The most important property of linear momentum is that it is conserved in all interactions. That means, in three dimensions, that each momentum component is separately conserved. Extending to four dimensions means that conservation of total energy is a consequence of conservation of the momentum 4-vector. Whereas energy and momentum obey separate conservation laws in classical mechanics they are now seen to be aspects of a single conservation law.

The components of this 4-vector will transform according to the Lorentz transformation equations. This gives us rules for the transformation of momentum and energy from one inertial reference frame to another. The prime in the equations below represents co-ordinates in a reference frame moving at velocity v along the positive x-axis of the unprimed frame.

$$\begin{aligned}
p'_x &= g\left(p_x - \frac{vE}{c^2}\right) & p_x &= g\left(p'_x + \frac{vE'}{c^2}\right) \\
p'_y &= p_y & p_y &= p'_y \\
p'_z &= p_z & p_z &= p'_z \\
E' &= g(E - vp_x) & E &= g(E' + vp'_x)
\end{aligned} \qquad (3.13)$$

The magnitude of the momentum 4-vector is an invariant. Its value is:

$$p^2 - \frac{E^2}{c^2} = p'^2 - \frac{E'^2}{c^2} \quad \text{etc.}$$

This is the invariant we constructed in section 2.14.2. Since an invariant has the same value in all inertial reference frames we can equate it to the value taken in a frame at rest with respect to the moving object, that is the frame in which $p^2 = 0$. In this frame the total energy of the moving object is simply its rest energy E_o.

$$E^2 - p^2c^2 = E_o^2 \tag{3.14}$$

This is one of the most useful relations in particle physics.

Example Prove that the magnitude of a particle's 4-momentum is an invariant.

To simplify this we choose axes so that the particle's velocity is parallel to the x- and x'-axes and then carry out a Lorentz transformation from an inertial frame S to an inertial frame S$'$ and show that the expression for magnitude has the same form in both.

$$p_4 = \left(p_x, \frac{iE}{c}\right) \qquad |p_4|^2 = p_x^2 - \frac{E^2}{c^2}$$

$$p'_4 = \left(p'_x, \frac{iE'}{c}\right) \qquad |p'_4|^2 = p'^2_x - \frac{E'^2}{c^2}$$

Lorentz transformation:

$$p'_x = \gamma\left(p_x - \frac{vE}{c^2}\right) \qquad \frac{iE'}{c} = \frac{i\gamma}{c}(E - vp_x)$$

$$p'^2_x - \frac{E'^2}{c^2} = \gamma^2\left(p_x - \frac{vE}{c^2}\right)^2 - \frac{\gamma^2}{c^2}(E - vp_x)^2$$

$$= \frac{\gamma^2}{c^2}\left(p_x^2c^2 - 2p_x vE + \frac{v^2E^2}{c^2} - E^2 + 2p_x vE - v^2p_x^2\right)$$

$$= \gamma^2\left[\left(1 - \frac{v^2}{c^2}\right)p_x^2 - \left(1 - \frac{v^2}{c^2}\right)\frac{E^2}{c_2}\right] = \underline{p_x^2 - \frac{E^2}{c^2}}$$

3.5.3 Interactions In Space-time.

The energy component of the 4-momentum means that a stationary particle has a non-zero 4-momentum:

$$p_4 = \left(0, \frac{iE_o}{c}\right) \qquad \text{(for a particle of rest energy } E_o \text{ at rest)}$$

A moving particle is represented by an inclined 4-vector, as shown in fig 3.24 below.

4-vectors provide a pictorial representation of space-time interactions. The resulting space-time diagrams can be analysed using simple trigonometry. These ideas are demonstrated in section 3.6.

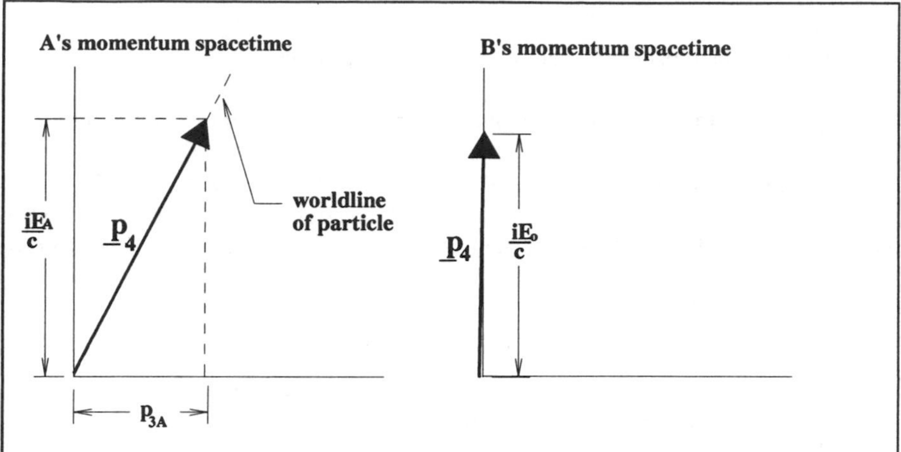

fig 3.24 The four-momentum of a particle relative to a laboratory frame (A) and in its own rest frame (B). The projection of four momentum onto the 'space axis' gives the component of linear momentum in that frame.

3.6 SPACE-TIME MECHANICS.

3.6.1 Collisions And Explosions.

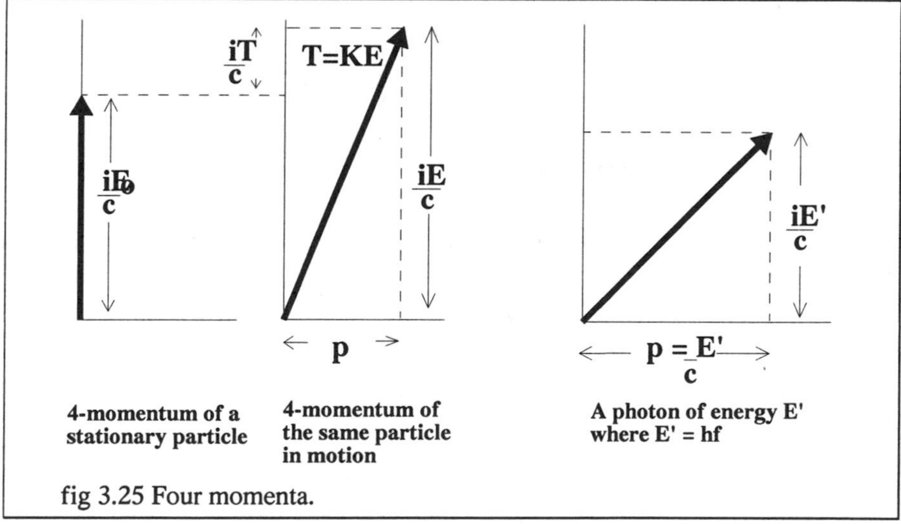

fig 3.25 Four momenta.

Figure 3.25 above shows the 4-momentum for stationary and moving particles and for a photon. In a collision or interaction the total momentum 4-vector will be conserved. This means each component is conserved separately. The 'time' component conserves total energy (that is rest energy plus kinetic energy of the particles) and the 'space' component conserves three-dimensional momentum (in its relativistic form). These values will change if we transform to another inertial frame but the invariant magnitude of the 4-momentum will be the same.

In an interaction involving more than one particle the total 4-momentum is the vector sum of the 4-momenta of each particle. Since the final 4-momentum of the system is unchanged by the interaction we can use simple trigonometric constructions and calculations to determine outcomes.

Consider an inelastic collision between two identical particles. Figures 3.26 and 3.27 below show this in two different inertial reference frames, the first is the rest frame of particle one (frame S), the second is the centre of mass frame (frame S'). We can analyse this collision to determine the rest mass of the composite body formed by collision. To simplify matters we use reference axes such that all relative motion occurs along the common x- and x'-axes of the frames involved. Thus 3-momentum can be represented simply by p (where $p = \gamma m_o v_x$) or p'.

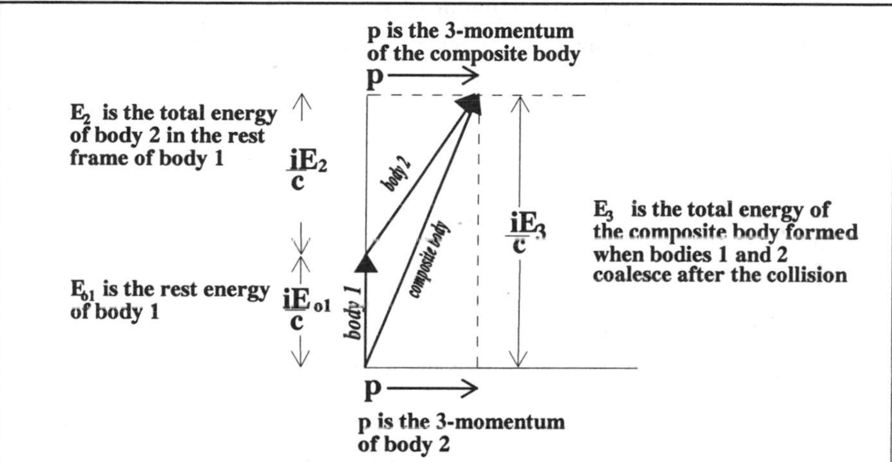

fig 3.26 4-momenta in rest frame of body 1. Since 4-momentum is conserved the 4-momentum of the composite body after the collision must equal the sum of 4-momenta before the collision. The 'time' and 'space' components are separately conserved and correspond to laws of conservation of total energy and linear 3-momentum.

This collision is analysed below. Notice that when $T<<E_o$, which is the condition for all classical collisions, we have $E_o(comp) \approx 2E_o$, and the rest mass of the composite is simply the sum of rest masses of the two colliding bodies. At higher incident energies the rest mass of the composite is greater than the sum of rest masses of the colliding bodies. However, it is clear from the equation that the increase in rest mass of the composite is less than the mass equivalent of T. This is because the composite must have linear momentum in the same direction as the original body 2, so it necessarily ends up with some of the collision energy remaining as kinetic energy.

In the frame of body 1 (S):

E_{o1}, E_{o2}, E_{o3} are the rest energies of the particles and $E_{o1} = E_{o2} = E_o$
E_1, E_2, E_3 are the total energies of the particles ($E_1 = E_o$)
p is the 3-momentum of the incident particle (particle 2)
T is the kinetic energy of the incident particle.

$$p_4(body\ 1) = \left(0, \frac{iE_{o1}}{c}\right) \qquad p_4(body\ 2) = \left(p, \frac{iE_2}{c}\right)$$

$$E_2 = E_o + T \qquad \text{where } T \text{ is the kinetic energy of body 2}$$

so $$p_4(comp) = \left(p, \frac{i(2E_o + T)}{c}\right)$$

we can use the invariant magnitude of $p_4(body\ 2)$ to express p in terms of energies. In the rest frame of particle 2 its energy is E_o and its 3-momentum zero so:

$$p^2 - \frac{(2E_o + T)^2}{c^2} = -\frac{E_o^2}{c^2}$$

giving $$c^2 p_2^2 = T(2E_o + T) \qquad (1)$$

we can now do a similar thing with the invariant magnitude of the 4-momentum of the composite body. That is we can equate its value in the frame of body 1 to its value in its own rest frame.

$$p^2 - \frac{(2E_o + T)^2}{c^2} = -\frac{E_o^2}{c^2}$$

using (1): $$T(2E_o + T) - (2E_o + T)^2 = -E_{o3}^2$$

giving: $$E_{o3}^2 = 2E_o(2E_o + T) \qquad (3.15)$$

If we are producing particle collisions in order to create new particles we really want as much as possible of the work done by the accelerator to be converted to rest mass in the collision. The example above is inefficient in this respect since the products of the collision retain some kinetic energy. It will be useful to compare result (3.15) with the rest mass of a composite body formed when the same two particles collide from opposite directions in the centre of mass frame. To make the comparison realistic we shall assume that each particle is given a kinetic energy $T/2$ by the accelerator. The total work done is then the same as in the example above.

In the CM frame (S′):

T/2 is the kinetic energy of each body.
p′ is the linear momentum of each body.

$$p_4(body\ 1) = \left(-p', \frac{i\left(E_o + T/2\right)}{c}\right) \qquad p_4(body\ 2) = \left(p', \frac{i\left(E_o + T/2\right)}{c}\right)$$

and $$p_4(comp) = p_4(body\ 1) + p_4(body\ 2) = \left(0, \frac{i(2E_o + T)}{c}\right)$$

but $$p_4(comp) = \left(0, \frac{iE_{o3}}{c}\right)$$

therefore $$E_{o3} = 2E_o + T$$

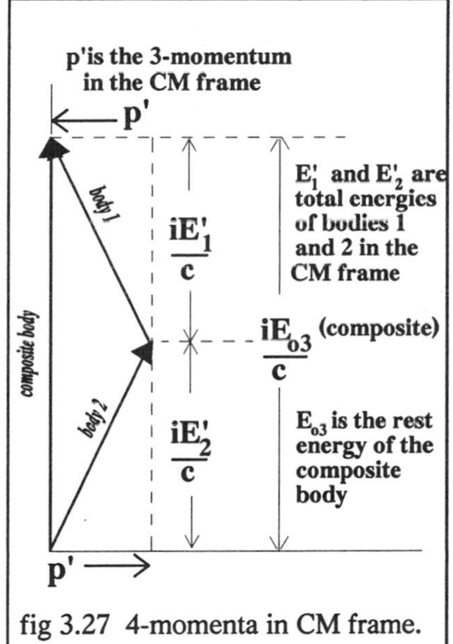

fig 3.27 4-momenta in CM frame.

All the input energy *T* is converted to rest mass in the composite body. (Note once again that, if *T* is small, the result agrees with classical physics.) We can see how significant this is if we compare the expressions for the total rest energy created in the two collision processes for an equal total input energy *T*:

$$\frac{E(\text{collider})}{E(\text{fixed target})} = \frac{2E_o + T}{\sqrt{2E_o(2E_o + T)}} = \sqrt{1 + \frac{T}{2E_o}} \qquad (3.16)$$

If $T = 100E_o$ this ratio is $\sqrt{51} \approx 7$. As the energy input increases so does the advantage in using a collider over a fixed target device.

An explosion is like an inelastic collision run in reverse, and can be analysed using the same approach as in the collision example illustrated in fig 3.27. Figure 3.28 shows the momentum 4-vectors for gamma-ray emission by a stationary nucleus. The sum of momentum 4-vectors after the emission must equal their sum before. This means that the momentum 4-vectors before and after the emission form a closed figure. The vectors are not to scale.

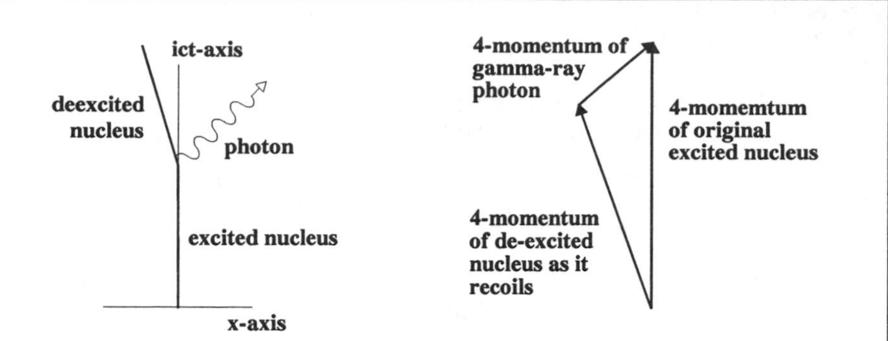

fig 3.28 Photon emission by a stationary excited nucleus. The diagram on the left is a spacetime diagram. On the right the 4-vector diagram shows how the sum of final 4-momenta must equal the 4-momentum of the original stationary nucleus.

3.6.2 The Compton Effect.

According to quantum theory the energy of a photon is $E = hf = \frac{hc}{\lambda}$ where λ is the wavelength of the radiation and h is Planck's constant. In the early 1920s Arthur Compton made a systematic study of how photons scatter from electrons in target atoms. He won the Nobel Prize for his work, in 1927.

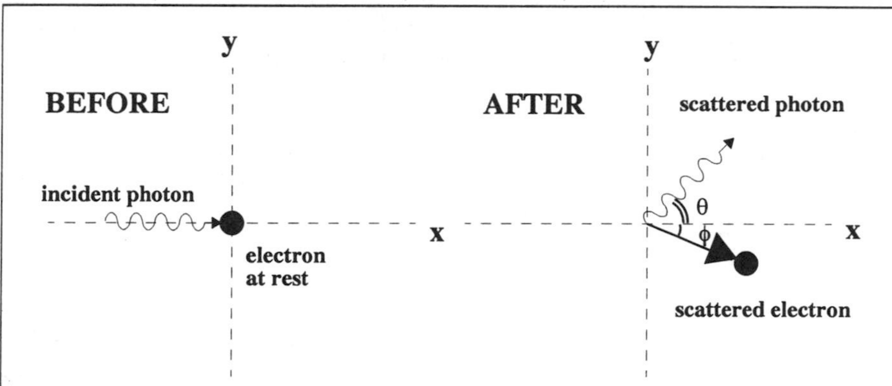

fig 3.29 Compton scattering. These two diagrams show the scattering in the xy-plane. The incident photon transfers energy and momentum to the electron.

Compton treated the incident photon as a mechanical particle and applied the laws for energy and momentum conservation to the collision in order to calculate how the energy of the scattered photon depends on the angle through which it is scattered. This is analysed below using conservation of 4-momentum and the energy-momentum invariant. Photon and electron energies before and after scattering are given by E_o, E, and e_o, e respectively. Momenta are E_o/c and E/c for the photon and p for the scattered electron. Angles are defined in fig 3.29.

Before: $$p_4(photon) = \left(\frac{E_o}{c}, 0, 0, \frac{iE_o}{c}\right) \quad (1)$$

$$p_4(electron) = \left(0, 0, 0, \frac{ie_o}{c}\right) \quad (2)$$

After: $$p_4(photon) = \left(\frac{E}{c}\cos\theta, \frac{E}{c}\sin\theta, 0, \frac{iE}{c}\right) \quad (3)$$

$$p_4(electron) = \left(p\cos\phi, -p\sin\phi, 0, \frac{ie}{c}\right) \quad (4)$$

Conservation of 4-vector components:

$$\frac{E_o}{c} = \frac{E}{c}\cos\theta + p\cos\phi \quad (5)$$

$$\frac{E}{c}\sin\theta = p\sin\phi \quad (6)$$

squaring and adding $p\cos\phi$ and $p\sin\phi$ from (5) and (6) gives:

$$p^2 = \left(\frac{E_o}{c} - \frac{E}{c}\cos\theta\right)^2 + \frac{E^2}{c^2}\sin^2\theta \quad (7)$$

using the invariance of the electron 4-momentum:

$$e^2 = p^2c^2 + e_o^2 \quad (8)$$

(8) can be substituted into $(1)^2$ to eliminate e and then the result can be used to eliminate p from (7):

$e^2 = (E_o + e_o - E)^2$ therefore $p^2c^2 = (E_o + e_o - E)^2 + e_o^2$

giving in (7): $(E_o - E\cos\theta)^2 + E^2\sin^2\theta = (E_o + e_o - E)^2 - e_o^2$

After a considerable amount of manipulation this reduces to:

$$\frac{(E_o - E)}{E_oE} = \frac{1}{e_o}(1 - \cos\theta)$$

Using $E = \frac{hc}{\lambda}$, $e_o = m_oc^2$, and $(\lambda - \lambda_o) = \Delta\lambda$ this becomes:

$$\Delta\lambda = \frac{h}{m_oc}(1 - \cos\theta) \quad (3.17)$$

3.7 EXAMPLES FROM PARTICLE PHYSICS

Example: Compton scattering, An X-ray of wavelength 10^{-11} m is scattered through 10° from a 'stationary' (i.e. low energy) electron. (a) What is the change of wavelength of the photon and (b) how much energy is transferred to the electron?

(a)

$$\Delta\lambda = \frac{h}{m_o c}(1-\cos\theta) = \frac{6.6\times10^{-34}\,\text{J s}}{9.1\times10^{-31}\,\text{kg}\times3.0\times10^{8}\,\text{m s}^{-1}}\left(1-\cos10^{\circ}\right) = \underline{3.7\times10^{14}\ \text{m}}$$

$$E = hf = \frac{hc}{\lambda} \qquad \frac{dE}{d\lambda} = -\frac{hc}{\lambda^2} \qquad \Delta E\,(\text{photon}) = -\frac{hc}{\lambda^2}\Delta\lambda$$

$$\Delta E\,(\text{electron}) = +\frac{hc}{\lambda^2}\Delta\lambda = \frac{6.6\times10^{-34}\,\text{J s}\times3.0\times10^{8}\,\text{m s}^{-1}\times3.7\times10^{-14}\,\text{m}}{10^{-22}\,\text{m}^2}$$

$$= 7.3\times10^{-17}\,\text{J} = \underline{460\ \text{eV}}$$

Example Show that it is impossible for a single γ -ray photon to produce an electron-positron pair in isolation.

Consider the CM frame, in which the electron and positron have equal but opposite 3-momenta and equal energies, E':

$$p_4\,(\text{photon}) = \left(\frac{E}{c}, \frac{iE}{c}\right)$$

$$p_4\,(\text{pair}) = \left(0, \frac{2iE'}{c}\right)$$

Conservation of 4-momentum means that the components of 4-momentum before and after pair production must be equal. This is not possible because the photon has non-zero 3-momentum in all frames (including the CM frame of the pair). However, it is possible for a photon to initiate pair production if another particle, e.g. a nucleus is involved - this can absorb the 3-momentum of the photon.

Example: Pion production. A collision between two protons can result in the creation of a positive pion and the conversion of one proton to a neutron:

$$p^+ + p^+ \rightarrow p^+ + n + \pi^+$$

Calculate the minimum kinetic energy (in MeV) for the proton(s) if this reaction is to happen (a) in a head-on collision between two protons of equal energy, (b) when a moving proton hits a stationary proton. (Take the rest energies of the p^+, n, and π^+) to be E_o, E_o, and 0.15 E_o.

(a) This reaction takes place in the CM frame and the threshold energy for pion production will be when the three product particles are at rest (no KE).

Initial 4 - momentum: $$p_4 = \left(0, \frac{2i\gamma E_o}{c}\right)$$

Final 4 - momentum: $$p'_4 = \left(0, \frac{2.15E_o}{c}\right)$$

4 - momentum is conserved so: $$\frac{2i\gamma E_o}{c} = \frac{2.15E_o}{c}$$

$$\gamma = 1.075$$

$\beta = \sqrt{1 - \frac{1}{\gamma^2}} = 0.37$ so both protons have velocities of $0.37c$ before collision.

$$KE = E - E_o = \gamma E_o - E_o = (\gamma - 1)E_o = 0.075E_o \approx 70 \text{ MeV}$$

(b) If one proton is fixed the collision is still identical to the one used in the calculation above, but is now being viewed from a different reference frame. We must transform the velocity in the CM frame to the laboratory frame and then use the new value of γ to calculate the kinetic energy.

$v' = \dfrac{2v}{1 + \frac{v^2}{c^2}} = 0.65c$ giving $\gamma' = 1.32$ and $KE = (\gamma - 1)E_o \approx \underline{300 \text{ MeV}}$

Note that the energy needed in the fixed target reaction is far greater than the combined energies of the two protons in the collider.

Example: Neutron absorption. A neutron of rest mass energy E_o and kinetic energy $2E_o$ is absorbed by a stationary boron-10 nucleus of rest energy $10m_o$. Find the rest mass of the boron-11 nucleus the instant after absorption of the neutron (before any other nuclear reactions occur).

For the neutron $$p^2c^2 = E^2 - E_o^2$$

and total energy $$E = E_o + 2E_o = 3E_o$$

these can be combined to give: $$p = \frac{E_o\sqrt{8}}{c}$$

The 4-momentum of the boron-11 nucleus in the laboratory frame is the sum of 4-momenta of the incident particles.

$$p_4 = \left(\frac{E_o\sqrt{8}}{c} + 0 \, , \, \frac{i(3E_o + 10E_o)}{c} \right) = \left(\frac{E_o\sqrt{8}}{c} \, , \, \frac{i(13E_o)}{c} \right)$$

The magnitude of this 4-vector equals the magnitude of 4-momentum in the rest frame of the boron-11 nucleus.

$$p_4(\text{boron}-11 \text{ rest frame}) = \left(0 \, , \, \frac{i\mathrm{E}_o}{c} \right)$$

(E_o is the rest energy of the boron-11 nucleus)

$$\frac{8E_o^2}{c^2} - \frac{169E_o}{c^2} = -\frac{\mathrm{E}_o^2}{c^2}$$

$$\mathrm{E}_o = 12.7E_o \qquad \text{so} \quad M_o(\text{boron}-11) = 12.7m_o(\text{neutron})$$

Note that this is not the rest mass of boron-11 in its ground state, this nucleus would be highly excited and highly unstable.

Example: photon absorption. Is it possible for a single isolated electron to absorb all the energy and momentum of a photon?

In the rest frame of the electron:

$$p_4(\text{electron}) = \left(0, \frac{iE_o}{c} \right) \qquad p_4(\text{photon}) = \left(\frac{E}{c}, \frac{iE}{c} \right)$$

so the final 4-momentum of the scattered electron would have to be:

$$p_4'(\text{scattered electron}) = \left(\frac{E}{c}, \frac{i(E_o + E)}{c} \right)$$

$$|p_4'|^2 = \frac{E^2}{c^2} - \frac{(E_o + E)^2}{c^2} = \frac{-E_o^2 - 2EE_o}{c^2} \neq |p_4|^2 = -\frac{E_o^2}{c^2}$$

The magnitude of a particle's 4-momentum is an invariant ($= -m_o$) in all inertial frames so this inequality shows that the proposed process is impossible.

Example: Pion decay. Stationary pions can decay to anti-muons and muon-neutrinos:

$$\pi^+ \rightarrow \overline{\mu}^+ + \nu_\mu$$

What is the kinetic energy of the emitted anti-muon?

Pion rest mass $M = 140$ MeV/c^2. Anti-muon rest mass is m_1=106 MeV/c^2.
Neutrino rest mass is m_2=0 MeV/c^2.

Initial 4 - momentum: $p_4(pion) = (0, iMc)$

Final 4 - momenta: $p_4(\text{anti} - \text{muon}) = \left(p_1, \dfrac{iE_1}{c}\right) \quad p_4(\text{neutrino}) = \left(p_2, \dfrac{iE_2}{c}\right)$

Conservation: $p_1 = -p_2 = p \quad (1) \qquad Mc^2 = E_1 + E_2 \quad (2)$

Invariants: $$p_1^2c^2 - E_1^2 = m_1^2c^4 \quad \text{(anti - muon)} \qquad (3)$$

$$p_2^2c^2 - E_2^2 = 0 \quad \text{(neutrino)} \qquad (4)$$

From (2): $E_1 = Mc^2 - E_2$

From (1) (3) and (4): $E_2 = -pc = -\sqrt{E_1^2 + m_1^2c^4}$

so $E_1 = Mc^2 + \sqrt{E_1^2 + m_1^2c^4}$

which leads to: $$\underline{E_1 = \frac{(M^2 + m_1^2)c^2}{2M}} = \frac{(140^2 + 106^2)}{280}\text{ MeV} = 110\text{ MeV}$$

Kinetic energy of anti - muon $= E_1 - E_o = 110\text{ MeV} - 106\text{ MeV} = \underline{4\text{ MeV}}$

This is in close agreement with experimentally measured values.

Example: Doppler effect. Derive the Doppler effect for a source moving at velocity v along the x-axis and transmitting a signal back to the origin of this reference frame by transforming the photon momentum from the source frame to the laboratory frame. The frequency of the emitted radiation in the rest frame of the source is f and the received frequency in the laboratory is f'.

$$p_4\,(\text{photon}) = \left(-\frac{E}{c}, \frac{iE}{c}\right)$$ in the source rest frame, and $E = hf$ and $p = \frac{h}{\lambda}$

Lorentz transformation of p_x component:

$$p'_{x'} = -\frac{hf'}{c} = \gamma\left(-\frac{E}{c} + \frac{vE}{c^2}\right) = \gamma\left(-\frac{hf}{c} + \frac{vhf}{c^2}\right)$$

$$f' = f\gamma\left(1 + \frac{v}{c}\right) = \frac{\left(1 - \frac{v}{c}\right)}{\left(1 - \frac{v^2}{c^2}\right)^{1/2}} f$$

$$f' = \frac{(1-\beta)^{1/2}}{(1+\beta)^{1/2}} f$$

The relativistic approach to the Doppler effect can be applied whatever the direction of relative motion between source and observer.

Example: Nuclear recoil. A nucleus has rest energy E_1 when in its ground state and $(E_1 + e)$ in a particular excited state. If it decays from this excited state to its ground state by emitting a single photon, find the frequency of the photon.

If recoil is neglected the photon energy is e and its frequency is $f = e/h$. However, the nucleus does recoil and this changes the frequency of the emitted photon (it now has an energy less than e).

Before: $$p_4 = \left(0, \frac{i(E+e)}{c}\right)$$

After: $$p_4\,(\text{nucleus}) = \left(p, \frac{i\gamma E_1}{c}\right) \qquad p_4\,(\text{photon}) = \left(\frac{E}{c}, \frac{iE}{c}\right)$$

$$p_4\,(\text{total}) = \left(p + \frac{E}{c}', \frac{i(\gamma E_1 + E)}{c}\right)$$

Conservation: $$p + \frac{E}{c} = 0 \qquad p = -\frac{E}{c} \qquad (1)$$

$$\frac{i(E_1 + e)}{c} = \frac{i(\gamma E_1 + E)}{c} \qquad \gamma = \frac{E_1 + e - E}{E_1} \qquad (2)$$

Energy - momentum invariant for nucleus in its ground state:

$$p^2 - \frac{\gamma^2 E_1^2}{c^2} = 0 - \frac{E_1^2}{c^2}$$

$$\frac{E_2^2}{c^2} - \frac{\gamma^2 E_1^2}{c^2} = -\frac{E_1^2}{c^2} \qquad \text{(after substituting for } p \text{ from (1))}$$

$$E^2 = (\gamma^2 - 1)E_1^2$$

$$\underline{E = \frac{e^2 + 2E_1 e}{2\,(E_1 + e)}} \qquad \text{(after substituting for } \gamma \text{ from (2))}$$

If the rest mass of the excited nucleus $(E_2 = E_1 + e)$ is used the expression can be rearranged to give:

$$E = e\left(1 - \frac{e}{2E_2}\right) \quad \text{and} \quad E \to e \quad \text{when} \quad e \ll E$$

3.8 SUMMARY OF IDEAS AND EQUATIONS IN CHAPTER 3

Space-time: If the laws of physics are referred to 4-dimensional space-time axes, relative motion is equivalent to a rotation and the principal effects of relativity have a simple geometric interpretation.

Event: A point in space-time having four co-ordinates.

4-vectors: Space-time vectors which have invariant magnitude and covariant co-ordinates.

Interval: Invariant magnitude of a displacement 4-vector.

$$s_4^2 = \Delta x^2 + \Delta y^2 + \Delta z^2 - c^2\Delta t^2 = \sum_{i=1}^{i=4} \Delta x_i^2 \quad \text{where } x_4 = ict$$

Worldline: Particle trajectory through space-time connecting all events that coincide with the origin of a set of space-time axes fixed in the particle.

4-momentum: The space components of the 4-momentum give the 3-momentum of the particle. The fourth component gives its total energy. The components are conserved separately in all inertial reference frames - this gives conservation of linear momentum, total energy and total mass. The magnitude of the 4-momentum is an invariant in all inertial reference frames.

$$p_4 = (p_x, p_y, p_z, \frac{iE}{c}) \quad \text{where } p_x = \gamma m_o v_x \text{ etc.}$$

Space-time rotation: $\tan\phi = \frac{v}{ic}$

4-displacement: $s_4 = (x, y, z, ict)$

4-velocity: $v_4 = (\gamma v_x, \gamma v_y, \gamma v_z, i\gamma c)$

Energy-momentum invariant: $E^2 - p^2c^2 = E_o^2$

Lorentz transformationof 4-momentum:

$$p'_x = \gamma\left(p_x - \frac{vE}{c^2}\right) \qquad p_x = \gamma\left(p'_x + \frac{vE'}{c^2}\right)$$

$$p'_y = p_y \qquad p_y = p'_y$$

$$p'_z = p_z \qquad p_z = p'_z$$

$$E' = \gamma\left(E - vp_x\right) \qquad E = \gamma\left(E' + vp'_x\right)$$

Compton effect: $\Delta\lambda = \frac{h}{m_o c}(1 - \cos\theta)$

3.9 PROBLEMS

The first questions in this section cover various aspects of special relativity discussed in both chapters 2 and 3. The later questions are best solved using the spacetime methods of chapter 3.

3.1 In the UK electricity consumers pay about 7.5p per kW h.
a. How much energy do they get per pound?
b. How much mass do they get per pound?
c. How much would 1 kg of energy cost?

3.2 The normal flux of solar radiation at the edge of the Earth's atmosphere is about 1400 W m^{-2} and the radius of the Earth's orbit is about 1.5×10^{11} m.
a. What is the total radiation flux from the Sun?
b. At what rate does the Sun convert rest mass to radiant energy?
c. The Sun's mass is approximately 2×10^{30} kg. Use this to suggest an upper limit to the lifetime of the Sun on the assumption that it will consume its total rest mass during its lifetime. Explain why this is an unrealistic estimate.
d. The main reaction that goes on in the Sun is the conversion of four hydrogen nuclei to helium. This process results in a reduction of rest mass of about 0.7%. Use this to revise your estimate of the Sun's lifetime. Is this realistic?
e. The Sun is thought to have been around for about 5 billion years (5×10^9 y) and is expected to continue in pretty much its present state for another 5 billion years. Comment on this in view of the values calculated above.

3.3 A 1 tonne car accelerates from rest to 30 m s^{-1}.
a. What is the mass of its kinetic energy?
b. Does the total mass of the car change? (Explain your answer clearly.)

3.4a. Write down an expression for the electrical potential energy of two electrons held a distance d apart.
b. Imagine a sphere of radius r in which there is a uniform number density n of electrons. Derive an expression for the total electrical potential energy of the electrons. (Hint: consider the increase in electrical potential energy as a thin layer of electrons is added to a complete inner sphere, and then integrate.)
c. Now consider a sphere of water of radius 10 cm. Estimate the number density of electrons in this sphere (density of water is 1000 kg m^{-3}, molar mass of water is 18 g, Avogadro's number is 6.0×10^{23}).
d. Use your expression to calculate the total electrical potential energy of an electron sphere with the number density and size of the water sphere in (c). (The electron charge is $e=1.6 \times 10^{-19}$C).
e. What is the mass equivalent of this electrical potential energy?
f. How does your answer to (e) compare with the actual mass of a 10 cm radius water sphere?

3.5 A torch emits 5 W of radiant energy in a narrow beam.
a. What is the recoil force on the torch?
b. Calculate the force this beam exerts when it strikes each of the following surfaces normally:

(i) A perfect absorber.
(ii) A perfect reflector
(iii) A surface that absorbs 40% and reflects 60% of incident light.

3.6 What are the values of (i) γ, (ii) total energy (in GeV) (iii) momentum (in GeV/c), and (iv) total mass (in GeV/c^2) for:

a. An electron accelerated through 50 GeV.
b. A proton accelerated through 50 GeV.
c. An electron moving at 0.99c.
d. A proton moving at 0.99c.

(electron mass = 0.511 MeV/c^2, proton mass = 938 MeV/c^2)

3.7 How much work must be done to accelerate and electron from rest to:
a. 0.5c.
b. 0.7c.
c. 0.9c.

3.8 What is the recoil velocity of a sodium atom in a vapour when it emits a photon of yellow light with wavelength 589 nm?

(atomic mass of sodium = 22.989771 u, 1 u = 1.66×10^{-27} kg)

3.9a. Show that if $v << c$ then $\gamma \approx 1 + \frac{v^2}{2c^2}$
b. Use this result to show that, to this approximation, the classical expression for kinetic energy is valid.
c. Suggest values of v for which the errors involved in using the classical expression are smaller than 5%.

3.10 Pions have a half-life of 2×10^{-8} s in their own rest frame. These are often produced when photons crash into fixed targets. In such an experiment 80% of the pions created survive at a distance of 25 m from the target. Assuming all pions are emitted with the same energy calculate the pion's total energy.
(m_π = 140 MeV/c^2)
3.11 A rocket heads away from the earth at a velocity of 0.5c. A radio operator on board the rocket transmits signals back to Earth once every 2 hours (rocket time) starting 2 hours after leaving Earth (ignore acceleration times).
a. How long after the rocket left Earth do the first three signals reach the Earth?
b. What is the interval between transmissions as calculated from Earth?

3.12 Our galaxy (the Milky Way) is about 10^5 light years across. How long would it take to cross the galaxy in a rocket:
a. Travelling at 0.90c?
b. Travelling at 0.99c?
In each case give the time elapsed for (i) an astronomer at rest in the galaxy and (ii) a passenger on the rocket.
c. What is the distance across the galaxy as measured by the rocket passengers in each case?

3.13 Is time dilation proof that time travel is possible?

3.14 Unstable particles with velocity v relative to the laboratory pass two detectors a distance d apart. The ratio of particles at the first detector to particles at the second is f. Derive an expression for the half-life of the particles.

3.15 The mean lifetime of a muon at rest is 2.2×10^{-6}s.
a. How far does light travel in this time?
b. Explain how it is possible for a muon to travel farther than the distance calculated in (a) before it decays.
c. How far would muons traveling at $0.999c$ go before decaying?
d. How far would a 200 MeV (total energy) muon go before it decayed? (Rest mass of muon = 105.6 MeV/c^2.)

3.16a. What is the period and frequency of the H_α-line (656.3 nm) in a stellar spectrum?
b. Calculate γ for a star moving away from the Earth at velocity 600 km s^{-1}.
c. What is the effect of motion alone (neglect time dilation) on the frequency of this radiation when it reaches the Earth?
d. What is the effect of time dilation on the emission frequency?
e. What is the frequency and wavelength of the radiation actually received?

3.17 A rod of rest mass m_0, length l and cross-sectional area A moves parallel to its own length along the x-axis of a laboratory at velocity v.
a. Derive an expression for its density in the laboratory.
b. Sketch a graph showing how its density in the laboratory changes as a function of its velocity.
c. What happens to its density as viewed by an observer moving with the rod?

3.18 A rocket passes through the centre of a spherical galaxy of proper radius r_0 at velocity $0.95c$. What shape does the galaxy appear to be?

3.19 A metal frame is an equilateral triangle with sides of length l. The triangular structure is passed by an observer moving at velocity v parallel to the base of the triangle. What are the lengths of the sides and angles in the triangle as seen by the moving observer?

3.20 Two atomic clocks are separated by 10^9 m and placed at rest with respect to one another. Clock A is constantly 1 second behind clock B. In what direction and at what speed would you have to move for the clocks to be synchronized in your reference frame?

3.21 Two stars are each 100 light years from Earth, but in opposite directions. One star goes supernova a year after the other as seen from Earth. Is there a reference frame in which the two explosions are simultaneous? Explain.

3.22 Use $E = \gamma m_o c^2$ and $p = \gamma m_o v$ to show that $E^2 - p^2c^2 = m_o^2 c^4$.

3.23a. What is the separation of points (1,2,3,4,5) and (5,4,3,2,1) in 5D Euclidean space?
b. How does 4D Minkowski spacetime differ from 4D Euclidean space?

3.24 Two points in 2D Euclidean space are at (1,2) and (5,5).
a. What is their separation?
b. Write down their co-ordinates in a reference frame whose x' and y' axes are rotated through 30° in a clockwise direction relative to those used to define the co-ordinates above.
c. Use the new co-ordinates to calculate the separation of the two points.
d. What does it mean to say that distance is an invariant with respect to rotation of 2D axes?
e. Give an example from relativity in which a similar transformation might be used.

3.25 Draw spacetime diagrams to show the following relative to a laboratory frame:
a. The decay of a stationary carbon-14 nucleus to a beta-particle and an anti-neutrino.
b. The absorption of a thermal neutron by a uranium-238 nucleus.
c. The annihilation of an electron-positron pair and the emission of two gamma-ray photons.

3.26 What is the condition that events at (x_1, t_1) and $(x_2\ t_2)$ are causally connected?

3.27 Use the definition of the metre and the second to write down as accurately as you can:
a. The frequency of radiation from the hyperfine levels of cesium-133 that is used in the definition of the second.
b. The wavelength of this radiation.

3.28 Penrose's suggestion that simply walking past someone in the street might result in a significant disagreement over the present moment in a distant galaxy seems rather counter-intuitive. This question looks at this claim quantitatively. Consider two people who pass with a relative velocity of 2 m s^{-1}. For the sake of argument we shall neglect the effects of the additional local velocities of our own and other galaxies (since these are the same for both observers).
a. Draw a spacetime diagram for the Earth and both walkers (assumed to have velocities $\pm$ 1 m s^{-1} relative to Earth). Draw a line of simultaneity for each of them at the moment they pass one another (the origin).
b. Explain how the diagram shows that although the two observers agree on the moment they pass one another they will disagree on the events that are

simultaneous with this moment and the amount of discrepancy will increase in direct proportion to the distance of the events.
c. Do they agree on the time order of distant events?
d. Calculate the time disagreement at the distance of Sirius (about 8.6 light years away).
e. Repeat this calculation for events in the Andromeda galaxy (about 2×10^{22} *m* away).
f. Penrose claims that for one observer the Andromedan spacefleet might have already set out to invade Earth whereas for the other the decision may not yet have been made. Does this mean that the future is pre-determined for at least one of these observers?
g. What happens to the two views of Andromeda held by the two walkers if they stop to discuss them?

3.29 A sets two clocks 10 m apart along the x-axis in his laboratory and synchronizes them using a light pulse from a point mid-way between them. The pulse is emitted at $t = 0$ in A's lab.
a. Draw a spacetime diagram showing the worldlines of the two clocks, and the light cone of the synchronizing pulse.
b. What are the co-ordinates of the synchronization events in this frame (i.e. the emission of the pulse and its reception at both clocks).
c. How can you tell from the spacetime diagram that the clocks are indeed synchronized?
d. B passes A moving in the positive x direction at velocity v and is opposite the light source at the instant it emits the synchronizing pulse. Call this time $t' = 0$ in B's reference frame. Now draw the spacetime diagram from B's point of view.
e. How can you tell from this diagram that the clocks are not synchronized in B's frame? Which clock starts first?
f. If the synchronization error is 10 ns in B's frame calculate B's velocity and the co-ordinates of all the synchronization events in this frame.

3.30 Two events occur in A's reference frame. One is at the origin and the other 2 years later at a distance of 10 light years along his x-axis. B passes A at a velocity v moving in the positive x direction and calculates that the two events occurred 1 year apart. How fast is B travelling and how far apart are the two events in B's frame?

3.31 In quantum theory physical processes are described by a wavefunction whose 'intensity' is related to the probability of particular events. For example, if a radioactive decay occurs at the origin emitting a single particle in a random direction then the wavefunction is spherical and uniform as it travels out from the source. However, if a geiger counter detects the particle at some point the probability there becomes one and the probability elsewhere drops instantaneously to zero. This 'collapse of the wavefunction' poses some problems for relativity.

a. Consider two points on opposite sides of the source at a distance of 1 m from it. if the particle is detected at one of these points the wavefunction falls to zero at the other. Assume this occurs instantaneously. What kind of interval separates these two events? Comment.
b. How would the time order of events in (a) be affected if the process was observed from a reference frame moving past the apparatus?
c. Draw a spacetime diagram with the source at the origin and the two events described in (a) marked in. Draw a lightcone centred on the event where the particle is detected. What is the relation of the other event to this lightcone and what is the significance of this construction?
d. Is it possible that the collapse of the wavefunction is triggered by some signal sent out from the point of detection?

3.32 Draw a spacetime diagram and draw 4-vectors of equal magnitude:
a. Parallel to the ict axis.
b. Parallel to an ict' axis for a particle moving at velocity (i) $0.1c$ (ii) $0.5c$ (iii) $0.9c$.

3.33 The proper time between flashes of a lighthouse beacon is 2.0 s.
a. What is meant by 'proper time'?
b. What is the measured interval between flashes for an observer moving at velocity $0.4c$ relative to the lighthouse?

3.34 Hermann Bondi invented another way to understand special relativity by using 'k-calculus'. The method exploits the symmetry between two inertial frames in relative motion. Consider two rockets (A and B) passing one another with a relative velocity v. A and B synchronize their clocks to $t = t' = 0$ as they pass. A radar operator on rocket A sends a short signal to B at time t_1 and it arrives at B at time t_2 (relative to A) and time t'_2 (relative to B). B immediately returns the pulse which arrives at A at time t_3 (relative to A) and time t'_3 relative to B. A immediately returns the signal to B and the process continues as they move apart.
a. Draw a spacetime diagram for A showing B's worldline and the events at which signals are transmitted and received, labelling times as described above on A and B's worldlines.
b. Write down an expression for t_2 in terms of t_1 and t_3.
c. By considering the motion of B relative to A and the time of flight of light show that $t_3 = \frac{(c+v)}{(c-v)} t_1$.
d. Since the two reference frames are symmetric the ratio of t'_2 to t_1 should be the same as the ratio of t_3 to t'_2. (In both cases a signal has been sent to a distant receiver moving away at velocity v). Call this ratio k. Write down these two ratios in terms of k.
e. Now substitute the ratios derived in (d) into the equations from (b) and (c) to eliminate all times and find k in terms of v and c.

f. Hence find t'_2 in terms of t_2. How does this relate to time dilation? What result would be obtained if we repeated the calculation based on a spacetime diagram drawn for B?

For more information about this method refer to 'Relativity and Common Sense' by Hermann Bondi, Heinemann 1970.

3.35 Another geometrical interpretation is developed by Lewis Caroll Epstein in his book 'Relativity Revisualised' (Insight Press, 1987). We have seen that the magnitude of a velocity 4-vector is always c. Epstein bases his interpretation around the idea that everything moves through spacetime at this same rate. This leads to the idea that when we observe a moving object its motion through our space is at the expense of its motion through our proper time, in 1 s of our proper time less than 1 s elapses for the moving object. His book is a delight to read and is highly original, but difficult to get hold of. This question introduces some of the simpler ideas.
a. Draw a spacetime diagram and label the vertical axis 'proper time' (measured as usual in metres) and the horizontal axis space.
b. Draw an arrow of length c up the proper time axis. This is not a worldline, it represents your displacement through space-proper time as 1 s of your proper time passes. Since you remain at the same place your 'motion' is entirely in the direction of proper time. What is your rate of motion through proper time?
c. Now draw an oblique line to represent an object moving away from you at velocity v. Make this line unit length as well because it moves through spacetime at the same rate you do. How far away from you is the end of its worldline? How much proper time has passed for it?
d. Draw a right-angled triangle on the diagram and use Pythagoras's theorem to calculate exactly how much proper time has passed for the moving object, call this t' and express it as a function of t. Impressed?
e. How far does light travel through space in 1 s? In which direction must a vector representing light be drawn on this diagram?
f. Your answer to (e) should show that light rays travel entirely in the space direction and not at all in the time direction, so photons do not age. Is this consistent with the equations of time dilation? Is it consistent with the idea that light has a finite velocity?
g. The diagram you have drawn is called a 'cosmic speedo' by Epstein. Describe what happens to motion through space and proper time as you turn the dial of this speedo in a clockwise direction.
h. How would you try to draw a line of simultaneity on this diagram?

3.36 Use a spacetime diagram to show that an isolated electron-positron pair cannot annihilate to a single photon.

3.37a. Convert the following to MeV:
i. The energy equivalent of 1 u (1.66×10^{-27} kg).

ii. A kinetic energy of 10^{-12} J.
iii. The rest energy of an electron (rest mass = 9.11×10^{-31} kg.
b. Show that a photon of energy 10 MeV has a momentum 10 MeV/c.
c. Show that a proton of kinetic energy 1 GeV has a momentum 1.7 GeV/c.
d. Convert the following to J or kg as appropriate:
i. 10 GeV.
ii. 10 GeV/c^2.
e. Express a momentum of 5 MeV/c in kg m s^{-1}.

3.38 A particular clock makes two short beeps 10 s apart. Calculate the interval, time and distance between these events as seen by an observer moving past the clock at: (i) 0.2c, (ii) 0.5c and (iii) 0.99c.

3.39a. Write down the components of the 4-momentum for a proton of total energy 10 GeV.
b. What is the magnitude of this 4-vector and why is this an important quantity?

3.40 A head-on collision between two particles each of rest mass m_o and kinetic energy E_K results in the creation of a pair of identical particles of rest mass M_o at rest. Find E_K in terms of M_o, and m_o.

3.41. A high energy particle has kinetic energy E_K and momentum p. What is its rest mass?

3.42. A nucleus of rest mass m_0 absorbs a photon of energy E.
a. Write down the components of 4-momentum for the resultant particle.
b. Write down the rest mass of the excited nucleus.

3.43 What are the 4-momentum components of a photon of energy E?

3.44 A stationary boron-10 nucleus absorbs a 50 MeV neutron.
a. Calculate γ for the neutron.
b. Write down the 4-momentum for the neutron and boron-10 nucleus before absorption.
c. Use the conservation of 4-momentum to determine the total energy and momentum of the boron-11 nucleus as it is formed.

3.45 A 100 MeV gamma-ray photon scatters off a stationary photon.
a. Calculate the initial and final photon wavelengths and hence the change in wavelength.
b. Use the Compton formula to calculate the photon's scattering angle.
c. By considering the energy transferred to the scattered proton calculate γ for the proton.
d. Calculate the proton momentum.

e. Use conservaton of momentum in 2D space to calculate the proton scattering angle (measured from the incident direction of the photon).
f. Find the angle between the final photon and proton paths.
g. Is there any limit to the energy that the photon can transfer to the proton? Explain.

3.46 An intense light source emits light equally in all directions. However, if this source is moving at high velocity through the laboratory (e.g. an electron in a synchrotron) the intensity of radiation is concentrated in a narrow beam in front of the particle - give a qualitative explanation of this.

4

GENERAL RELATIVITY AND COSMOLOGY

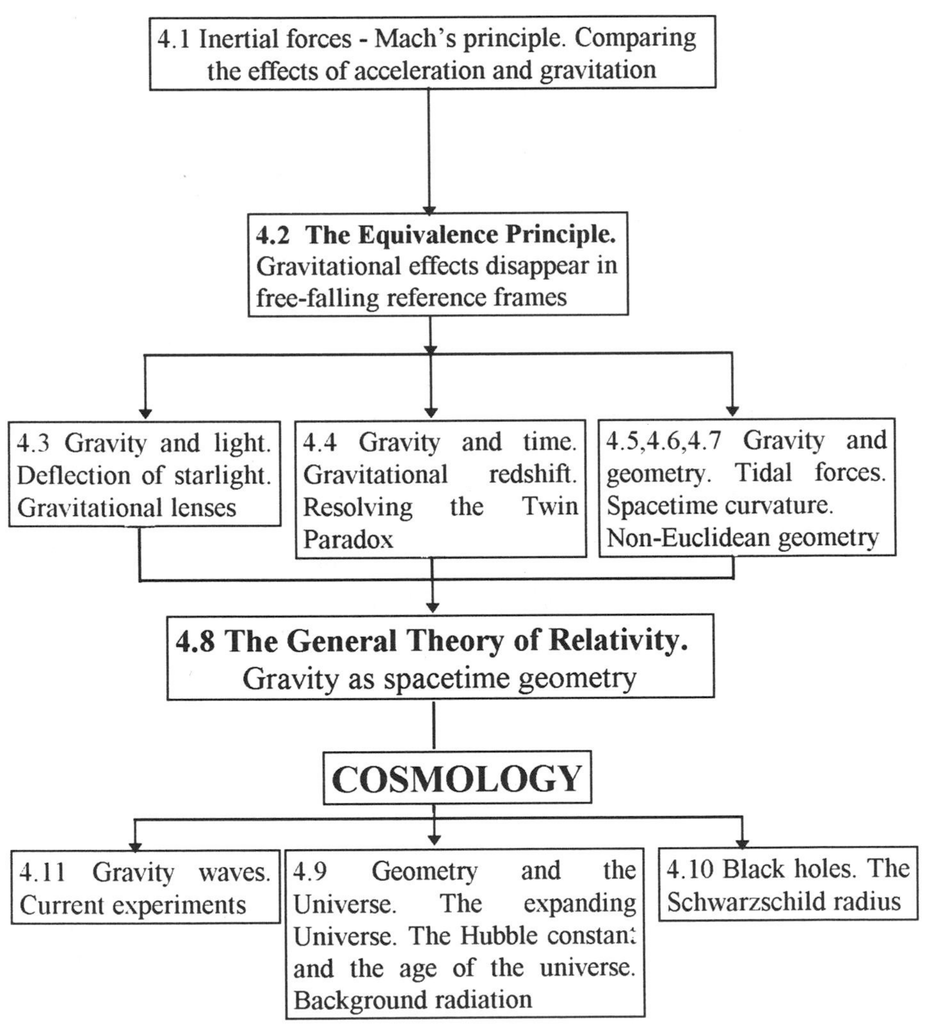

4.1 INERTIAL FORCES.

4.1.1 Why Special Relativity Is Special.

Special relativity is 'special' in the sense that it is restricted. The principle of relativity states that the laws of physics are the same in all inertial reference frames. In such a reference frame a 'free' body will continue at rest or at constant velocity until acted upon by a resultant force. However, there also exist reference frames in which this does *not* appear to be the case. In these frames bodies spontaneously accelerate when released. Three obvious examples are: (i) that coffee tends to slop from a cup when the cup rests on a table in an accelerating train; (ii) that we experience a 'centrifugal force' pushing us to the outside of a vehicle when it corners; and (iii) that objects released close to the earth have a free-fall acceleration directed toward the Earth's centre. Thus acceleration, rotation and gravitational fields all produce 'non-inertial' reference frames. In these non-inertial frames the spontaneous accelerations can be accounted for using normal Newtonian mechanics if we introduce new force fields which exert 'inertial forces' on otherwise free bodies. Centrifugal force is an example of an inertial force.

The classical approach to acceleration and rotation is to describe the forces (at least in (i) and (ii)) as 'fictional', arising only because of an inappropriate choice of reference frame. If we transform to an inertial reference frame then the inertial forces are seen to be illusions. It is not that the coffee slops from the cup, the cup accelerates away from the coffee; we are not thrown outwards as the car corners, it turns inwards toward us. Gravitational fields however, are treated rather differently. The acceleration of a free-falling body is said to be 'caused' by a gravitational force arising from the real attraction of massive bodies. These contrasting explanations of apparently similar phenomena are described in more detail below.

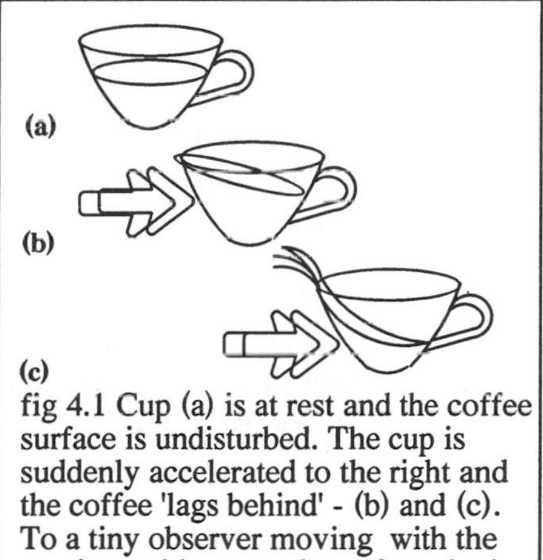

fig 4.1 Cup (a) is at rest and the coffee surface is undisturbed. The cup is suddenly accelerated to the right and the coffee 'lags behind' - (b) and (c). To a tiny observer moving with the cup it would appear that a force had been applied to the coffee.

Seen from an external inertial reference frame (e.g. by a spectator outside the train looking in through a window as it pulls out of the station) the coffee is just obeying Newton's first law. The coffee is in contact with the cup only at its outer surfaces and as the cup accelerates this applies a force to one side of the liquid but not to the liquid in bulk. If the cup contained some rigid solid then the whole would be accelerated with the cup via this contact force, but liquids can flow and the net effect is that the centre of mass of the liquid does not accelerate as quickly as the cup and so spills over its rear lip.

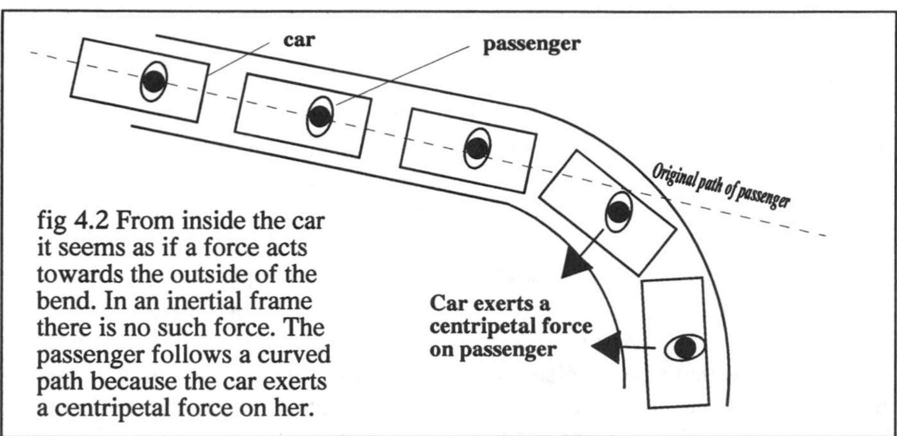

fig 4.2 From inside the car it seems as if a force acts towards the outside of the bend. In an inertial frame there is no such force. The passenger follows a curved path because the car exerts a centripetal force on her.

Viewed from the sidewalk a passenger in a moving car is also obeying Newton's first law. The vehicle experiences a resultant force from the road (as a result of its tyres pressing back on the road) which causes it to change direction. This force is not applied directly to the passenger who, because of her inertia, continues to move in a straight line at constant speed. The relative motion of the passenger to the vehicle causes her to slide towards the outside of the turning circle. Seen from inside the car it appears that an inertial force has acted to push her outwards (a centrifugal force). Viewed from the sidewalk this is clearly not the case. There is no resultant force on the passenger until she interacts with the side of the car and then experiences a centripetal force that changes her direction of motion in the same sense as the car's. Centrifugal force arises when we try to describe the physics of a non-inertial reference frame as though it were an inertial frame. (The example has been distorted slightly to emphasize the nature of the motion, in practice the car seat exerts a centripetal force on the passenger and prevents her hitting the door! This is why seats are usually contoured around the passenger for comfort).

The classical treatment of gravity is very different from the inertial forces described in the previous examples. With gravity we usually assume we are in an inertial reference frame and that the acceleration of a falling body is due to some real gravitational force acting on it. The gravitational field is then simply a region of space in which such forces would be exerted on any massive body that happened to be there. Einstein realized that it *is* possible to treat gravitational forces in a similar way to inertial forces. To do this he needed to observe the effects of gravity from an inertial frame inside the gravitational field. He discovered that the way to do this is to compare gravitation with acceleration, a link he formalized in the 'equivalence principle' which we shall meet very soon.

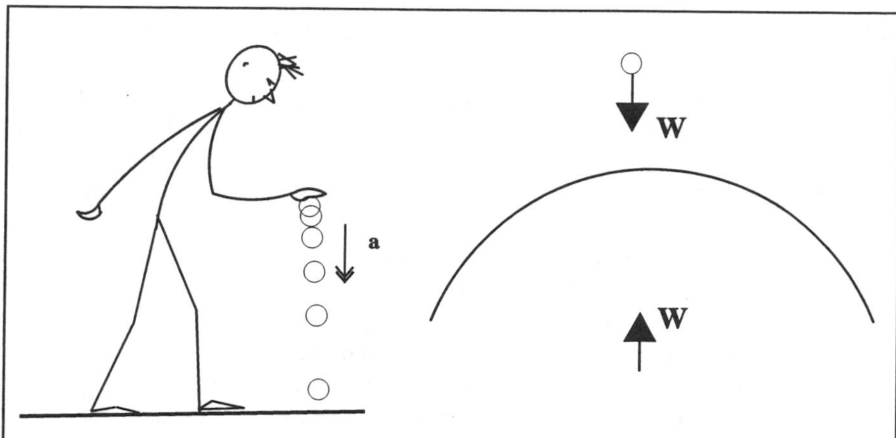

fig 4.3 'Free' bodies accelerate towards the Earth. This is explained in Newtonian mechanics by introducing the gravitational interaction. The Earth and ball exert equal attractive forces on one another.

4.1.2 Mach's Principle.

'When .... we say that a body preserves unchanged its direction and velocity in space, our assertion is nothing more or less than an abbreviated reference to the entire universe.'
(Ernst Mach, The Science of Mechanics, 1883)

Although it is unnecessary for the development of our ideas here, it is worth digressing slightly to consider what is now known as 'Mach's Principle'. This is a speculation on the origin of inertia and the relativity of acceleration. Mach's ideas strongly influenced Einstein but were never clearly incorporated into general relativity. However, the origin of inertia is still obscure and Mach's ideas are simple, compelling and utterly relativistic!

The existence of inertial forces in accelerated reference frames suggests that acceleration takes place against some background of absolute space. This is the conclusion Newton drew from his bucket experiments. The surface of the water in the bucket becomes concave whenever the water rotates with respect to the fixed stars. Rotation relative to the bucket itself is irrelevant (once the bucket has been rotating for a while it will be at rest relative to the water although the water surface remains curved).

Bishop Berkeley responded to Newton's ideas by arguing that rotation relative to empty space is a nonsense. Think of an otherwise empty universe containing Newton's bucket. If there is nothing to measure the rotation against then the concept of rotation is meaningless. However, we do not live in an 'otherwise empty universe', so the curvature of the water surface (and the existence of all inertial forces caused by rotation) must be generated by rotation relative to 'the fixed stars'. This view is supported by the oscillation of a Foucault pendulum as discussed in chapter 1.

Ernst Mach made rotation relative by suggesting that the same forces would result if we rotate an object relative to the fixed stars as if we considered the object to be at rest and the stars in motion rotating about it. In this model the inertial forces in the rotating reference frame are the result of an interaction of sorts with the rest of the matter in the entire universe! The idea can be extended. The reluctance of a body to change its state of motion relative to the fixed stars, that is its inertia, may be caused by the interaction with distant matter. In an empty universe a body would have no inertia and acceleration would be a meaningless concept.

Up to about 1920 Einstein tried very hard to make Mach's Principle a foundation of general relativity alongside the principle of relativity and the principle of equivalence. It also influenced his early attempts to construct a unified field theory but was never completely and successfully incorporated. However, the bold idea that the inertia of a body is wholly or partially caused by an interaction with the entire universe has an obvious appeal and should not be dismissed lightly.

4.2 THE EQUIVALENCE PRINCIPLE.

4.2.1 Everything Falls With The Same Acceleration.

'In contrast to electric and magnetic fields the gravitational field exhibits a most remarkable property which is of fundamental importance for what follows. Bodies which are moving under the sole influence of a gravitational field receive an acceleration which does not in the least depend on the material or the physical state of the body.' (Einstein)

The key to a general theory of relativity that could include accelerated reference frames was intimately linked to gravitation. Think of the forces acting on a charged particle in an electric field. The size of the force depends on the size of the charge, but its subsequent acceleration depends inversely on its mass. For this reason different charges in the same electric field will generally have different accelerations. With gravity the situation is rather different. The size of the force depends on the mass of the body but its acceleration is inversely proportional to its mass. Therefore a body of greater mass will experience a larger force but also has greater inertia. The ratio of force to mass is constant so all bodies have the *same* acceleration in the same gravitational field.

Galileo drew attention to this remarkable property. According to physics legend he climbed to the top of the leaning tower of Pisa and dropped two spheres of similar radius but made of materials of different density. It is claimed that they both hit the ground at the same time. A similar experiment was carried out by the Apollo astronauts on the Moon. They dropped a hammer and a feather. Both struck the Moon's surface at the same time (the Moon was an ideal laboratory for this experiment since it lacks an atmosphere). These crude experimental results have been confirmed by very accurate laboratory experiments to compare the free-fall accelerations of objects made from different materials. Interest in these experiments intensified in the 1980s when there was a suggestion from some that small differences had been detected for some objects. For a while rumours circulated that a 'fifth force' of nature had been discovered, but these were unfounded. More accurate measurements again confirmed the equality of free-fall accelerations in the gravitational field.

Many modern experiments designed to test this are based on an arrangement first used by Baron Roland von Eötvös around the turn of the century. The idea is to balance the gravitational and inertial forces on a pair of different bodies of equal mass suspended from a torsion balance. Eötvös used the weight of the bodies in the Earth's gravitational field and the centrifugal force on them due to the Earth's rotation on its axis. If the gravitational force is greater on one than the other then the orientation of the balance will change as the two masses are interchanged. Later experiments used the centrifugal force due to the Earth's orbital motion and the gravitational attraction of the Sun as it passes overhead. This removes the need to rotate the apparatus and allowed the equality of accelerations to be checked to an accuracy of about 1 part in 10^{11}! The diagrams below show the experimental arrangement.

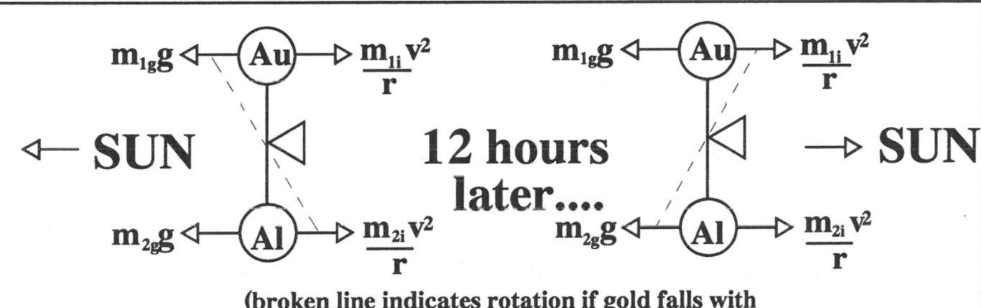

fig 4.4 The Dicke experiment.The Earth is in free fall around the Sun. This should eliminate inertial forces in the Earth's frame. Another way to see this is to say that centrifugal force is exactly balanced by the gravitational attraction to the Sun. If different bodies have different free-fall accelerations in the same field this balance will not always occur. In this experiment Dicke looked at a rotation of a torsion balance connected to one aluminium and one gold sphere on a 12 hour cycle as the Earth rotates. The spheres have equal inertial masses.

4.2.2 Gravitational And Inertial Mass.

The difference between a mass in the gravitational field and a charge in the electric field is that mass determines *both* the size of force that the field exerts on the body *and* the response to that force, the acceleration of the body. We can emphasize this by labelling two kinds of mass: a gravitational mass m_g and an inertial mass m_i. The former determines the gravitational force exerted on the body in a field of strength g and the latter determines the inertia of the body in Newton's second law.

$$F = m_g g$$

$$a = \frac{F}{m_i} = \left(\frac{m_g}{m_i}\right) g$$

From these equations we can see that the equality of accelerations for all masses in the same field is telling us that the ratio of gravitational to inertial mass must be a constant, or that they are directly proportional to one another. By choosing to measure both in kilograms we set the constant of proportionality to one and emphasize their equivalence. This is often glossed over in elementary physics books, but it is extremely significant. In classical physics there is *no reason at all* that these two quantities should be equivalent.

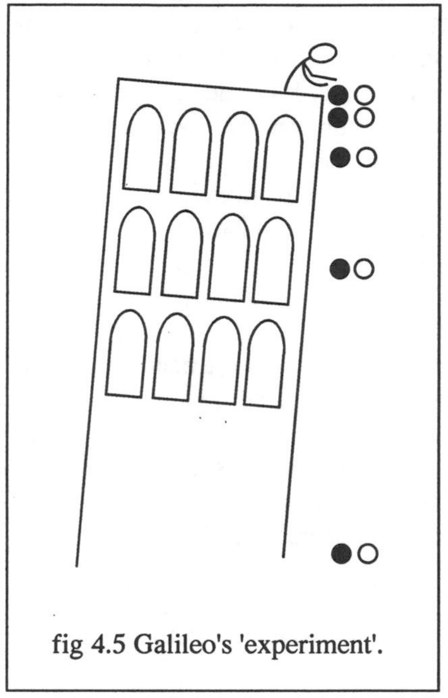

fig 4.5 Galileo's 'experiment'.

4.2.3 Acceleration And Gravitation.

Einstein realized that the equivalence between gravitational and inertial mass must arise from some fundamental physical connection between gravitation and acceleration. To illustrate this connection he devised a thought experiment. Imagine two observers, both confined in closed laboratories, such that they each have no knowledge of the state of motion of their own laboratory with respect to the rest of the Universe. Each observer releases a ball and watches it closely. Unknown to the observers, laboratory A is being accelerated upwards at a rate equal to the acceleration of freely falling objects close to the Earth's surface whilst laboratory B is at rest on the surface of the Earth itself. In both cases the observers record that the ball bearing falls vertically downwards relative to their own frame with an acceleration equal to 9.81 m s^{-2}. Their observations are *equivalent*. This is reminiscent of Galileo's thought experiment below decks on a uniformly moving ship, but now we are dealing with acceleration and gravitation. If the observers go on to carry out any other mechanical experiments within their laboratories they will again find they get identical results. The laws of mechanics in a uniform gravitational field are equivalent to the laws of mechanics in a uniformly accelerated reference frame. No mechanical experiment carried out in either laboratory can inform the inhabitant that they are in a gravitational field rather than an accelerating laboratory or vice versa.

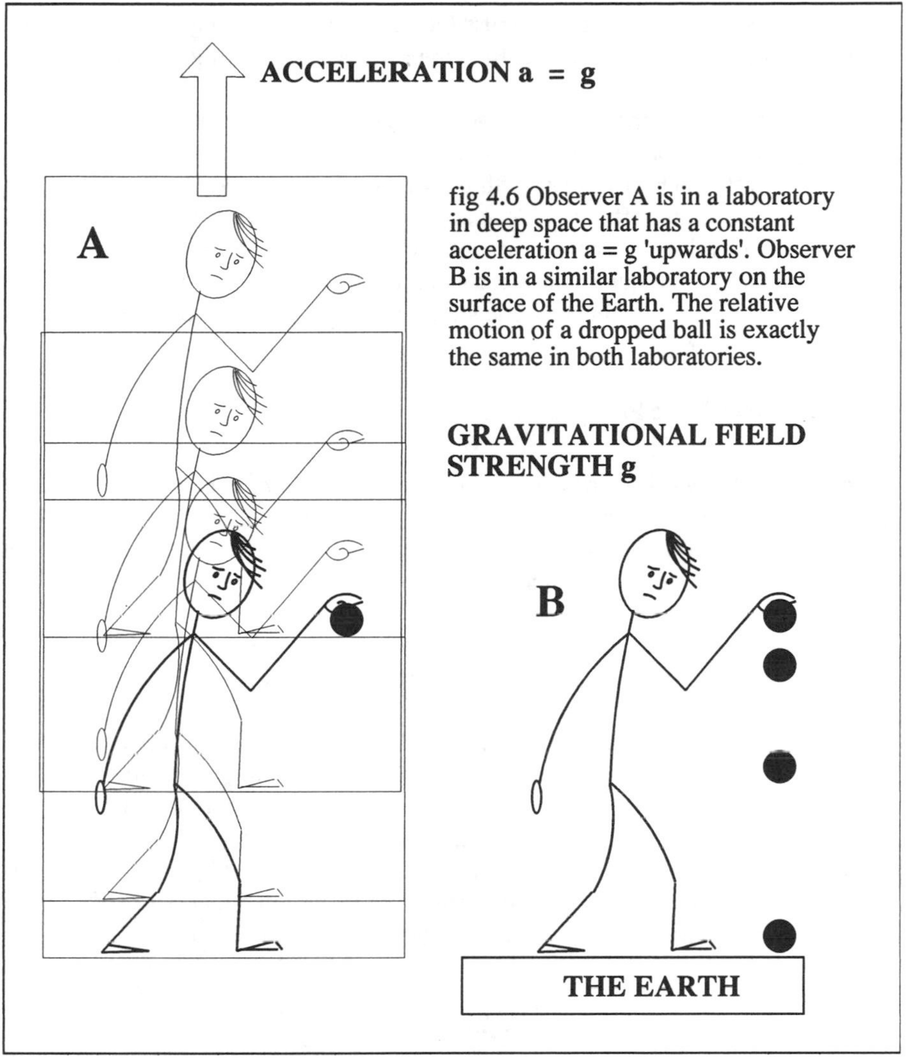

fig 4.6 Observer A is in a laboratory in deep space that has a constant acceleration a = g 'upwards'. Observer B is in a similar laboratory on the surface of the Earth. The relative motion of a dropped ball is exactly the same in both laboratories.

Could it be that gravitational forces, like the forces experienced in an accelerated reference frame, are actually inertial forces and have a relative rather than an absolute claim to existence? This is an interesting question since it is possible to remove inertial forces by transforming to an inertial reference frame. Is it also possible to remove gravitational forces in a similar way? The answer is 'yes', there is a reference frame in a gravitational field in which there are no gravitational forces. Einstein's thought experiment shows us what this frame must be. If we consider the ball released by observer A above and observe it from outside the laboratory from a frame in which the laboratory has an upward acceleration of 9.81 m s^{-2} we shall see it stay suspended in space as A accelerates past it. What is the equivalent frame from which to observe the falling ball in B? We must remain at rest with respect to the ball so we must fall with it! If we make our observations from freely falling reference frames then the effects of uniform gravitational fields disappear.

This has a practical consequence. When NASA trains astronauts to prepare them for 'weightlessness' part of the training involves flying up to high altitude in a large jet aircraft which then dives with a downward acceleration g. The astronauts fall at g inside the aircraft and so lose all reaction forces with their surroundings. They feel 'weightless'. Whether or not we regard this as true weightlessness depends on what reference frame we choose to inhabit! The film Apollo 13 used the same technique to obtain authentic shots and required the actors and film crew to undertake about 600 dives.

4.2.4 The Equivalence Principle.

In special relativity Einstein extended Galilean relativity to include *all* laws of physics, not just mechanics. The equivalence principle plays the same crucial role in general relativity as the principle of relativity in the special theory. Einstein described it as 'the happiest thought of my life':

- *The laws of physics are the same at each point in a uniform gravitational field as in a reference frame undergoing uniform acceleration.*

Think back to Galileo's experiment at Pisa. If he had jumped with the balls and looked only at them he would have seen them hover in front of him all the way to the ground. There would have been no need to invent a gravitational force to describe their motion, they would continue at rest in the absence of any external resultant force. By extending the principle to all the laws of physics Einstein was asserting that light must behave in the same way in a gravitational field as if it were observed from an accelerating reference frame - this led to some startling new predictions.

4.2.5 An Experiment With Antimatter.

The creation of anti-hydrogen atoms at CERN in January 1996 will soon allow the equivalence principle to be tested for antimatter. Antiatoms produced by LEAR (the Low Energy Antiproton Ring) will be slowed down and trapped using electric and magnetic fields before combining them with positrons to form slow moving anti-hydrogen atoms. All that will remain to be done is to measure their free-fall acceleration in the Earth's gravitational field. The expectation is that antimatter, like matter, will fall downwards with an acceleration equal to the local gravitational field strength.

4.3 GRAVITY AND LIGHT.

4.3.1 Gravity Deflects Light.

If the equivalence principle applies for all the laws of physics then it will apply to electromagnetism as well as mechanics. This means that no experiment in optics can distinguish a reference frame at rest in a uniform gravitational field from a reference frame undergoing uniform acceleration. The path of a light ray in both frames will be identical.

Einstein considered another version of the thought experiment discussed previously. A light ray travelling 'horizontally' through our inertial reference frame also passes through the accelerating frame. Clearly the path of the ray in the accelerating frame will be parabolic. To the internal observer it will appear to 'fall' just like any other projectile in that frame.

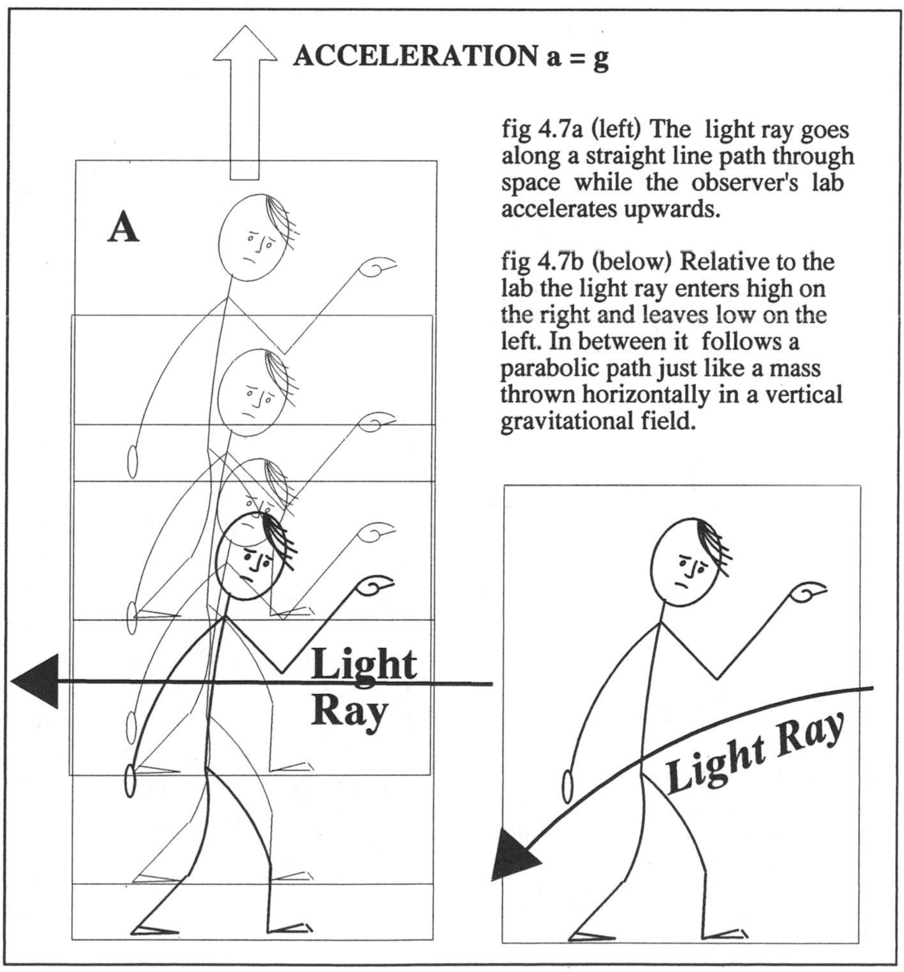

fig 4.7a (left) The light ray goes along a straight line path through space while the observer's lab accelerates upwards.

fig 4.7b (below) Relative to the lab the light ray enters high on the right and leaves low on the left. In between it follows a parabolic path just like a mass thrown horizontally in a vertical gravitational field.

fig 4.8 If a uniform gravitational field is equivalent to a constantly accelerating reference frame, light rays should be deflected by gravity.

Applying the equivalence principle, a light ray should also bend in a uniform gravitational field. This is a *testable* prediction from the equivalence principle: gravity bends light, so observed images will be displaced if the light rays pass through a gravitational field on their way to us.

We can also interpret the bending of light in a gravitational field in terms of the equivalence of mass and energy we met before. Light is a form of energy and so has mass. Mass creates gravitational fields and experiences forces from them. Light falls just like anything else.

Yet another way to look at this is to say that light always follows the most direct route through spacetime and that massive bodies distort the geometry of spacetime. In this interpretation the gravitational field *is* the geometry of spacetime. Furthermore, gravity is transformed away at each point in space when we observe events from a freely falling reference frame. For example, light bends near the surface of the Earth, but if we observe it from a freely falling reference frame it will go straight through our laboratory.

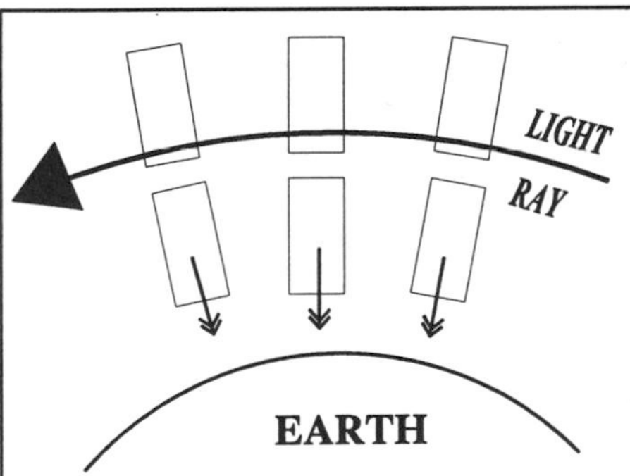

fig 4.9 Gravity bends light but it can be transformed away inside a freely falling reference frame. Each of the free-falling laboratories above is a local inertial frame. Inside them light travels at velocity c in a straight line.

4.3.2 Testing The Deflection Of Starlight.

Einstein suggested that the deflection of light by gravity could be used experimentally as a test of General Relativity. He calculated how much a ray of light passing close to the surface of the Sun would be deflected due to the Sun's gravity. This angular displacement would mean that the apparent position of stars behind the Sun would change as the Sun passed across our line of sight to them. This deflection would be most noticeable as they reappeared from behind the occluding Sun. Unfortunately the presence of the Sun itself would usually prevent us from seeing the far dimmer stars. However, if the Sun's disc was blocked out, as it is during a total eclipse, it might just be possible to measure the apparent shift and test the theory.

fig 4.10 Deflection of starlight. The apparent displacement of stars close to the disc of the Sun can be measured during a total eclipse.

In 1911 Einstein published a paper predicting a deflection of 0.875 seconds of arc for a ray passing at grazing incidence to the surface of the Sun. His calculation was made on the basis of the equivalence principle, but can also be obtained from Newtonian theory if light rays fall in a gravitational field like other projectiles. (It is interesting that the German astronomer Johann Georg von Soldner (1776-1833) had derived exactly this result from classical mechanics in 1801, but his work was forgotten until it was rediscovered in 1921.)

The early attempts to measure this deflection during a total eclipse were unsuccessful, the deflection was close to the limit of detection and the experiment was difficult to carry out accurately. This was in some respects fortunate for Einstein. In 1915 he realized that the deflection due to the equivalence principle is only one contribution to the deflection of light in a gravitational field. The other is a purely relativistic effect due to the distortion of geometry itself. It turned out that this additional deflection was equal to the original prediction so that the total prediction became 1.75 seconds of arc at grazing incidence. The deflection of light rays passing some distance from the Sun's surface falls in inverse proportion to their angular displacement from the surface.

The crucial experiments were carried out in May 1919 by a team led by Sir Arthur Eddington. This eclipse was particularly good for testing the theory since it occurred when the Sun was surrounded by a field of many bright stars. Even so, it was a difficult experiment to carry out accurately enough to distinguish between the three possible outcomes:

- no deflection at all would indicate that light is not affected by gravity;
- a deflection of 0.875 seconds is consistent with a Newtonian model in flat spacetime;
- a deflection of 1.75 seconds would support general relativity.

Problems of cloud cover and poor weather conditions meant only two images involving five stars could be used, but these gave a result of 1.60 seconds deflection with a likely error of 0.31 seconds. A second experiment by a team led by Andrew Crommelin during the same eclipse also supported Einstein's theory.

4.3.3 Calculating The Deflection Of Starlight.

The calculation below treats the deflection of light in a semi-classical way and gives Einstein's 1911 result. A ray travelling close to the surface of the Sun is deflected by the component of gravitational force acting perpendicular to its path as any mechanical particle would be. We assume that the deflection is small and neglect the effect of the gravitational component parallel to the path. It is easiest to picture this as if the light consisted of discrete particles having energy E, but this is unnecessary since the mass itself will cancel from the expression for deflection (light 'falls' at the same rate as everything else). See fig 4.11.

$$\frac{x}{R} = \tan\theta \quad \therefore \quad \frac{dx}{d\theta} = R\sec^2\theta \quad dx = R\sec^2\theta\, d\theta$$

$$r = \frac{R}{\cos\theta}$$

$$E = mc^2 \quad \therefore \quad m = \frac{E}{c^2}$$ (we shall not substitute for m)

$$\delta t = \frac{\delta x}{c}$$ (time for light to move along element δx)

$$F = \frac{GMm}{r^2} = \frac{GMm\cos^2\theta}{R^2}$$ (gravitational force on mass m, distance r)

$$F_\downarrow = \frac{GMm\cos^3\theta}{R^2}$$ (component of F perpendicular to light ray)

$$\delta p = F_\downarrow \delta t$$ (impulse perpendicular to light ray)

$$= \frac{GMm\cos^3\theta}{R^2}\cdot\frac{\delta x}{c} = \frac{GMm\cos\theta\delta\theta}{Rc}$$

$$p = mc$$ (momentum of light of energy E/c^2)

$$\delta\phi = \frac{\delta p}{p} = \frac{GM\cos\theta\delta\theta}{Rc^2}$$ (deflection during path length δx)

$$\phi = \frac{GM}{Rc^2}\int_{-\pi/2}^{\pi/2}\cos\theta\, d\theta = \frac{GM}{Rc^2}[\sin\theta]_{-\pi/2}^{\pi/2}$$

$$\phi = \frac{2GM}{Rc^2}$$ actual value is $$\phi = \frac{4GM}{Rc^2} \quad (4.1)$$

For grazing incidence to the Sun:

$R = 7\times10^{8}$ m

$G = 6.7\times10^{-11}$ N m^{2} kg^{-2}

$M = 2\times10^{30}$ kg

$\phi = 4.2\times10^{-6}$ rads $= 0.88$ seconds of arc

We shall return to this problem and discuss why this result is *wrong* later.

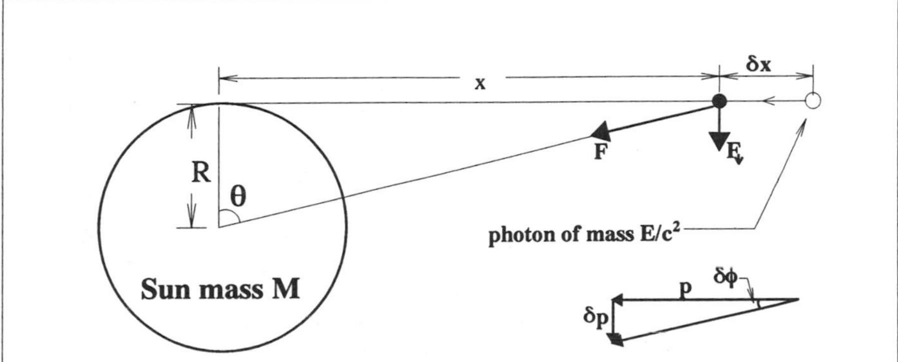

fig 4.11 The deflection of a photon as it passes the Sun at grazing incidence can be calculated as the sum of small deflections over short sections of path. The component of attraction acting perpendicular to the grazing path gives the photon an impulse and change of momentum perpendicular to its original momentum, this results in a small angular deflection.

4.3.4 Gravitational Lenses.

Glass lenses form images by refraction. As light passes through the lens its velocity changes and it is deviated from its original path. Light from distant stars or galaxies may be deflected by the gravitational field of intermediate objects and in some cases this produces a lensing effect. Astronomers have detected a number of gravitational lenses and the photograph in plate 4.2 shows one example. In this case multiple images of a quasar have been formed as light on its way to Earth is deflected around a massive galaxy.

The idea of a gravitational lens was first proposed by Fritz Zwicky in the 1930s but astronomers were not particularly excited because the predicted focal length for these lenses would be comparable to the scale of the observable universe. However quasars, discovered in the 1960s, are very distant and very luminous so they are visible from Earth. The first example of a gravitational lens was discovered in 1979 with quasar QSO957+561 which appeared as a double image. The two images were shown to come from one object by their near identical spectra and the fact that they both changed in intensity in the same way (though not at exactly the same time since light travelling the longer route is delayed - in this case the delay was about

14 months). Measurement of these delays can give an idea of the distance of the quasar and this, combined with the quasar's redshift, can be used to calculate the Hubble constant (see section 4.9).

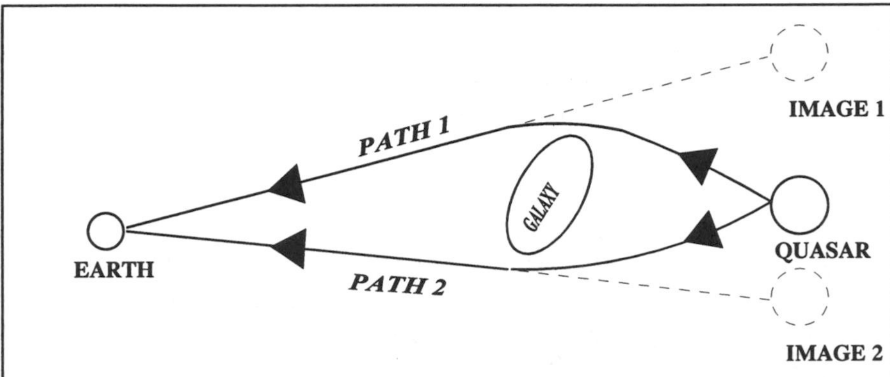

fig 4.12 A gravitational lens. The light paths have different lengths so the light takes longer to reach Earth by one path than the other. If the quasar changes in brightness this change is seen at different times in each image and the delay can be used to measure the Hubble constant.

More recently, astronomers have become interested in gravitational microlenses, formed by individual stars as they pass in front of a quasar. The optical effects of such microlenses will vary on a much shorter timescale than macrolenses because of the smaller scale of the lens. What is observed is a fluctuation in the brightness of the quasar over a period of days or weeks (the corresponding changes when a galaxy drifts in front of a quasar occur in hundreds of thousands of years). The obvious place to look for microlensing is in one of the multiple images formed by a macrolens (i.e. a galaxy). The variation over a short time would indicate the movement of a single star across the line of sight of the quasar. This variation is typically an increase in intensity as the star focuses the quasar's light into an 'Einstein ring' (see fig 4.13). If the star is slightly out of alignment with the quasar this separates into a pair of images formed on either side of the lens.

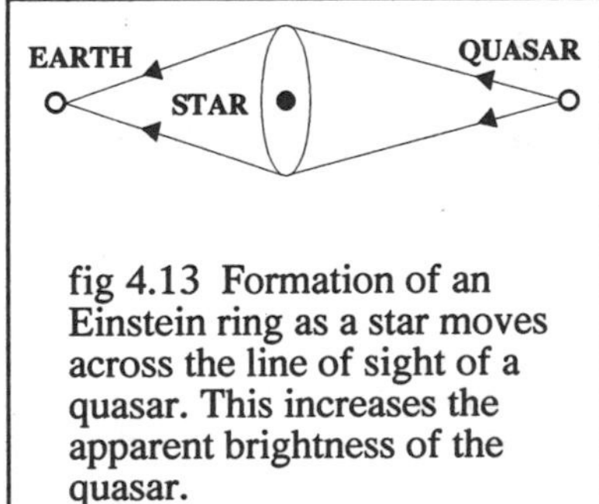

fig 4.13 Formation of an Einstein ring as a star moves across the line of sight of a quasar. This increases the apparent brightness of the quasar.

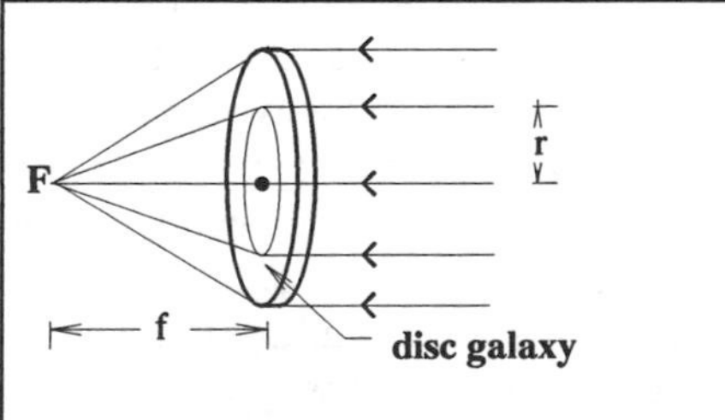

fig 4.14 A simple gravitational lens formed by a disc-shaped galaxy of constant surface density. F is its focal point.

The focal length of a gravitational lens can be calculated from the deviation of light as it passes the mass of the star or galaxy that causes the lensing. In the calculation below we assume that the lens is a disc perpendicular to the quasar light and consider the convergence of parallel rays by the lens. There are a number of simplifications used in the calculation, for example, the surface density of the galactic disc is assumed constant, but the calculation is still useful to give an idea of the scale on which such lenses may be important. The resulting focal length, billions of light years, means that the most likely objects we will observe with gravitational lenses are very bright and very distant. This emphasizes the importance of exotic objects like quasars (which are extremely luminous and yet also very far away) to relativistic cosmology.

Angular deflection for light grazing surface of lens of radius r and mass m is α

$$\alpha = \frac{4Gm}{rc^2}$$

Rays parallel to the 'principle axis' of the lens will intersect this axis at a distance f from the lens such that:

$$\alpha = \frac{r}{f} = \frac{4Gm}{rc^2}$$

If the galaxy is a disc lying perpendicular to the incident light and has constant surface density σ then the rays passing at distance r from the centre are deflected by a lens of mass :

$$m = \pi r^2 \sigma$$

$$\frac{r}{f} = \frac{4G\pi r^2 \sigma}{rc^2}$$

so

$$f = \frac{c^2}{4\pi G\sigma} \tag{4.2}$$

Massive galaxies have a surface density of around 10 kg m^{-3}.

This gives a focal length of about 10^{25} m or about a billion light years.

4.4 GRAVITY AND TIME.

4.4.1 Comparing Clocks.

Once again we shall consider weak field cases. These are simpler to analyse than the full general relativistic theory but yield results that are useful in many real situations. To compare clock rates between observers at different points in a gravitational field we must first decide what will constitute comparable clocks and then ask how one observer's clock will appear to the other observer. Let us assume each observer uses the international agreed standard for the second by calibrating local clocks against the oscillations of radiation emitted by cesium atoms. Comparing clocks is now equivalent to comparing the frequency of radiation emitted by atoms in each reference frame.

fig 4.15 The frequency of light falls as it climbs out of the gravitational field. If both use local sources to govern their clocks A will see B's clock run slow and B will see A's run fast.

In an experiment to compare clock rates observers A and B (above) transmit photons emitted from excited cesium atoms in their own frame to each other. They then compare the frequency of the arriving photons with the frequency of their own transmission. According to the photon theory the energy associated with a photon of frequency f is $E = hf$. Photons have mass, because of their energy, and so work must be done in 'climbing out of the field' from A to B. Similarly work is done on the photons by the gravitational field as they 'fall' from B to A.

Let a photon have frequency f on leaving the stronger gravitational field at A. On arrival at B it has a lower energy and frequency f'. One second is 9 192 631 770 periods of this radiation so it will take longer in B's frame for this number of periods of the received radiation to pass than for the same number of oscillations from her own cesium atoms. B concludes that one second lasts longer in A's frame than in her own. Time passes more slowly in stronger gravitational fields.

Unlike special relativistic time dilation, this is not a symmetric effect. Photons arriving in A's frame from B have gained energy and so oscillate at a higher frequency. A concludes that one second passes more quickly in B's frame and that time passes more quickly in the weaker field. A and B *agree* on gravitational time dilation.

4.4.2 Time Dilation In A Weak Field.

It is not difficult to calculate the amount of gravitational time dilation between two points in a weak field. We shall consider two points A and B at gravitational potentials φ_1 and φ_2 respectively. Assume B is at the higher potential. A transmits a photon of frequency f to B.

$E = hf$ so if E changes by δE then f changes by δf and $\delta f = \dfrac{\delta E}{h}$

photon mass $m = \dfrac{E}{c^2} = \dfrac{hf}{c^2}$

$\delta E = -m\delta\phi$ since photon energy falls as gravitational p.e. rises.

$$\delta f = \frac{-m\delta\phi}{h} = \frac{-hf\delta\phi}{hc^2} = \frac{-f\delta\phi}{c^2}$$

$$\underline{\underline{\frac{\delta f}{f} = -\frac{\delta\phi}{c^2}}} \qquad (4.3)$$

If we wish to compare time periods in the two frames then:

$T = \dfrac{k}{f}$ that is, any time interval is a multiple of the period $\dfrac{1}{f}$.

$$\frac{dT}{df} = -\frac{k}{f^2} = -\frac{T}{f} \quad \text{so that} \quad \underline{\underline{\frac{\delta T}{T} = -\frac{\delta f}{f} = \frac{\delta\phi}{c^2}}} \qquad (4.4)$$

Example This effect is very small in the weak gravitational field around the Earth. The height of Mount Everest is about 8 km. This gives:

$$\frac{\delta T}{T} = \frac{gh}{c^2} \approx \frac{9.8 \times 8000}{9 \times 10^{16}} = \underline{9 \times 10^{-13}}$$

That is, clock rates differ by only about 1 part in 10^{12}. Atomic clocks at the base and summit of the mountain would develop a difference of 1 second in 35000 years! The clock at the summit would gain with respect to the clock at the base.

4.4.3 The Gravitational Red Shift.

The last section showed that the frequency of a photon falls when it is transmitted from a lower to a higher gravitational potential. The observer at higher potential will therefore associate it with a different part of the electromagnetic spectrum. It is shifted to a lower frequency and longer wavelength. This is the gravitational red shift. Its effect is similar to the Doppler red shift caused by the relative motion of a receding source, but the origin of the effect is now due to gravitation. The gravitational red shift for a photon moving through a gravitational potential difference $\delta\phi$ is calculated as follows:

$$c = f\lambda \quad \text{giving} \quad \frac{df}{d\lambda} = -\frac{c}{\lambda^2} \qquad (1)$$

$$\frac{df}{f} = -\frac{\delta\phi}{c^2} \quad \text{(previous section)} \qquad (2)$$

now combine (1) and (2):

$$\frac{\delta\lambda}{\lambda} = \frac{\delta\phi}{c^2} \qquad (4.5)$$

Waves reaching us from distant stars will be red shifted as they 'climb out' of the gravitational field of the source (they will also be slightly blue shifted as they 'fall into' the gravitational field of our solar system). In practice these gravitational red shifts are just one component of the total spectral shift caused by gravitation, relative motion and the expansion of the Universe and it is not always clear how to separate these components. The most accurate tests of the gravitational red shift have been terrestrial ones where other factors can be controlled or eliminated.

4.4.4 Measuring The Red Shift.

Between 1960 and 1965, Pound, Rebka and Snider carried out a series of experiments in which the gravitational red shift was measured for photons moving up or down the Jefferson Physical Laboratory Tower at Harvard University. The source was excited iron-57 nuclei which emit gamma-rays as they decay. The detector contained iron-57 nuclei in their ground state and so acted as a very finely tuned receiver. Any shift in frequency of the arriving photons and the detector cannot absorb them. It is like a radio station and a radio receiver, the receiver must be tuned to the carrier frequency in order to receive a strong signal.

Unfortunately, when photons are emitted from a nucleus the nucleus recoils and there is a slight change in frequency. In a similar way a free iron-57 atom is unable to absorb a photon at its resonant energy because it cannot satisfy both the conservation of energy and momentum required for the resulting recoil of the absorbing atom. This is a problem for experiments to measure gravitational red shifts since it is essential that the emitting and detecting atoms are 'tuned' to one another. It is also important that the spread of frequencies present in the gamma ray emission is extremely narrow otherwise it will mask the effect of a small additional change due to gravity.

Pound, Rebka and Snider's experiment was made possible by a discovery in 1958 by R.L. Mössbauer. If the iron-57 atoms are embedded in a solid, as they are in this case since the excited iron-57 originates from a decay of cobalt atoms, then a significant proportion of them behave as if they are locked into the extended structure and do not recoil like free atoms. This has the effect of increasing the effective mass of the emitter or absorber many times and makes them resonate at precisely the same frequency.

One further complication is the natural line width of the gamma ray emission. For iron-57 this amounts to about 3 parts in 10^{13}, but is still of considerably greater magnitude than the expected gravitational red shift:

$$\text{tower height} \quad h = 22.5 \text{ m}$$

$$\frac{\delta f}{f} = -\frac{\delta\phi}{c^2} = -\frac{gh}{c^2} = \underline{-2.5\times10^{-15}}$$

In the experiment the detecting atoms absorb photons from the emitter and then, a short time later, re-emit them in random directions. This *'resonant scattering'* process reduces the number of photons at the resonant frequency that are re-emitted in their *original* direction. A photon counter placed behind the detector will therefore register a reduction in photon arrival rate at this frequency. If the detector does not absorb them then more will pass through and arrive at the counter.

Consider photons emitted from the top of the tower. These move from higher gravitational potential to lower gravitational potential so they will have an increased frequency as they reach the detector and so will not be so strongly absorbed as if they had the resonant frequency. However, if the detector is moved away from the source then a Doppler shift will occur and the frequency of the arriving photons will be reduced. At a certain velocity the arriving photons will resonate with the detector and the counter will give a minimum reading. The speed at which the detector has to move (which is very low for such a tiny frequency shift) can be used to calculate the gravitational shift (by using the Doppler formulae). The results agreed with the prediction to within about 1%.

A more accurate experiment was carried out in 1976. This agreed with the predictions of the equivalence principle to better than 0.01%. The experiment was carried out by R. Vesot and M. Levine (once again from Harvard) and measured the shift of microwaves transmitted to earth from a hydrogen maser atomic clock carried up to 10 000 km on a Scout D rocket.

A contemporary version of Eddington's 1919 experiment uses radio interferometers to pinpoint the position of a distant radio source very accurately. Quasars are very intense and very distant radio sources that appear to the radio astronomer a little like the stars of optical astronomy. The pair of quasars, 3C273 and 3C279 are conveniently close to the Sun and so are used in a radio version of the experiment to measure the deflection of starlight. In practice what happens is that the two galaxies are observed for several days either side of 8 October (when they are closest to the Sun as seen from the Earth) and their separation is seen to vary as the radio waves from the quasar closer to the Sun are deviated more than those from the other quasar. Since it is not necessary to wait for a solar eclipse, observations can be made every year and the variation of deflection with proximity to the Sun can be studied. Results from these experiments are in agreement with general relativity to about 1%.

4.4.5 The Shapiro Time Delay.

Another accurate test of general relativity comes from the time delay of radio echoes as they pass through the Sun's gravitational field. There are two ways of thinking about this delay, we can say that the radio waves travel through regions of curved space in which time is dilated or we can say that the speed of light relative to us varies. Either way, a round trip journey for light or radio waves from the Earth to a point beyond the Sun and back takes longer if the Sun is there than if it is not.

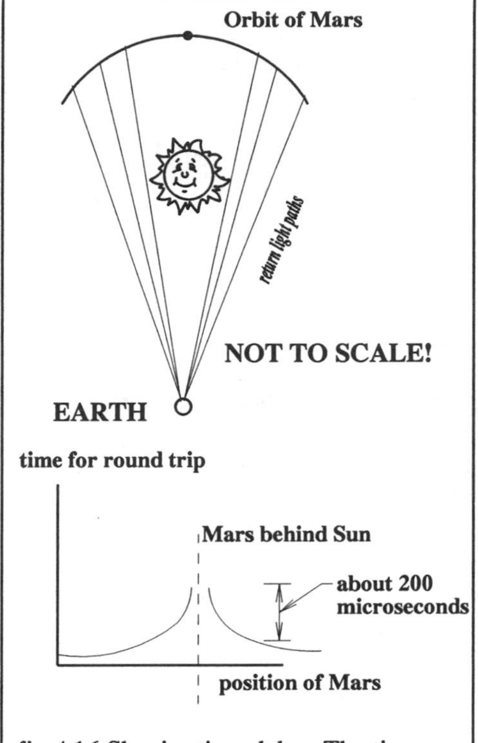

fig 4.16 Shapiro time delay. The time for light to make a round trip to a point behind the Sun increases for rays that pass close to the Sun's surface where the gravitational potential has a larger negative value.

Shapiro was the first to study this delay. He calculated that it would be about 250 μs for a round trip to Mars when Mars was at its farthest point from the Earth on the far side of the Sun. Of course, there is an obvious problem here: we cannot remove the Sun to find out how long the radio waves actually take in its absence! We also need a very powerful terrestrial radio source if, after such a long journey, we expect a strong enough return pulse to detect. These two problems were both solved. The first was tackled by calculating the distance from the Earth to Mars based on Newton's theory of gravity which is valid in the weak field region where these planets orbit. The strong radio source was provided by the Haystack radar antenna in Massachusetts which emits 500 kW pulses.

The initial results were obtained in 1967 and were within 20% of the predicted value. This was soon improved to 5%. Reflected pulses from the Mariner spacecraft gave agreement to 3% but dramatic improvements were obtained after the Viking spacecraft made a Mars landing. Using this to return radar pulses gave agreement with theory to 0.1%. The great advantage here is that the point of reflection is sharply defined. Earlier planetary reflections have a considerable uncertainty because of the mountainous nature of planetary surfaces. (On Earth for example it would take about 30 μs to travel from the summit of Mount Everest to sea level). The main disadvantage of a freely moving spacecraft is that it will be subject to the buffeting of radiation and meteorites and so its precise trajectory is uncertain to a greater degree than the position of the landed spacecraft.

4.4.6 Calculating The Shapiro Time Delay.

We can estimate the Shapiro time delay by considering how gravity affects the rates of free-falling clocks at each point along the light path. This is based on the equivalence principle but gives an underestimate of the actual value because it neglects the effect of spacetime curvature (as did our derivation of the deflection of starlight). However, it does give the correct order of magnitude and is a genuine gravitational effect. The Viking test involved a reflection from the surface of Mars when it was almost opposite the Earth behind the Sun. In our approximation we shall calculate the time delay by considering a hypothetical ray passing through the centre of the Sun. This sounds rather crude, but the radius of the Sun is less than 0.5% of the radius of the Earth's orbit (and an even smaller percentage of the radius of Mars's orbit). We then integrate the small contributions to time delay along each small radial element δr along the light path. The total delay will be the sum of delays for rays traveling to and from the Earth and to and from Mars.

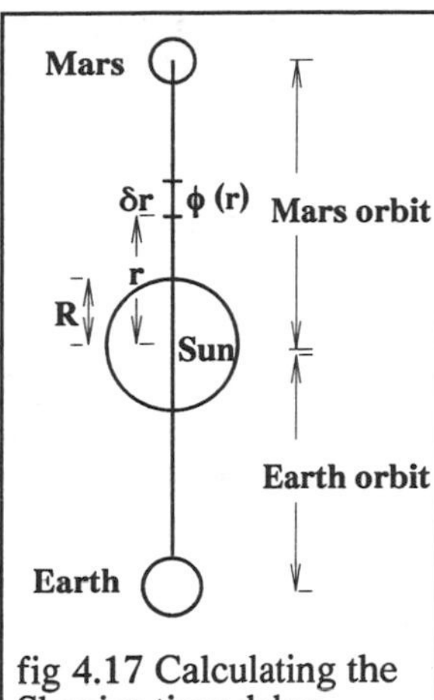

fig 4.17 Calculating the Shapiro time delay.

$\delta\tau =$ infinitesimal delay during section δr

$t =$ time to move through $\delta r \approx \dfrac{\delta r}{c}$

$$\frac{\delta\tau}{t} = -\frac{\phi}{c^2} \qquad \text{(gravitational time dilation)}$$

$$\delta\tau = -\frac{\phi\delta r}{c^3} = \frac{GM\delta r}{rc^3}$$

$$\tau = \int_R^{r'} \frac{GMdr}{rc^3} = \frac{GM}{rc^3}[\ln r]_R^{r'} \qquad (r' \text{ is the planet's orbital radius})$$

$$R \approx 7\times10^8 \text{ m} \qquad r_E \approx 1.5\times10^{11} \text{ m} \qquad r_M \approx 2.5\times10^{11} \text{ m}$$

This gives a total delay for the reflected rays of:

$$\tau_{tot} = \frac{2GM}{c^3}\left\{\ln\frac{r_E}{R} + \ln\frac{r_M}{R}\right\} \approx 110\ \mu s \qquad \text{about half the actual delay}$$

This argument is quite subtle. What we are saying really is that the light passes through an infinite number of reference frames in which the rates of clocks are determined by the gravitational potential. If we judge the progress of the ray from a distant frame (in this case the Earth) then time passes more slowly in the regions of lower potential. However, light moves through these frames with constant velocity c. Relative to us it must therefore *slow down*, accumulating a delay relative to its expected time of arrival in the absence of gravitational effects. The total delay is calculated by summing the infinitesimal delays along each small section of its path.

4.4.7 The Twin Paradox Again!

We are now in a position to understand the asymmetry between the two twins in the famous twin paradox and use it to calculate the time difference that develops between them as a result of their different trajectories in spacetime. The spacetime diagram below emphasizes the unavoidable accelerations experienced by twin B. During these periods that twin is effectively constrained in a gravitational field of strength $g = a$ (where a is the actual acceleration) while the other twin, who experiences no inertial forces, is effectively in a free falling reference frame in a uniform gravitational field of strength g. This is reminiscent of Einstein's thought experiment that led him to the equivalence principle. Twin B sees time speed up in the accelerating frame (which is at the higher potential) and this gravitational time dilation accounts for the 'gain' of time by that twin on their reunion. By including the gravitational effects we are able to analyse the situation from either twin A's or twin B's point of view and reach exactly the same conclusion about the relative times that have elapsed during their separation. This is very satisfying.

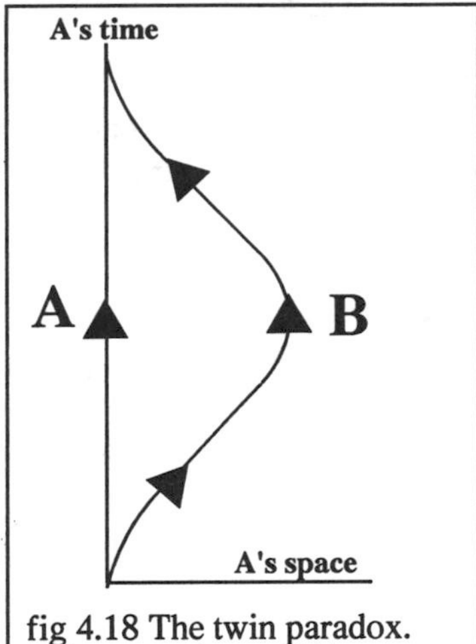

fig 4.18 The twin paradox. Curved sections of B's worldline represent periods of acceleration.

When we first derived the gravitational time dilation effect we did so by considering the work done by a photon as it 'climbs out' of a gravitational well. This implies that there will be no difference in clock rates between clocks that are not separated in space even if one of them is accelerating. In this case that means the time difference will be entirely due to the acceleration of the travelling twin at the distant star and not to the initial and final accelerations which take place in close vicinity of the Earth. (There is of course a slight approximation introduced by neglecting these completely.) The derivation that follows makes a number of simplifying assumptions, but a full treatment also gives agreement on the expected outcome.

Consider the journey from B's point of view:

acceleration lasts a time T therefore $g \approx \dfrac{2v}{T}$

A is in a gravitational field of strength g at distance d from B so

$$\text{A's clock gains by } \delta T = -\frac{T\delta\phi}{c^2} = \frac{Tgd}{c^2} = \frac{2vd}{c^2}$$

But $d = \dfrac{vt}{2}$ where t is the total time for the round trip $(t_A \approx t_B \approx t)$

$$\therefore \quad \delta T \approx \frac{v^2}{c^2}t$$

Now, during the period of constant velocity B will see A's clock run slow, so we can apply special relativistic time dilation to this part of the motion:

$$t_A = t_B\sqrt{1-\frac{v^2}{c^2}}$$

The total loss during the journey will therefore be:

$$\delta t = t_B\left(1-\left(1-\frac{v^2}{c^2}\right)^{½}\right) \approx \frac{v^2 t}{2c^2}$$

Relative to B, twin A's clocks therefore have a net *gain* of:

$$\delta t = \left(\frac{v^2}{c^2} - \frac{v^2}{2c^2}\right)t = \frac{v^2}{2c^2}t$$

This is exactly equal to the time loss calculated by A for twin B's clocks on the basis of special relativity alone. The paradox is resolved.

4.5 THE GEOMETRY OF CURVATURE.

4.5.1 Introduction.

Is the universe infinite or finite in extent? Does it have a beginning and end in time? Does it have an edge or boundary? Before general relativity it was impossible to give a convincing answer to any of these questions. Whilst the concept of an infinite space challenges our imagination, the idea that space stops somewhere leads to the question of what lies 'beyond' the end of space. This challenges our very notion of space. In a similar way, if we trace the universe back to some originating event then the question arises as to what existed *before* that event. This leads to an infinite regress. If the universe is not eternal then what will come after? These questions force classical science into a difficult corner and some then resort to a different kind of solution, often mystical or religious. General relativity does not solve all of these problems but it does show how they might be approached and how science does have something very important to say about the fundamental questions of existence. The ideas are radical and in order to appreciate them we must first consider the geometry of space-time itself.

A full mathematical treatment of the general theory is beyond the scope of this book but the conceptual basis can be understood in terms of weak field and low speed approximations. In practice, since the gravitational field we inhabit and the objects we usually observe satisfy these criteria, it is in this range that most experimental tests of general relativity are carried out.

4.5.2 Worldlines In A Gravitational Field.

Free bodies in special relativity move along straight worldlines. The presence of a gravitational field changes this. For example, a falling body accelerates relative to a reference frame 'at rest' in the gravitational field (e.g. the Earth's surface) and its worldline in this frame is curved. On the other hand an observer falling with the object will disagree. In his reference frame, which is locally an inertial frame, the body has a straight worldline and the laws of mechanics apply with no need to introduce a uniform gravitational field. For this observer the paths of the particles correspond to the shortest paths through space. However, for the observer 'at rest' in the field this is not the case, the paths are curved. Einstein's idea was that the curvature is a clue to the nature of gravitation. The gravitational field is a distortion of space-time geometry and free bodies move along the shortest available path through it. In an inertial frame this will correspond to the straight lines of familiar Euclidean geometry. In a gravitational field these will be 'geodesics', the shortest paths in the distorted space-time analogous to the great circle routes over the surface of the Earth which provide the shortest links (on the surface) between two places. The example above shows that the question of whether a particular trajectory through space is straight or curved is itself a relative judgment affected by the choice of reference frame.

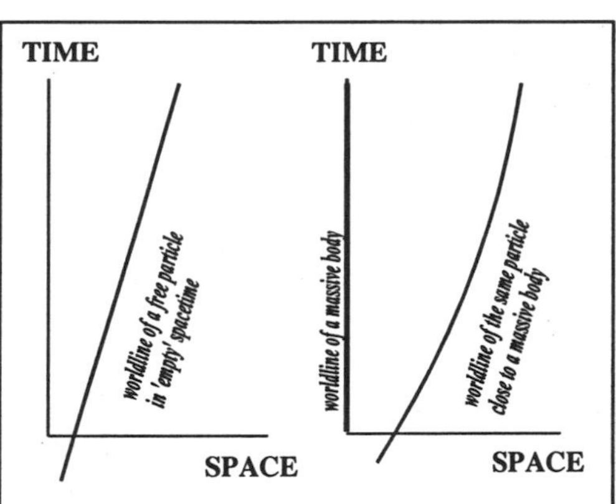

fig4.19 Worldlines appear curved in the spacetime of a massive body.

If it is true that free bodies move along geodesics in space-time then we should expect the curvature of all worldlines to be the same at the same point in space. This does not *appear* to be the case. Consider three projectiles launched from the same point in a uniform gravitational field. The first is a ball, the second a bullet fired from the same point in the same direction, the third is a ray of light. All three are projected in such a way that their start and finish positions in the laboratory are at the same height. The three paths appear to have very different curvatures. They all follow parabolic paths but the ball's path is sharply curved whereas the bullet's is nearly straight and that of the light beam is so close to a straight line that its deflection on this scale is undetectable.

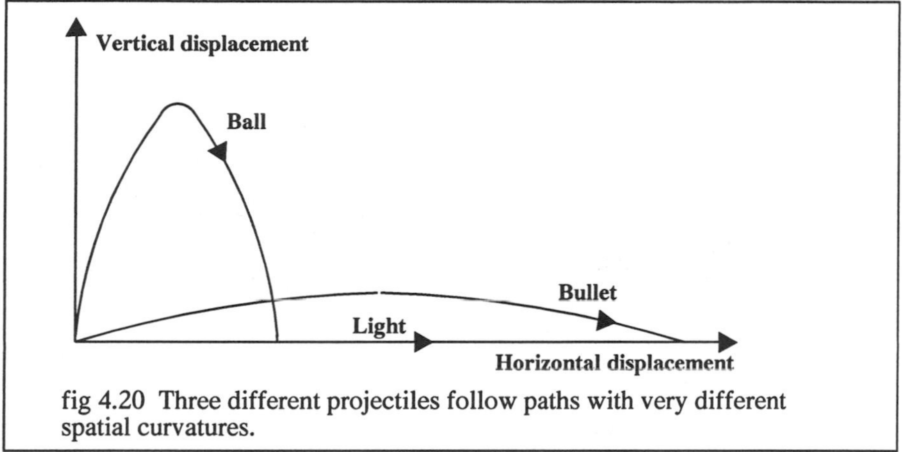

fig 4.20 Three different projectiles follow paths with very different spatial curvatures.

However, we are looking in the wrong way. Worldlines are curved in *time* as well as space. We should compare the curvature of the *space-time worldlines*, not just the spatial trajectories. When we do this an interesting picture emerges. Sections of all three worldlines have the same curvature and can be superimposed on one another.

The reason we find it difficult to detect the curvature of the bullet's worldline is because we see it for such a short time. Its velocity vector has barely turned from its original direction during this time so it appears to be going straight. However, the amount of turning per unit of its path is the same as that of the ball, and of the light beam and is determined by the strength of the local gravitational field.

This gives us a glimpse of the mathematics of curvature. We can measure the degree of curvature at a point by considering the rate at which a vector changes direction in the vicinity of that point. This is determined in Einstein's theory by the strength of gravity at that point. We can go beyond this statement and say that the curvature of space-time *is* the strength of the gravitational field at a point. The general theory of relativity can be paraphrased as:

Matter tells space how to bend;
Space tells matter how to move.

This is obviously simplistic but it captures the essence of the theory. The presence of matter distorts space-time geometry so that the curvature at any point is determined by what we previously interpreted as the gravitational field. A freely moving object passing through this non-Euclidean space-time follows a geodesic path that will appear to be curved relative to a distant observer. With this interpretation planetary orbits are geodesics within a curved space-time around the Sun.

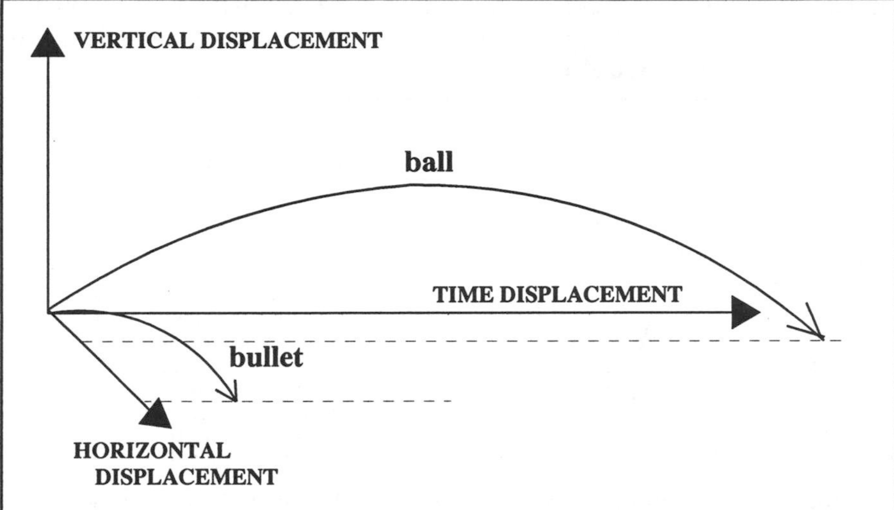

fig 4.21 The worldlines of ball, bullet and light beam have the same curvature in spacetime. However, their projections onto the space plane do not. The light ray is not shown because it lies so close to the horizontal displacement axis.

4.5.3 Curves In Space.

To understand Einstein's theory we must first get to grips with curved lines and curved surfaces in our own space. Consider an arc of a circle. The tangent to this arc changes direction as we move along the arc and one way to measure the curvature is to use the rate at which the direction of the tangent changes.

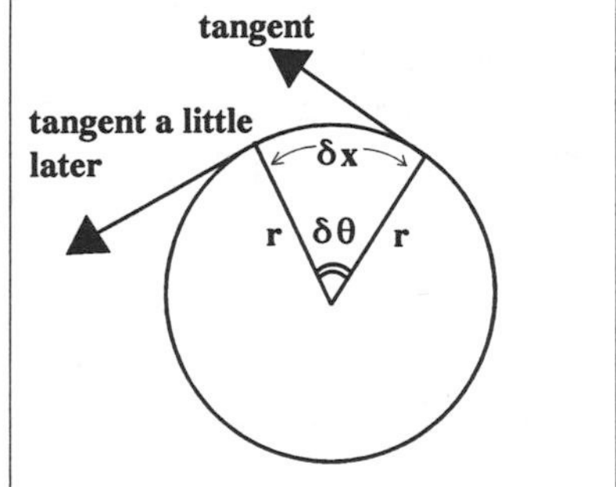

fig 4.22 If a path is curved, the tangent to the path changes direction as it moves along the path.

$$\delta x = r\delta\theta$$

$$\frac{\delta\theta}{\delta x} = \frac{1}{r}$$

The change in direction per unit path is equal to the reciprocal of the radius of the circle. This distance r is the radius of curvature for the path. We can define a radius of curvature at each point along an arbitrary path by measuring or calculating the rate of change of direction of a tangent to the path with distance

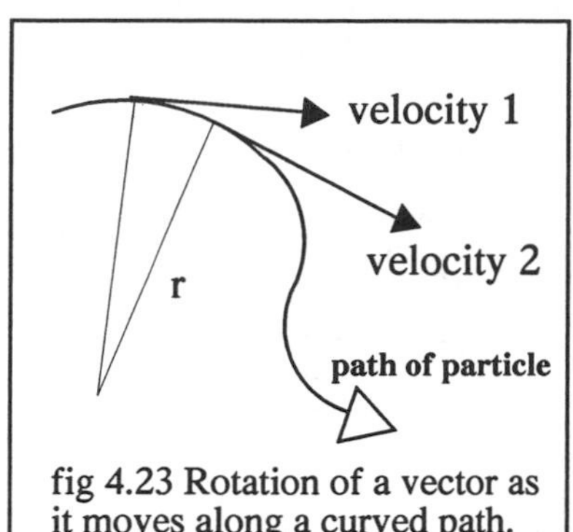

fig 4.23 Rotation of a vector as it moves along a curved path.

moved along the path. The idea here is one of 'parallel transport'. The tangent is a vector parallel to the path at each point and we measure the curvature in terms of the rate of angular rotation of this vector as it is transported along the path. The important point about such parallel transports is that the vector concerned remains in the same relation to the path at every point. A trajectory with a small radius of curvature is sharply curved whilst one with a very large radius is much straighter. A truly straight line would have $r = \infty$.

4.5.4 Flatland.

The curvature of a two-dimensional surface can be measured in a very similar way, but there is a complication because the radius of curvature can be measured in different directions. We shall avoid this complication for the time being and consider the surface geometry of a sphere (the radius of curvature for a sphere's surface is the same in whichever direction it is measured). Of course, if the sphere is viewed from the outside, that is from a Euclidean three-dimensional reference frame, then its curvature can be interpreted as arising because its surface is not restricted to one Euclidean plane in the three-dimensional space. This misses an essential point. When we are considering our own world's space-time geometry we do not have access to a higher (five dimensional) view and there may be no higher dimensional view to be had (although recent versions of 'string theory' propose many higher dimensions). For this reason we must look for the intrinsic properties of curvature that would be apparent to creatures confined to the surface or space we are considering. A sphere's surface is 'obviously curved' when seen from the outside, but would we be able to measure this curvature if we were two-dimensional creatures living in the surface?

The idea of two-dimensional creatures, 'flatlanders', goes back to a famous book ('Flatland') written by Edwin Abbott in 1854. These creatures have no perception of the third spatial dimension and so can only interpret the geometry of their world in terms of two-dimensional measurements made within the surface of that world. Could they tell that the surface was curved? Could they measure its curvature? Would the curvature influence the laws of physics in Flatland? The answers to all of these questions turn out to be 'yes'.

Imagine that the radius of the sphere on which flatlanders live is very large compared to the size of the flatlanders themselves. This will mean that, as long as they do not wander far across its surface, they will think the surface geometry is Euclidean: parallel lines remain parallel, the sum of angles in a triangle is 180 degrees and the area of a circle is πr^2. All of these results turn out to be approximations that fail if geometry is explored over a distance comparable to the radius of the sphere. This can clearly be seen in the diagrams below.

GEOMETRY IN FLATLAND

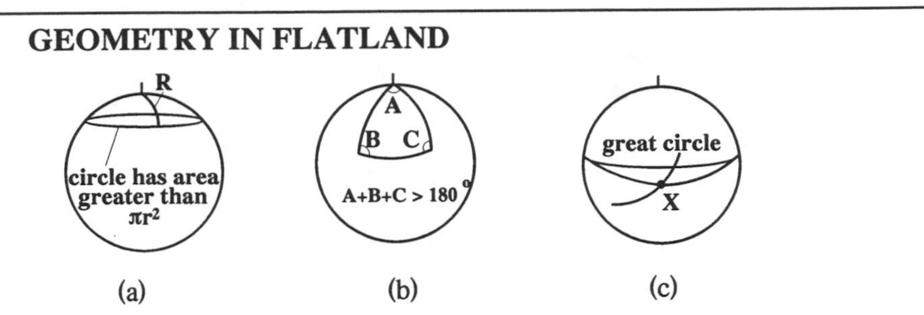

fig 4.24 (a) The area of a circle on the surface of a sphere is greater than that of a circle of the same radius on a plane surface. (b) The sum of the angles inside a triangle formed from three intersecting geodesics is greater than 180 degrees. (c) All geodesics through X will eventually intersect the great circle above them - there are no lines through X that remain parallel to the arc of the great circle.

The straightest lines in flatland are geodesics on the surface that are great circles around the sphere. All great circles intersect so lines drawn parallel in a local reference frame will intersect far from the origin. No local Euclidean reference frames are adequate for the global geometry. This should ring some bells - local inertial reference frames in a gravitational field appear to be accelerating with respect to a distant observer. (Some people object to this idea of intersecting parallels by arguing that we should adopt 'lines of latitude' as our parallels instead of the great circles of longitude. The problem with this is that any two points connected by a line of latitude other than the equator (which *is* a great circle) can be joined by a shorter line following a great circle. Lines of latitude are not lines of minimum distance on a spherical surface.

It is also clear that the surface area inside a circle is greater than would be expected in Euclidean geometry. We can see that this is due to the bulging outwards of the surface contained within the circle, the fact that the circle is not flat. The flatlanders do not have access to our privileged view. For them the geometry of their world obeys non-Euclidean rules which only approximate to our Euclidean geometry as the size of the circle considered is made smaller and smaller. Similarly with the triangles - the sum of angles in large triangles is more than 180°.

An even more surprising result concerns the topology of Flatland. It is finite in extent but has no boundaries! Forget the three-dimensional edge *we* can see perpendicular to their surface, *their* world has a limited area but, because of the way it curves around to join with itself, it has no edge or boundary. A flatlander can move as far as it likes in any direction without hitting a wall or falling off!

4.5.5 The Curvature Of A Spherical Surface.

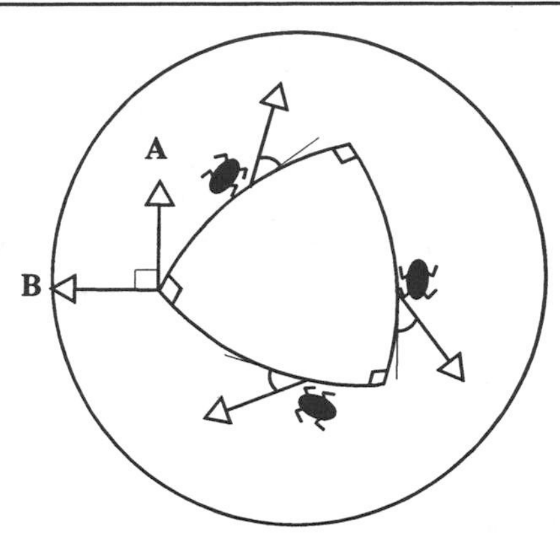

fig 4.25 Vector A is carried around a closed triangular loop in spherical flatland. At all points the vector is at a fixed angle to the motion but has rotated by 90 degrees when it completes the loop (at B). Note also that the triangle covers exactly an eighth of the sphere's surface and that the sum of its internal angles is 270 degrees.

Can Flatlanders measure the curvature of their surface even if they cannot leave it? Clearly they can. They could measure it in terms of how quickly the area of a circle deviates from πr^2 as the radius is increased, or how the angle between two parallel lines changes as they are extended. There are many geometric effects that reveal the intrinsic curvature of their world. *The third spatial dimension does not have to exist, it is quite possible for a two-dimensional surface to be intrinsically curved.* In the same way it is possible for our four-dimensional space-time geometry to have an *intrinsic* curvature and we do not need access to any fifth dimension in order to measure and detect it.

The curvature of a two-dimensional curved surface can be measured by the rotation of a vector transported around a closed loop. This is illustrated in the diagram above. The flatlander carries the vector around a large triangle on the surface of the sphere. It is quite clear that the vector has rotated on its return. In this simple case the angle of rotation is $\pi/2$ radians (90°). The curvature is calculated as the ratio of the angle turned through to the area enclosed by the loop used to transport it. In this case we can see that the triangle covers exactly one-eighth of the surface so we can calculate the curvature as:

Angle turned through $\theta = \dfrac{\pi}{2}$

Area enclosed $A = \dfrac{1}{8} \times 4\pi r^2 = \dfrac{\pi r^2}{2}$

Curvature $= \dfrac{\theta}{A} = \dfrac{1}{r^2}$

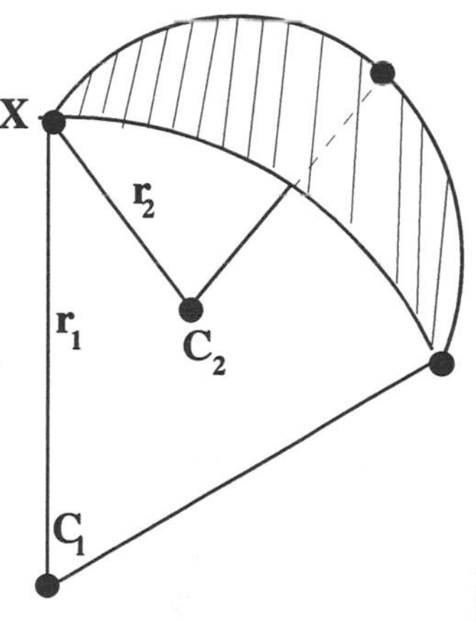

fig 4.26 The curvature at X can be described by giving the radius of curvature along two intersecting axes.

It is interesting to note that the flatlanders can calculate the length of the radius which, in our view, is perpendicular to their entire world. Furthermore, even if the third dimension of space did not exist this value would still represent the radius of curvature of their two-dimensional world. In a similar way we can calculate the radius of curvature of the four-dimensional space-time world in which we live. Of course, the spherical model we have chose for flatland is a very special case. A more arbitrary two-dimensional world would require two radii of curvature at each point to define its geometry. In four-dimensional space-time there are many more components and this is the reason, at least in part, for the mathematical complexity of the theory.

4.6 GRAVITY AND GEOMETRY.

4.6.1 Another Thought Experiment.

One way to create an 'artificial gravity' in space is to use the inertial forces created by rotation. A large cylindrical space station rotating about its long axis would exert an inward force on inhabitants in contact with its inner wall to provide a centripetal force for their circular motion. Viewed from within the rotating reference frame the force from the wall would feel like a reaction to a gravitational force acting *away* from the axis, a centrifugal force.

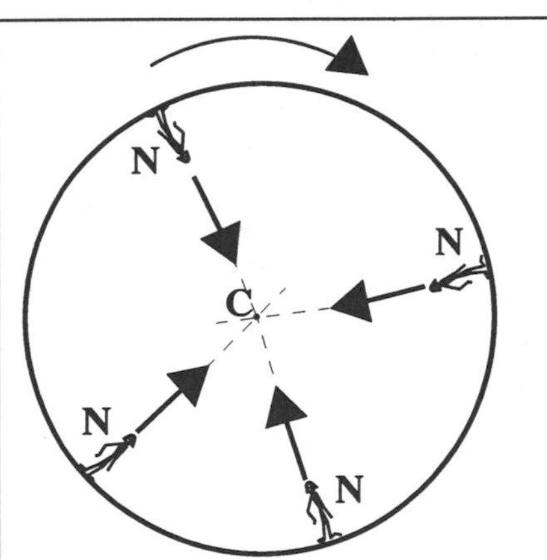

fig 4.27 End view of a cylindrical rotating space station. The contact force N with the 'ground' provides centripetal force to make occupants move in circular motion. This creates artificial gravity in the frame of the space station.

According to the equivalence principle a uniform gravitational field is equivalent to an accelerating reference frame. The centripetal acceleration of the space station inhabitants relative to a distant inertial observer should therefore give the same physical and geometric effects as a gravitational field of equivalent strength ($a = g$). Consider an attempt by the inhabitants to measure the radius and circumference of the space station. Assume they would use 'rigid' measuring rods placed end to end along the radius and the circumference. According to the distant observer the radial rods are transported sideways in a direction perpendicular to their length. This should be no problem. However, the rods along the circumference are moving parallel to their lengths and so ought to suffer a length contraction. This will lead to the inhabitants of the space station concluding that it has a non-Euclidean geometry. The circumference of a circle is greater than

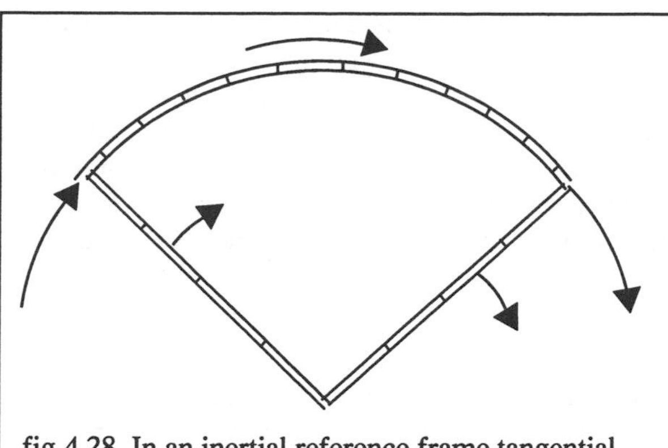
fig 4.28 In an inertial reference frame tangential rods undergo a Lorentz contraction but radial rods do not.

$2\pi r$. This effect will get greater as radius increases since the tangential velocity increases with radius. The other thing that increases with radius here is the centripetal acceleration ($a = r\omega^2$), and this is equivalent to a gravitational field strength. It seems that spacetime is more severely curved where the field strength is greater. The non-Euclidean geometry of this particular example, in which curvature increases with distance from the axis, is rather different from what we would expect near a massive body. In this case the strongest field and greatest curvature should be close to the body itself.

4.6.2 The Tidal Nature Of Gravitation.

The results we derived for the deflection of starlight and the Shapiro time delay in section 4.4 were both about half the actual values. There we used the equivalence principle alone and ignored the effects of spacetime curvature. The reason we should not do this is shown below. The equivalence principle allows us to transform to a freely falling inertial reference frame at each point in a gravitational field. However, if we extend the axes of any one of these free falling reference frames to any significant extent they will soon pass into regions in which they are no longer valid. This is easy to see. Imagine a free-falling room in New York. Now expand the room so that it is larger than the Earth. Freely moving objects will fall in all directions inside this super room and so it is no longer a valid inertial reference frame. On a smaller scale, the fact that all objects are attracted towards the centre of the Earth means that worldlines which start parallel to one another are bound to converge. Two free particles at extreme ends of a falling laboratory will

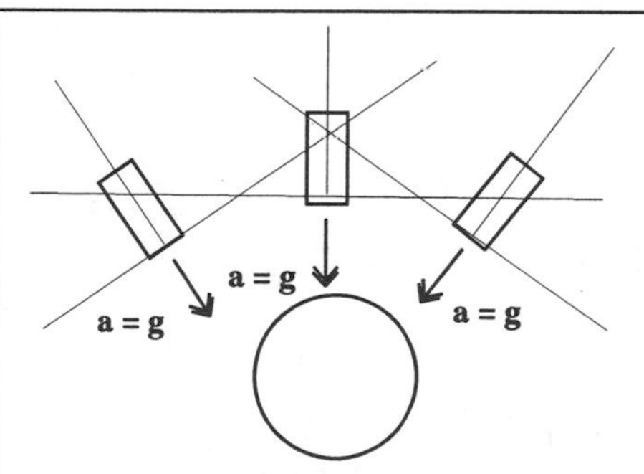

fig 4.29 Free-falling reference frames define local inertial frames, but there is no global inertial reference frame.

move together as the laboratory falls. In a sense this 'tidal effect' is the true gravitation, there is *no* global reference frame that can remove it.

The calculations of section 4.4 assumed that we could replace a global gravitational field by a series of free-falling reference frames in which measurements are made relative to local straight lines. Unfortunately these local straight lines are geodesics which appear curved to the external observer. This gives an extra deflection to the starlight and a longer delay to the reflections.

4.6.3 Calculating The Radius Of Curvature Near Earth.

We can calculate the radius of curvature in the Earth's field by transporting a *4-vector* parallel to itself around a closed loop in spacetime in a free-falling reference frame in the gravitational field. The curvature is then equal to the angle turned through divided by the area of the loop, as it was in the case of Flatland. Because the field is non-uniform we should choose a small loop. We will use a loop consisting of two space-like sides of length 1 m and two time-like sides of length 1s (we shall have to convert time to metres using c as before). Once the curvature has been calculated we can calculate a value for the radius of curvature at that point in the gravitational field (in that orientation in spacetime). The diagram below shows the spacetime loop in both a uniform (left) and a non-uniform (right) gravitational field.

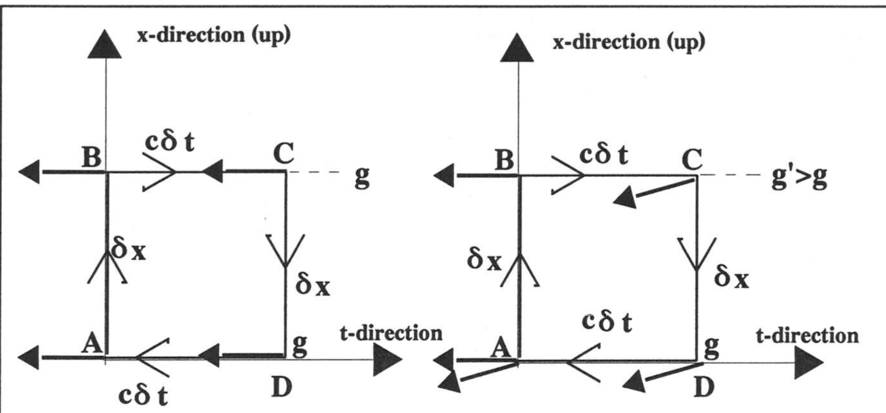

fig 4.30 The diagrams show the transport of a 4-velocity vector for a particle initially at rest (at A) around a closed spacetime loop in a freely falling reference frame. In the left hand diagram there is no rotation (uniform field). In the right hand diagram the vector rotates as a result of the timelike transport CD from the region of stronger gravitational field strength g'.

The right hand diagram could represent the situation in either the Earth's or the Sun's field. The positive x-direction is 'up', that is away from the center of the gravitating body. The field strength $g(x)$ is approximated by the Newtonian inverse-square law and so is less at $x + \delta x$ than at x. The vector chosen to transport around the loop is the 4-velocity of a particle initially at rest at the origin. This is a vector of length c directed parallel to the 'time' axis at A. The stages in moving around the loop are described below.

- AB: a *space-like* transport parallel to the x-axis. The vector remains parallel to the time axis. The field strength at B is different to A. At B it is $g(x + \delta x)$.
- BC: a *time-like* transport is equivalent to watching the particle begin to accelerate relative to the origin because of the change in field strength. It does this for a time of 1 second. At the end of this time it is at C and the velocity vector has been rotated by the extra velocity component.
- CD: another *space-like* transport. No time passes so the velocity vector does not rotate any further relative to this reference frame.
- DA: a *time-like* transport back to the origin. The field here is $g(x)$ so there is no rotation relative to this reference frame.

The difference in g as a result of transport through δx is given by:

$$\delta g = \frac{dg}{dx} \cdot \delta x$$

$$g = -\frac{GM}{x^2} \quad \text{where} \quad M = \text{mass of the Earth } x = \text{distance from Earth's centre}$$

$$\frac{dg}{dx} = \frac{2GM}{x^3}$$

$$\delta g = \frac{2GM\delta x}{x^3} \quad \text{in the } x\text{-direction}$$

If the time-like side is of duration δt seconds

the velocity vector changes by:

$$\delta v = \frac{2GM\delta x \delta t}{x^3}$$

the angular deflection is approximately:

$$\delta\theta = \frac{\delta v}{c} = \frac{2GM\delta x \delta t}{cx^3}$$

$$\text{and the curvature} = \frac{\delta\theta}{\delta x c \delta t} = \frac{2GM}{cx^3} \quad \text{(ratio of angle of rotation to loop area)}$$

Equating this to the reciprocal of the radius of curvature gives:

$$\frac{1}{r^2} = \frac{2GM}{c^2 x^3} \tag{4.6}$$

$$r = \sqrt{\frac{c^2 x^3}{2GM}} = \underline{\underline{1.7 \times 10^{11} \text{ m}}} \quad \text{if A is at Earth's surface and the } x\text{-axis is vertical.}$$

In effect the rotation of vectors in a gravitational field is a measure of the tendency for separate particles in an extended free-falling reference frame to drift together or apart as a result of the non-uniformity of the gravitational field. It is a manifestation of the tidal nature of gravitation.

This calculation gives a very large radius of curvature compared to the dimensions of the Earth itself. It is about 25 000 times the Earth's radius and is greater than the radius of the Earth's orbit around the Sun. The scale on which we survey our local geometry is hardly likely to show up this relatively tiny curvature. Local spacetime appears to obey the rules of Euclidean geometry.

We can use the equation above to calculate the radius of curvature close to the Sun's surface. It comes out to about double this value, 3.4×10^{11}m. This value gives an idea of the rotation of a vector as it moves along a worldline in a local free-fall reference frame near the Sun. If it moves a distance δx along a trajectory grazing the surface of the Sun then it will be deflected by an angle of about $\delta x/r$ radians. That is about 3×10^{-12} radians per metre of its path as it grazes the Sun. (However, it is important to realize that this is a deflection in spacetime and so we need to be careful about how we interpret it.) This deflection is in addition to the deflection we calculated previously on the basis of the equivalence principle applied to locally straight lines in free-falling reference frames. Contributions like this summed along the path of a light ray at grazing incidence give the extra deflection due to spacetime 'stretching' or the non-Euclidean geometry of spacetime near the Sun.

4.6.4 The Deflection Of Starlight.

In section 4.3.3 we showed that the equivalence principle alone leads to a deflection of about 0.88″ of arc relative to local straight lines for light grazing the surface of the Sun. We can now estimate the additional deflection due to spacetime curvature in the non-uniform gravitational field. Consider a free-falling reference frame at some point on the light path. If it extends over a path length δx then particles at its ends will be in slightly different strengths of field and will gain a relative velocity δv in the direction of the field as the light passes. This will produce a rotation $\delta v/c$ radians in spacetime. Since light travels at c in the free-falling frame it will deflect by this angle in space. By summing these small deflections in all adjacent frames we can find the total deflection of light as it grazes the Sun. To simplify the calculation we have assumed that the deflection will be close to the value calculated along a line passing from the Sun's surface to infinity.

The derivation below is interesting, but it should be remembered that this is based on several rather dubious assumptions and the fact that the correct value emerges should not be taken too seriously! The point is that spacetime curvature is caused by the non-uniform character of extended gravitational fields and results in a greater deflection than the equivalence principle alone. This 'extra' deflection must be added to Einstein's 1911 result to give the total deflection.

$$g = -\frac{GM}{x^2}$$ at distance x from the Sun's centre

$$\frac{dg}{dx} = \frac{2GM}{x^3}$$

$$\delta g = \frac{2GM\delta x}{x^3}$$ between opposite edges of the reference frame

$$\delta t = \frac{\delta x}{c}$$ time for light to cross reference frame

$$dv = \frac{-2GM\delta x\delta x}{cx^3}$$ velocity difference developed between particles

$$\delta\theta = \frac{\delta v}{c} = \frac{-2GM\delta x\delta x}{c^2x^3}$$

$$\theta = 2\int_R^\infty\int \frac{-2GMdxdx}{c^2x^3} = 2\left[\frac{GM}{c^2x}\right]_\infty^R = \frac{2GM}{c^2R}$$

Putting in values for the Sun we get $\theta = 0.88''$ of arc.

This gives a total deflection of $1.76''$, in agreement with Einstein's 1915 prediction.

The weak field approximation to general relativity gives a deflection of:

$$\frac{1}{2}(1+\gamma)\times 1.75'' \text{ of arc}$$

for a ray grazing the surface of the Sun. The parameter γ is a measure of the curvature and is equal to one in general relativity. Accurate measurement of the deflection of electromagnetic waves gives a way to measure γ. This is important since some of the competing theories of gravitation predict slightly different values. Experimental results to date support general relativity. The Shapiro time delay also depends on curvature and when this is taken into account the expected delay for a reflection from Mars is about 250 μs. This also is in agreement with experimental values.

4.6.5 The Advance Of Perihelion.

According to Newtonian theory a planet will follow a closed elliptical path with the Sun at one focus. By the mid nineteenth century it had been established that this seemed to be the case for all of the planets except Mercury. Mercury's point of closest approach to the Sun is advancing around the Sun by about 574 arcseconds per century. This was not necessarily a problem for the Newtonian theory, Mercury is subject to more than just the attraction of the Sun - it is perturbed by all the other planets as well. In 1859 Le Verrier analysed the effects of other planets and showed that they would produce an advance of perihelion, but not enough to account for the observations. Even taking account of the oblateness of the Sun there remained an unexplained advance of about 43 arcseconds per century. For a while there was speculation that this must be evidence for the existence of an undiscovered planet (Vulcan) between Mercury and the Sun, but there was no observational evidence to support this and calculations on the orbits of other planets did not reveal the additional perturbations this would cause.

In 1915 Einstein showed that general relativity predicts an advance of perihelion of 42.98 arcseconds per century! A group from MIT compiled results between 1966 and 1976 based on radar reflection measurements of Mercury's position and measured the advance of perihelion as $43.11 \pm 0.21''$ per century.

It is not difficult to see why general relativity leads to this effect. Firstly the weak field approximation to general relativity approaches Newtonian gravitation so the basic shape of all planetary orbits will be an ellipse. However, relative to a distant observer, spacetime near the Sun is distorted and the planet's mass will increase due to its velocity. Both effects make the trajectory of the planet bend more in a stronger field than expected in Newton's theory. This extra deflection prevents it from repeating its orbit and causes the axis of the ellipse to rotate.

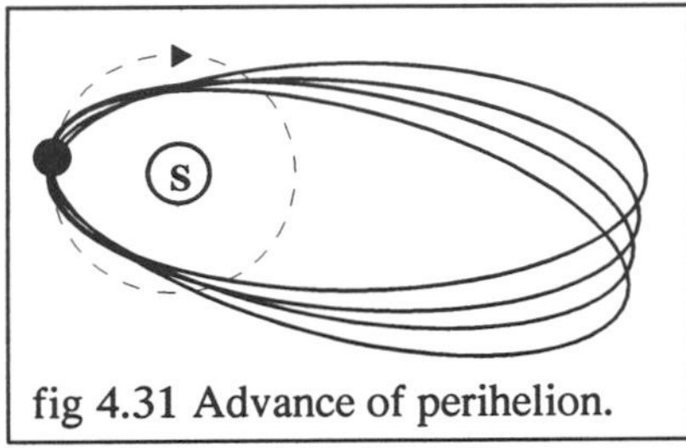

fig 4.31 Advance of perihelion.

The advance of perihelion is most obvious for Mercury since its orbit is closest to the Sun. The Earth, which orbits in a weaker field, where the geometry is much closer to that of Euclid, has a predicted advance of just $3.8''$ per century. Measurements of the advance of the Earth's perihelion have relatively greater errors but they are in agreement with this prediction.

4.6.6 Gyroscope Precession.

A gyroscope spinning about its axis will conserve angular momentum about that axis and so (if it is isolated from external torques) defines a constant direction in space. The essence of space-time curvature is that vectors transported around a closed path will end up with a rotation from their starting position. If a gyroscope rotates inside an orbiting satellite then space-time curvature should cause the axis to tip and the gyroscope to precess as the satellite orbits. The precession can be measured relative to the distant fixed stars and used to test the predictions of general relativity. The expected rotation is about $7''$ per year due to curvature and a further $0.05''$ due to the 'dragging' of space-time by the rotating Earth. Experiments to measure this effect are being carried out at the time of writing.

4.7 NON-EUCLIDEAN GEOMETRY.

4.7.1 Beyond Euclid.

Euclid, in the 'Elements', asserted that all the theorems of geometry derive from a small number of self-evident axioms. The truth of the theorems obviously depends on the truth of the axioms, which explains why the axioms themselves must be self-evident. Euclidean geometry is the geometry we study at school, it is the geometry of straight lines on flat sheets of paper. Triangles have angles whose sum is 180°, circles have circumferences equal to $2\pi r$ and parallel lines never meet. For over two thousand years this seemed to be the only conceivable form for geometry and appeared to describe the geometry of space in our universe very well. However, Euclid himself had some doubts about one of the axioms even before he issued the Elements. These doubts were shared by several nineteenth century mathematicians who recognized that the axiom of the parallels was based on an assumption. They asked whether a logically consistent alternative to Euclidean geometry might be constructed by changing this axiom:

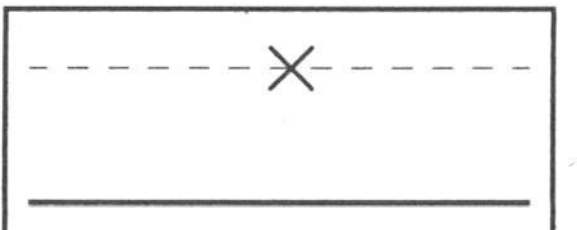
fig 4.32 The axiom of the parallels.

Given a straight line and a point which does not lie on the line there is one and only one line parallel to the first that can be drawn through the point.

Lobachevski, Bolyai, Gauss and Riemann considered the possibilities that *any number* of parallels can be drawn through the point, or that *no* parallels can be drawn through it (that is that an infinite number of lines through P do not intersect the extension of AB or that all lines intersect AB somewhere). Surprisingly, the new 'Non-Euclidean Geometries' were self consistent. They also generated some strange results. For example, the angles in a triangle no longer add up to 180° (it is less in the geometry of Bolyai and Lobachevski and more in Reimann's) and the area of a circle is not πr^2. Bolyai announced:

'I have made such wonderful discoveries that I am myself lost in astonishment: Out of nothing I have created a new and another world.'

If alternative geometries are equally valid how do we decide which geometry applies to the space we inhabit? Gauss suggested that this could be determined *experimentally* by measuring the angles in a real triangle. He even carried out the experiment by positioning three observers on mountain peaks and getting them to measure the angle between their lines of sight to each of the other two observers. The result was inconclusive (and also depended on the assumption that rays of light follow straight line paths). Of course, this does not prove that our space is Euclidean, the scale of the experiment may have been too small to detect deviations from Euclidean results.

4.7.2 Visualizing Non-Euclidean Geometries.

Non-Euclidean geometries, like Euclidean geometry, can represent a space of any number of dimensions. It happens, however, that in two dimensions they can all be visualized as the intrinsic geometry of a curved surface. The diagrams below compare the three cases and illustrate how the parallel postulate and the geometry of a triangle appear in each geometry. The fact that the non-Euclidean surfaces appear to curve through a higher dimensional space is irrelevant, the internal surface properties could be measured by a creature (e.g. a flatlander) entirely confined to the surface itself. The extra dimension is simply an aid to our visualization.

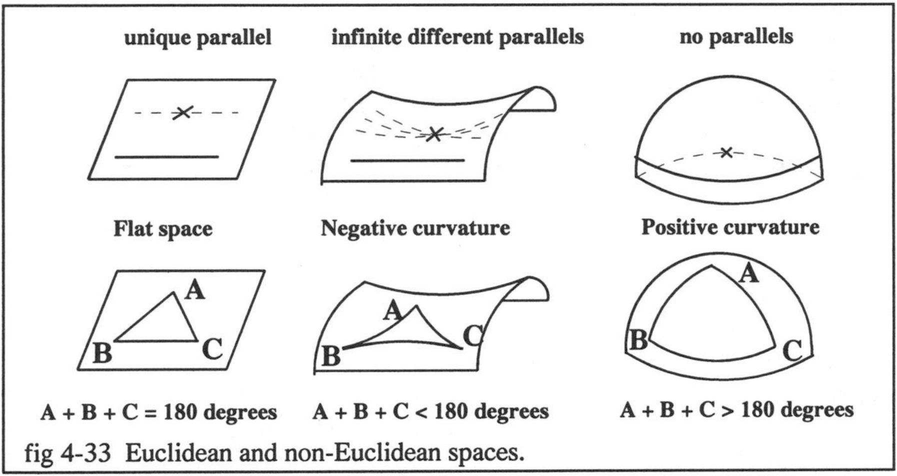

fig 4-33 Euclidean and non-Euclidean spaces.

Notice that 'straight lines' in curved space appear curved. They are actually the shortest paths between points in the surface and are called '*geodesics*'. Once again their apparent curvature is because we are viewing them from a higher dimensional viewpoint. If we were confined to the surface itself they would be the obvious direct routes between places. The geodesics on the positively curved two-dimensional surface of a sphere are great circles.

4.7.3 The Gravitational Field.

In the absence of matter, space-time has a neo-Euclidean geometry (the neo- refers to the addition of an imaginary time axis). This is the space-time geometry of special relativity. The worldlines of free particles in this space-time are straight lines. However, if a massive body is present the worldlines are curved due to gravitation. Newton's explanation was that masses exert mutually attractive forces on one another causing accelerations in an otherwise inertial reference frame. Einstein's interpretation is different. The effect of massive bodies is to distort space-time geometry so that freely falling particles will then follow geodesics in this non-Euclidean geometry. In this theory there is no need to introduce an additional gravitational force, the same trajectories arise geometrically. Once again recall the paraphrase of Einstein's theory:

- *Matter tells space how to curve and space tells matter how to move.*

The general theory of relativity is a theory of mechanics in which the effects of gravity emerge as a natural consequence of geometry and do not have to be introduced as an additional force field in otherwise flat space-time. Newton's first law of motion is reinterpreted in terms of geodesic paths through space-time.

A simple 2D analogy is provided by the surface of a stretched rubber membrane. If this is kept flat and horizontal marbles will roll across it in straight lines at constant speed. If it is depressed at some point then the marbles deflect from their previous straight-line paths and follow curved trajectories around the depression. Slow moving marbles may fall into the depression and become bound to it, others may even orbit it. Creatures confined to the surface might conclude that some mysterious force of attraction acts toward the centre of the depression, perhaps even a 'gravitational' force (of course there *is* a force acting in this analogy and it is gravitational, but the picture is quite useful to visualize the effects of massive bodies on their local space-time continuum).

4.8 THE GENERAL THEORY.

4.8.1 Bits And Pieces.

It took Einstein ten years to go from the special theory of relativity to the general theory. During this time he struggled to find a mathematical form which would incorporate general covariance and the equivalence principle and which would go smoothly over to the special relativistic form in flat spacetime in the limit of weak gravitational fields. In the end he needed help from his friend, the mathematician Grossmann who alerted him to the work of Reimann and others on non-Euclidean geometry. A number of general considerations enabled him to adapt Reimann's mathematics to his own physical principles.

- In the absence of gravity spacetime is flat and the laws of nature transform according to the Lorentz transformation equations. The general theory must go smoothly to this limit as g approaches zero.
- Since gravity distorts spacetime geometry there are no global inertial reference frames. There are therefore no privileged sets of co-ordinates in which the laws of nature should be written. The laws of nature must therefore take the same form with respect to any generalized co-ordinates. This is general covariance.
- The equivalence principle shows that the gravitational field at a point can be transformed away in a local freely falling reference frame. This means that the

allowable generalized systems of co-ordinates must all be realized by continuous transformations of Minkowski spacetime.

- The mass-energy relation and the energy-momentum 4-vector show that mass is associated not only with particles in space but also with the energy and momentum associated with the gravitational field itself. This implies that masses generate a field which is itself a source of gravitational effects. This makes gravity 'non-linear' and accounts for some of the complex technical structure of the theory.

It is beyond the scope of this book to delve deeply into these technical complexities, but it is interesting to consider some of the preliminary steps towards what is generally regarded as the most elegant of all physical theories.

4.8.2 Generalizing Pythagoras.

Pythagoras's theorem gives us a simple method for calculating the separation of adjacent points in Cartesian space. If we have two orthogonal axes x_1 and x_2 and want to calculate the length of a small displacement δs in this two-dimensional space we use the familiar equation:

$$\delta s^2 = \delta x_1^2 + \delta x_2^2$$

Now imagine that the regular Cartesian mesh is actually drawn on a rubber sheet which is distorted by some external influence. This will change the angle between the axes and the separation of intersections on the mesh so that both angle and measure are a function of position. We can still locate a point on the mesh by giving its co-ordinates in this generalized scheme, but the meaning of the co-ordinates is not clear and the separation of two points cannot be calculated using the simple version of Pythagoras's theorem above. We have generated a non-Euclidean two-dimensional space which can be described by giving a set of numbers representing the scaling factor (measure) and angle of intersection at each point in the space.

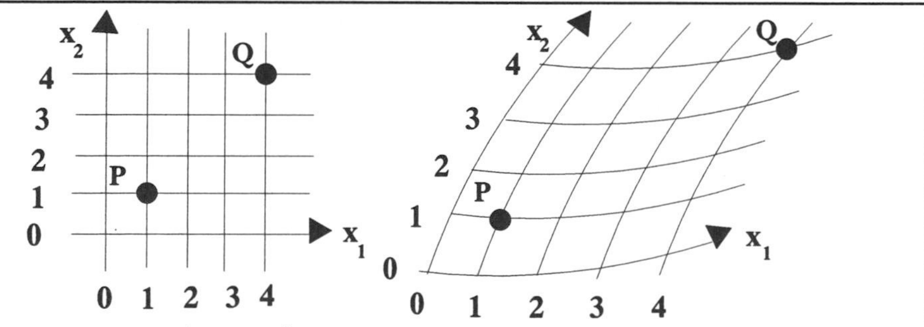

fig 4.34 The (gaussian) co-ordinates of events P and Q above are the same in both co-ordinate systems, but the second is distorted in both the direction of the axes and the scale along those axes.

How do we calculate the separation of points in this new space? Consider a small displacement δs at some place where the angle of intersection is θ and the scaling factors for each axis are a and b respectively. Referring to fig 4.35 below:

$$\delta s^2 = a^2 \delta x_1^2 + b^2 \delta x_2^2 - 2ab\cos\theta\,\delta x_1 \delta x_2$$

This is a generalized version of Pythagoras's theorem. Its significance and relation to the more familiar version becomes clearer if we adopt a matrix representation for the calculation.

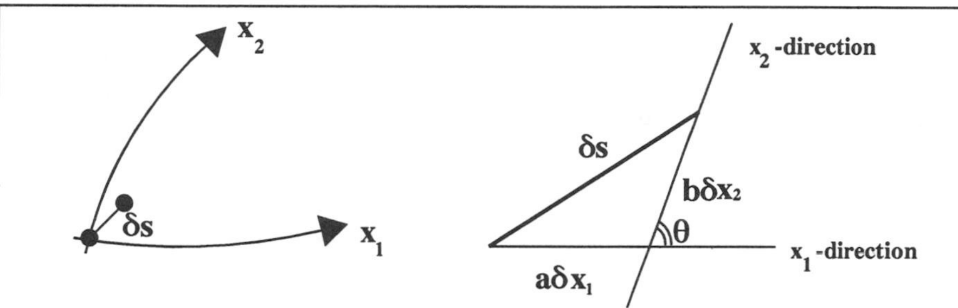

fig 4.35 The generalized version of Pythagoras's theorem in 2 dimensions can be derived with the cosine rule by taking account of the angle between axes and the local scaling factors along them.

We can write the original as:

$$\delta s^2 = (1 \times \delta x_1^2) + (0 \times \delta x_1 \delta x_2) + (0 \times \delta x_2 \delta x_1) + (1 \times \delta x_2^2)$$

The set of coefficients 1,0,0,1 can be written in the form of a matrix. These numbers define the 'metric' g of the two-dimensional surface.

$$g = \begin{bmatrix} 1 & 0 \\ 0 & 1 \end{bmatrix}$$

The 1s pick out the combinations of a single co-ordinate difference, the zeros exclude 'cross-products' of co-ordinate differences. Also, unit value for the scaling factor tells us that the mesh is not stretched or compressed in any way, a unit along the axis has measure one as well. This matrix can be generalized to describe an arbitrarily distorted surface:

$$g = \begin{bmatrix} g_{11} & g_{12} \\ g_{21} & g_{22} \end{bmatrix}$$

This will involve cross-products and scaling factors as occurred naturally in the example given above. If we return to that example we can identify the four coefficients in this metric.

$$g_{11} = a^2$$
$$g_{12} = g_{21} = ab\cos\theta$$
$$g_{22} = b^2$$

The equality of the off-diagonal elements g_{12} and g_{21} makes this a *symmetric tensor*. This is important in general relativity since it reduces the number of independent components of the metric from 16 (a 4 by 4 tensor) to 10. If the g_{ij} are defined as a function of the co-ordinates then the geometry of the surface is determined. Einstein's theory of gravitation gives a set of differential equations which determine the g_{ij} from the distribution of mass-energy-momentum. In the weak field limit this metric becomes the Minkowski metric of flat spacetime:

$$\begin{bmatrix} 1 & 0 & 0 & 0 \\ 0 & 1 & 0 & 0 \\ 0 & 0 & 1 & 0 \\ 0 & 0 & 0 & -1 \end{bmatrix}$$

The arrays of ones show that there is no spacetime stretching. The minus sign distinguishes the time co-ordinate in the same way that the square root of minus one (squared) did in our previous representation.

In 4-dimensional spacetime the metric is a symmetric sixteen component tensor having the form shown below:

$$\begin{bmatrix} g_{11} & g_{12} & g_{13} & g_{14} \\ g_{21} & g_{22} & g_{23} & g_{24} \\ g_{31} & g_{32} & g_{33} & g_{34} \\ g_{41} & g_{42} & g_{43} & g_{44} \end{bmatrix}$$

The shorthand way to write this is g_{ij} where it is understood that the subscripts i and j have the range 1 to 4. (Strictly speaking we need to distinguish between raised and lowered suffixes, but we shall skip this for the sake of this argument.) The generalized Pythagoras theorem now takes the form:

$$\delta s^2 = g_{11}\delta x_1^2 + g_{22}\delta x_2^2 + g_{33}\delta x_3^2 + g_{44}\delta x_4^2 + 2g_{12}\delta x_x\delta x_2 + 2g_{13}\delta x_1\delta x_3$$
$$+ 2g_{14}\delta x_1\delta x_4 + 2g_{23}\delta x_2\delta x_3 + 2g_{24}\delta x_2\delta x_4 + 2g_{34}\delta x_3\delta x_4$$

I wonder if Pythagoras would have recognized this? Although this will simplify in many practical situations, it does give some idea of the technical complexity of the theory. For our purposes we have come far enough, a continuously curved four-dimensional spacetime continuum is defined once we have expressions for the g_{ij} at each point. How do we determine these expressions?

4.8.3 What Matter Does To Space.

In special relativity the laws of nature have the same form in all inertial reference frames. In a gravitational field the laws of physics will take the same form at each point in a freely falling reference frame (the equivalence principle). However, no extended global inertial reference frame can be constructed, so a transformation law is needed between different local frames. This generalized transformation must leave the *form* of the laws of nature unchanged regardless of the set of curvilinear co-ordinates used to measure phenomena (this is *general covariance*).

Einstein's law of gravitation is a set of differential equations for the g_{ij} that occur in the spacetime metric. Solving these equations subject to some distribution of mass energy and momentum determines spacetime geometry in the vicinity of this distribution. This is equivalent to determining the resultant gravitational field for some distribution of masses in Newtonian theory. The set of differential equations can be written in tensor form and gives another symmetric tensor which obeys the same transformation rules as the g_{ij} themselves. This tensor can be written as R_{ij} and the distribution of mass-energy-momentum as T_{ij}. Einstein's equations of general relativity are then:

$$R_{ij} - \frac{1}{2} g_{ij} R = T_{ij} \tag{4.7}$$

or: *spacetime curvature ↔ mass-energy-momentum distribution*

In Newtonian theory we use the gravitational field to calculate the trajectories of particles moving subject to gravitational forces. In the Einstein theory there are no 'gravitational forces' and particles move freely through curved spacetime.

4.8.4 What Space Does To Matter.

Particles moving through spacetime follow geodesics. These are determined by the intrinsic geometry of spacetime. For example, light rays in special relativity move along a null worldline, that is the magnitude of the interval between events on the lightcone is always zero. This must also be true in general relativity since we can observe a small section of a light ray from a local free-falling reference frame and then transform to a different frame. Light therefore follows a null geodesic through spacetime $\delta s^2 = 0$.

The trajectory of the Earth around the Sun appears circular to us but could be regarded as a helical worldline whose axis of rotation in spacetime is the Sun's own worldline (this is adopting a heliocentric reference frame). The path followed by the Earth is the straightest path it can follow in the curved spacetime around the Sun. This can be expressed in the following way:

$$\delta s^2 = \text{minimum}$$

Subject to the momentum possessed by the Earth, its path will be such as to minimize the interval between successive positions. This minimum principle replaces the Newtonian law of motion.

4.9 THE GEOMETRY OF THE UNIVERSE.

4.9.1 Newtonian Cosmology.

Newton used the theory of gravitation to support the idea of an infinite unbounded universe. The quotation below is from a letter to a clergyman, Richard Bentley:

'....it seems to me that if the matter of our Sun and planets and all the matter of the universe were evenly scattered throughout all the heavens, and every particle had an innate gravity toward all the rest, and the whole space throughout which this matter was scattered was but finite, the matter on the outside of this space would, by its gravity, tend toward all the matter on the inside and, by consequence, fall down into the middle of the whole space and there compose one great spherical mass. But if the matter was evenly disposed throughout an infinite space, it could never convene into one mass; but some of it would convene into one mass and some into another, so as to make an infinite number of great masses scattered at great distances from one another throughout all that infinite space.'

Forces are transmitted instantaneously in Newton's theory so masses everywhere contribute to the gravitational force on a mass at any point. Unless matter is distributed perfectly uniformly throughout the universe there will be an infinitely large imbalance in the force acting on individual particles, so the infinite Newtonian universe is also unstable. In general relativity gravity is a distortion of spacetime geometry that propagates at the finite speed of light. This means that the field at a point is determined by the positions of distant masses at some past time, not at the moment the field is measured. This gives a way out of the 'gravity paradox'. An infinite universe can survive as long as it has not been around for an infinite time. Distant masses would not then have had time to interact and the infinite unbalanced forces are avoided.

This gives us a static, infinite universe of finite age, but this is not the only Newtonian universe. In 1934 two British cosmologists, Edward Milne and William McCrae, showed that dynamic models could be derived from Newton's equations. To understand their models (which reproduce many of the relations obtained in general relativity) think of a projectile thrown upwards from the surface of a planet. It will have an escape velocity v_{esc} given by:

$$v_{esc} = \sqrt{\frac{2GM}{r}}$$

There are three alternatives. We can throw the projectile faster or slower than this or we can throw it at exactly this speed. If it is thrown faster than v_{esc} it will escape from the field and retain some kinetic energy. At the escape velocity it will also escape but only just, its velocity will slow towards zero and after an infinite time it will come to rest infinitely far from the planet. Below the escape velocity it slows, stops and eventually falls back to the planet's surface. These expanding and collapsing universes played no part in Newton's cosmology but they correspond to very important models based on general relativity and in some cases can be used to derive the same equations.

4.9.2 The Einstein And de Sitter Universes.

Like Newton, Einstein expected a static universe. In 1917 he published a paper entitled 'Cosmological considerations on the general theory of relativity' in which he proposed a model of the universe with positive curvature in which the geometry of space closed on itself and remained static for all eternity. Time was treated differently and the universe can be considered as analogous to a cylinder with the time axis running up its centre. This was a finite but unbounded universe, but it was constructed at a price. All static finite universes tend to collapse. To prevent this Einstein introduced a new term into the equations of general relativity, a 'cosmological constant' that had the effect of a long range gravitational repulsion to oppose the attractive forces causing collapse. Later Einstein described this as the greatest mistake of his life (not all cosmologists would agree, and whilst the Einstein static universe is no longer a serious contender many current models still incorporate a cosmological constant).

In the same year de Sitter showed that the equations of general relativity *do* allow the construction of a universe in which the geometry of spacetime is flat. Unfortunately this universe contains no matter at all and its spacetime is in a state of cosmic expansion.

4.9.3 The Expanding Universe.

Meanwhile the astronomers had been busy. Since 1912 Vesto Slipher had been measuring the wavelengths of spectral lines emitted by visible galaxies. By 1923 he had discovered that 36 of the 41 galaxies he had studied had red shifts - the observed wavelengths were slightly longer than those from similar sources on Earth. If this was caused by a Doppler shift due to relative motion it implied that the majority of the observed galaxies must be moving away from the Earth. It was as if the Earth is situated at the centre of some cosmic explosion. (Later we will re-interpret the concepts of 'centre' and 'motion of the galaxies').

According to the theory of the Doppler shift the fractional change in wavelength due to relative motion is given by:

$$\frac{\delta\lambda}{\lambda} = \pm\frac{v}{c}$$

While Slipher measured red shifts, Edwin Hubble had been developing and improving techniques to measure the distances of galaxies (he was the first to show that Andromeda is outside our own galaxy). Combining Hubble's distances and Slipher's red shifts, Howard Robertson showed that the red shift increased in proportion to distance and interpreted this as evidence that the galaxies are moving away from us with speeds which increase in proportion to their distance. Hubble continued to refine the data and is remembered for the 'Hubble Law':

$$\text{recession velocity} \propto \text{distance} \quad \text{or} \quad v = H_o d \tag{4.8}$$

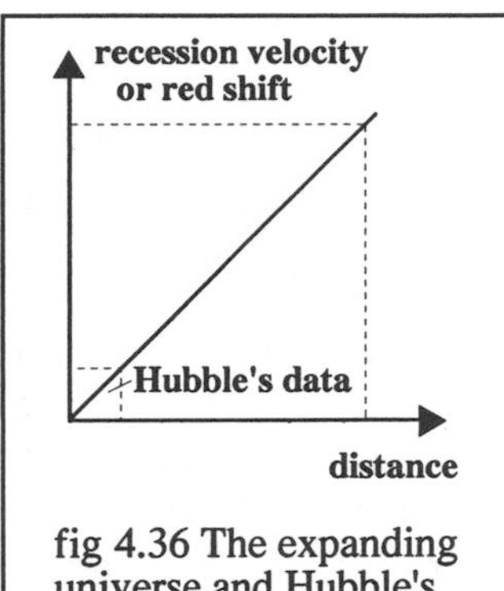

fig 4.36 The expanding universe and Hubble's law.

where H_o is known as the Hubble constant. The value of this constant can be derived from the red shifts and distances of galaxies. Unfortunately the distances are difficult to measure and so there is considerable uncertainty (perhaps a factor of 2) in the value of the Hubble constant. Since the Hubble constant determines the rate of expansion and hence the age of the Universe this has led to controversy over the status of the Big Bang model. In 1995 new measurements of the constant increased its value to such an extent that newspaper articles proclaimed that astronomers had proved the Universe was younger than some of the stars it contains (the model for stellar evolution is fairly well-established). If such a claim could be substantiated it would lead to a reassessment of fundamental theory, but it would be unlikely to deal the body blow to cosmology predicted by some journalists.

Robertson and Hubble both calculated a value for H_o of around 150 km s^{-1} per million light years. At that time the distances to galaxies were all greatly underestimated. The modern value for H_o lies in the range 15 to 30km s^{-1} per million light years. We can use this value to estimate the age of the universe. A first approximation assumes that the expansion has taken place at a constant rate since the Big Bang. This gives the 'Hubble period' T_h:

$$T_h = \frac{1}{H_o} \tag{4.9}$$

If we take H_o as 20 kilometres per second per million light years this gives an age of about 15 billion years (15×10^9 years). Since the value of H_o has such uncertainty it is not worth trying to refine this value.

If all the galaxies are now rushing apart, they must have been much closer together in the past. The Big Bang is the moment when they were packed together in a point of infinite density called a singularity. Of course, the known laws of physics are not adequate to describe these asymptotic conditions but predictions based on a hot Big Bang model produce consistent physical ideas right back to within a tiny fraction of a second of the event itself. Since then the scale of the universe has increased and we can introduce a scaling factor R that allows us to compare the size of the universe at different epochs. One way of visualizing what R represents is to imagine an elastic ruler marked in arbitrary units with two coins resting on it at different positions. As the ruler is stretched the coins maintain their own size but separate from one another. Their 'co-ordinate distance' from one another is exactly the same as it was before but their actual separation has increased. This is because the interval between co-ordinates has increased by a factor R, where this is the proportional change in length of the ruler. If the ruler were to stretch more and more in a steady continuous way then the two coins would move apart from one another at an increasing rate. The reason the rate of separation increases is that as their separation increases the same fractional increase in a larger separation results in greater movement in the same time, hence a faster recession velocity. This model reproduces the Hubble law.

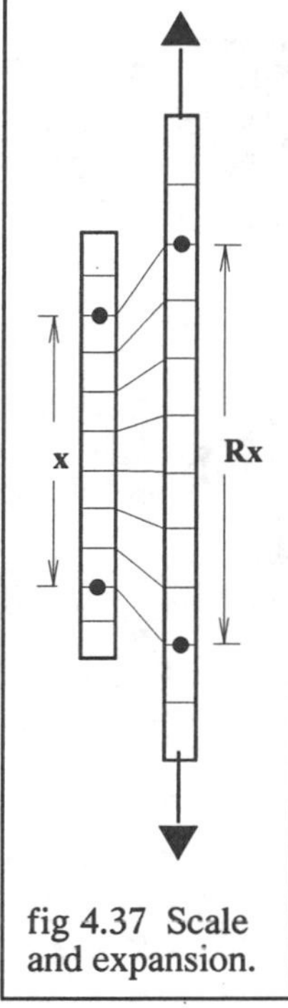

fig 4.37 Scale and expansion.

The 'actual separation' of points on our expanding ruler is given by the co-ordinate distance multiplied by the scaling factor R. Since the co-ordinates are 'co-moving', that is they are unaffected by the expansion, we can rewrite the Hubble law in terms of the scaling factor as follows:

$$v = \frac{ds}{dt} = H_o s \quad \text{(Hubble law)} \qquad s \text{ is the actual distance}$$

$$s = Rx \qquad (1) \qquad x \text{ is the constant co-ordinate distance}$$

$$v = x\frac{dR}{dt} \qquad (2)$$

substituting (1) and (2) into the Hubble law gives:

$$x\frac{dR}{dt} = H_o Rx$$

$$H_o = \frac{\dot{R}}{R} \qquad \text{where } \dot{R} \text{ is the time derivative of } R \qquad (4.10)$$

This gives the Hubble law in a form that is independent of the actual co-ordinates and shows that it represents a uniform universal expansion. Our one-dimensional expanding ruler model can be developed to a two-dimensional infinite sheet or even the expanding curved surface of an inflating balloon. In all cases co-moving co-ordinates expand with the sheet so that the separations of points on the expanding sheet are given by constant co-ordinate differences. In the case of the balloon universe the galaxies might be represented by small rigid discs stuck onto the surface and the co-ordinates by lines of 'latitude and longitude' drawn on it. This emphasizes an important point that is also true for our universe, the space between the galaxies expands but the galaxies themselves do not. The three distinct classes of universe described by this scaling law are:

$$\dot{R} > 0 \quad \text{expanding}$$

$$\dot{R} = 0 \quad \text{static}$$

$$\dot{R} < 0 \quad \text{contracting}$$

It is possible to develop this description of an expanding universe to consider the rate at which expansion slows due to gravitational attraction between the receding galaxies. This introduces a deceleration term dependent upon the second derivative of R. We can also derive equations for the variation of areas and volumes, and hence densities. This is useful in considering the physics of the universe at different epochs.

fig 4.38 The Virgo Cluster. This is the largest of the nearby clusters and is about 70 million light years from the Milky Way. The red shift of this cluster is about 0.016. The large scale structure of the Universe is dominated by objects that are bound by their own gravity: stars, nebular clusters, galaxies, galaxy clusters and super-clusters. As the universe expands these objects maintain their own scale as their separation increases.

4.9.4 The Expansion Of Spacetime.

In 1922 Alexander Friedmann discovered that it is not necessary to introduce a cosmological constant in order to produce a model of a universe containing matter which is consistent with the equations of general relativity. He proved that there are three basic solutions resulting in three physically distinct models for the geometry of the universe as a whole. In all three models the universe begins with a Big Bang after which space undergoes a period of expansion. However, the mutual gravitation of matter in the universe slows this rate of expansion and may even reverse it. In 1927 George Lemaître also published work on the expansion of the universe. He rediscovered many of Friedmann's results and considered other possible solutions to the Einstein equations involving non-zero values for the cosmological constant. For this reason the works of Friedmann and Lemaître are often regarded together and we describe the models as 'Freidmann-Lemaître universes'. The three basic models with zero cosmological constant are:

- Gravity dominates. The deceleration term is large and the universe will eventually recollapse. Space is everywhere positively curved. This leads to a closed universe, one that is finite and unbounded like the Einstein universe.

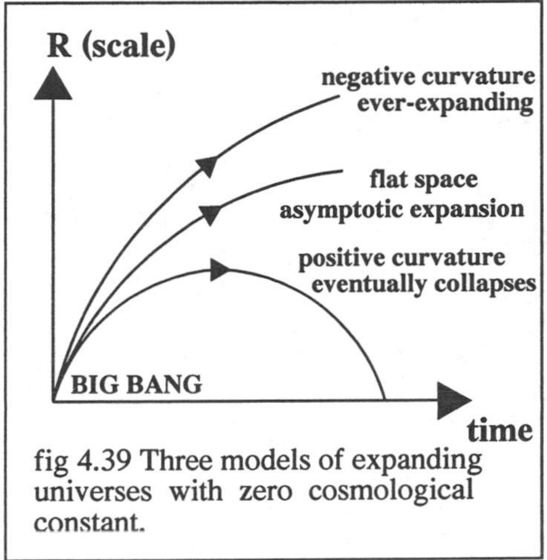

fig 4.39 Three models of expanding universes with zero cosmological constant.

- If the deceleration term is very small then the galaxies are effectively receding from one another with enough kinetic energy to eventually escape to infinity. This universe expands, at an ever slowing rate, forever. Space is everywhere negatively curved.

- Poised between the extremes above is a universe that would just cease to expand after an infinite time. This model is sometimes referred to as the Einstein-de Sitter universe since it combines features of both models. Space is everywhere flat in this universe.

The history of these ideas is quite interesting. Friedmann's work was not recognized when it was published, despite the fact that he sent it to Einstein for comment. Lemaître rediscovered many of Friedmann's results in 1927 but even his work went unnoticed by most astronomers until 1930 when it was publicized by Eddington. Perhaps the discovery of red shifts made the ideas seem more acceptable.

In addition to the three models described above, Lemaître produced a number of alternative pictures by adjusting the value of the cosmological constant. These included models in which the universe underwent rapid expansion followed by a long pause before further expansion. More recent cosmological models have introduced a super-rapid period of *inflation* prior to the steadier Hubble expansion. This has been necessary in order to generate the tiny variations in density that eventually led matter to collapse into the galaxies and galaxy clusters we see today. However, the inflationary model is still controversial and may well be revised in the future.

It is important to point out that the theoretical expanding universe has a rather different interpretation to the exploding universe often inferred from the Hubble

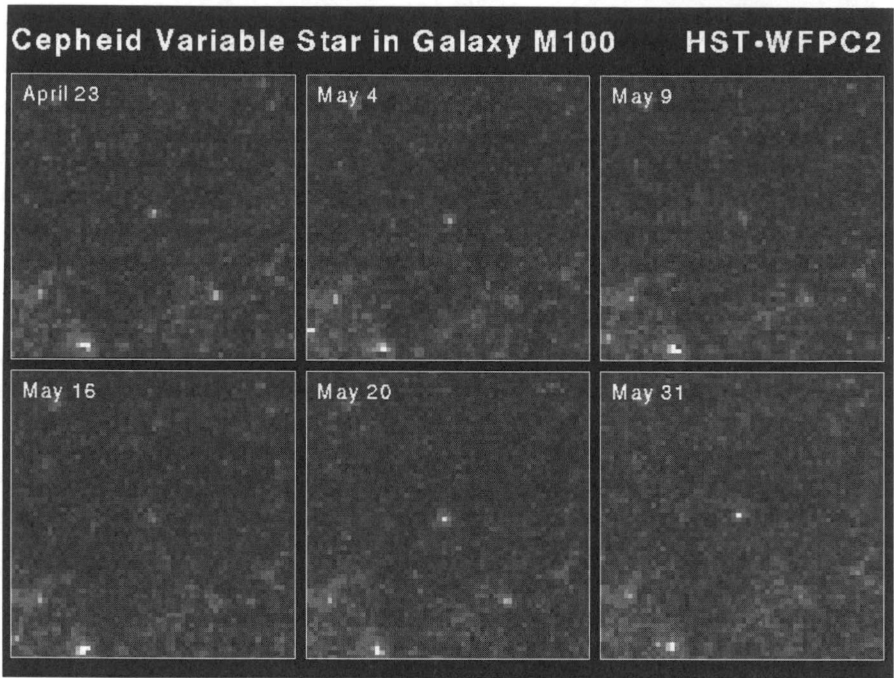

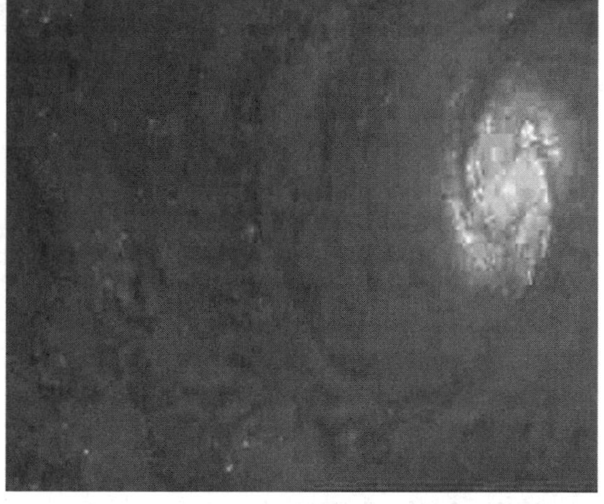

figs 4.40 and 4.41 The Hubble constant. To calculate H_o from the Hubble Law we need the distance and recession velocity of distant galaxies. However, measuring distances accurately is a real problem. Nearby stars can be measured by parallax, but this method is very limited and other distance markers must be used for distant galaxies. One of the most successful methods uses var-iable stars. Cepheid variables fluctuate in intensity with a period that depends on their absolute magnitude. Once this is known their distance can be calculated using the inverse-square law and their apparent magnitude when viewed from the Earth. The Hubble Space telescope can resolve individual stars in quite distant galaxies, so these can be used as distance markers. The photograph above (fig 4.40) shows the variation of a Cepheid in the galaxy M100 (shown in fig 4.41). This is a spiral galaxy like our own Milky Way, it is in the Virgo Cluster of about 2500 galaxies at a distance of about 56 million light years. The large uncertainties in the distance scale affect our estimates of H_0 and therefore of the age of the universe.

Photograph credits: W.Freedman (CIW), R.Kennicutt (UA), J.Mould (ANU), J.Ttrauger (JPL) and NASA, HST.

law. Slipher and others interpreted the red shifts as caused by the *motion* of galaxies *through space* in some cosmic explosion. This is not the picture we obtain from general relativity. Here *space expands* and the galaxies are *carried with it*, they do not move through it. In this respect cosmic expansion is exactly like the expanding rubber models described in the previous section. The galaxies themselves do not expand because they are bound together by relatively strong fields while space as a whole expands. The red shift is explained because the waves get 'stretched' as they move across the expanding surface. We can derive the red shift from the fact that the scale of the universe is increasing.

Let the wavelength emitted by a distant galaxy be $\lambda = \lambda_o R$

Let the actual distance of this galaxy be d

The time of flight of the light is thus $t = \frac{d}{c}$

The rate of change of wavelength is $\frac{d\lambda}{dt} = \lambda_o \dot{R}$

The red shift is therefore: $\delta\lambda = \frac{\lambda_o \dot{R} d}{c}$

the fractional change is: $\frac{\delta\lambda}{\lambda_o} = \frac{\dot{R}d}{Rc}$ (4.11)

Red shift is proportional to distance as Slipher observed.

The ratio $\frac{\dot{R}}{R} = H_o$, the Hubble constant.

The equation for the expansion red shift derived above can be rearranged to tell a slightly different story.

$$\frac{\delta\lambda}{\lambda} = \frac{\dot{R}}{R}\cdot\delta t$$ where δ t is the time between emission and absorption

$$\frac{\delta\lambda}{\lambda} = \frac{\delta R}{R} \qquad (4.12)$$

The red shift is a measure of the fractional increase in the size of the universe between the emission and absorption of the light. This is perhaps a more significant result than the previous version since the value d/c used there does not specify exactly what d is.

Object	Comment	Red shift
Virgo	Cluster of galaxies	0.016
0957+561	Galaxy (gravitational lens)	0.36
3C9	Quasar	2.01
PC1158+4635	Quasar	4.73

Quasars (quasi-stellar-objects) provide the largest values of red shifts (up to about 5). If these are due to cosmic expansion they tell us that the universe was only 20% of its present size and age when these quasars emitted their radiation. However, there is no reliable independent means to verify that quasars really are very distant and there is the possibility that the observed red shift is caused, at least in part, by some other mechanism.

fig 4.42 Quasar 0051-279. This faint star-like object was one of the most distant known objects in the universe (before the HST). It is a quasar with a red shift of 4.43 and the light recorded on this plate left the quasar when the scale of the Universe was less than a quarter of its present value. The original plate was taken with the UK Schmidt telescope in Australia.

4.9.5 Critical Density.

For uniform homogeneous models of the universe the Newtonian equations lead to exactly the same equations as general relativity. This is because local and global physics is the same in such a universe. We shall use this approach to calculate the critical density ϱ_c for the Friedmann-Lemaître universe of flat space balanced between eternal expansion and ultimate recollapse. Imagine a universe filled with dust of density ϱ in a state of expansion. If we identify a small mass m of this dust on the surface of an imaginary expanding sphere of radius r we can use Newton's equations to calculate the trajectory of this 'galaxy'. The gravitational effect of the homogeneous external universe will cancel to zero as far as the galaxy is concerned (this is a consequence of Gauss's theorem, but is reasonable on the grounds of symmetry alone). It will move as if attracted to a point mass M at the centre of the sphere (M is equal to the total mass of dust in the imaginary sphere). The three Friedmann universes - expanding, contracting and critical - are distinguished by their rates of expansion. Effectively, we are asking the question: 'is this galaxy moving away from the centre fast enough to escape?' In the Newtonian model we can answer the question in the same way we would calculate escape velocity for a projectile leaving the surface of a planet - by considering its total energy.

$$GPE = \frac{-GMm}{r^2}$$

$$KE = \frac{1}{2}mv^2$$

$$v = H_o r \qquad \text{(Hubble's law)}$$

$$TE = GPE + KE = \frac{-GMm}{r} + \frac{1}{2}mv^2 = \frac{-GMm}{r} + \frac{1}{2}mH_o^2 r^2$$

For escape we need $TE = 0$ therefore:

$$\frac{GMm}{r} = \frac{1}{2}mH_o^2 r^2$$

Now substitute for M in terms of the critical density ϱ_c

$$M = \frac{4}{3}\pi r^3 \varrho_c$$

after simplifying this gives:

$$\varrho_c = \frac{3H_o^2}{8\pi G} \qquad (4.13)$$

Current estimates put H_o in the range 50 to 100 km s^{-1} per megaparsec. (One megaparsec is 3.26 million light years, so this is about 15 to 30 km s^{-1} per million light years.) These values give a critical density in the range 5-10 $\times$ 10^{-28} kg m^{-3}. This is equivalent to 3 to 6 hydrogen atoms per cubic metre of space. This doesn't sound very much, but is about ten times *greater* than the average density of matter we observe through telescopes. One of the most intriguing unanswered questions in cosmology is whether there is a great deal of invisible or 'dark matter' in the universe and if so whether it consists of exotic new particles or unexpected objects formed from the matter we are already familiar with. Apart from the aesthetic appeal of a universe balanced on the knife edge of expansion and collapse, the observed structures and motions of galaxies also suggest the existence of unseen masses in 'halos' surrounding them.

4.9.6 The Cosmic Background Explorer (COBE).

The energy density of the early universe was enormous. If we look back in time the temperature of the universe and therefore the average particle energy rises as its scale shrinks. Galaxies formed from hot gas clouds between one hundred million and one billion years after the Big Bang. Between one hundred thousand and one million years after the Big Bang the temperature was about 3000 K and the universe was dominated by radiation. Photons were so energetic that atoms could not exist, they would be ionized by collisions with photons. This was an important era, before it radiation and matter interacted continually, afterwards radiation decoupled from matter and travelled freely through the universe. However good our telescopes become we never 'see' back before the radiation decoupled because the photons emitted then were continually scattered and the universe was opaque to light. The existence of this 'radiation era' was predicted in 1948 by George Gamow who was modelling the nuclear reactions that took place in the early universe. The radiation would have the spectrum of an ideal hot body (black-body radiation) and the effect of expansion would be to stretch the wavelengths towards the red end of the spectrum and lower the radiation temperature. In 1965 Arno Penzias and Robert Wilson were trying to eliminate unwanted sources of radio noise from a large horn-shaped antenna at Holmdel New Jersey. However, a low intensity radio hiss persisted despite all their efforts. Eventually they concluded that this was a weak radio signal coming almost equally from all directions in the sky. Moreover, the radiation had the characteristic black-body spectrum. Dicke and others at Princeton realized that this was the 'electromagnetic echo' of the Big Bang predicted by Gamow - microwave radiation at 2.8 K above absolute zero. Penzias and Wilson received the Nobel Prize for Physics in 1978.

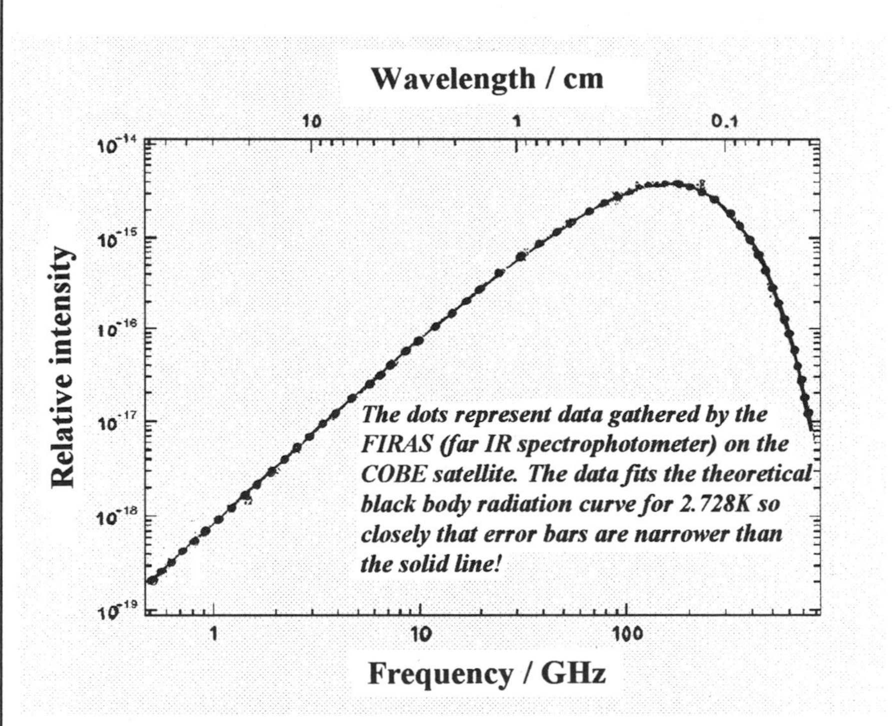

fig 4.43 The cosmic background radiation fits the theoretical spectrum for black-body radiation at 2.728 K almost perfectly. The COBE experiments were designed to measure the spectrum and to look for small intensity fluctuations from different regions of the sky. These were caused by local anisotropies in the distribution of matter in the early universe and are needed to explain the subsequent evolution and distribution of galaxies.

The fact that the radiation seems to come equally from all directions poses a problem. If the background radiation really is homogeneous then it follows that the early universe must also have been fairly uniform, with a constant density and temperature everywhere. However, matter clumped into stars, galaxies, and clusters of galaxies in the period following the end of the radiation era. On the large scale there is now a great deal of inhomogeneity, regions of space with a lot of matter and huge voids. Presumably this clustering was controlled by gravitational forces pulling matter into regions of above average density and the evidence for these 'bumps' should be imprinted on the background radiation. We should see slight irregularities in the spectrum and intensity of radiation coming from different parts of the sky.

In 1989 COBE, the Cosmic Background Explorer, was launched. Its job was to search for these irregularities in the microwave sky. The first data was announced in 1992 and, after correcting for other effects which could have caused apparent fluctuations (e.g. the Doppler shift of radiation due to the earth's local motion) fluctuations of about ±30 μK began to show up. Tiny though these are, they seem to be genuine non-uniformities and are just (only just!) large enough to account for the structures we see in the present universe. Earth and balloon based receivers are now being used to verify these results at other frequencies, but the COBE results have been hailed by cosmologists as among the most significant this century.

4.10 BLACK HOLES.

4.10.1 Early Ideas.

We tend to think that black holes are an entirely relativistic phenomenon. It is surprising therefore to discover that at least two eighteenth century physicists speculated that thcy might exist as a consequence of Newtonian theory. John Mitchell and Pierre Laplace both argued that sufficiently massive astronomical bodies might generate such an intense gravitational field that even light rays would be unable to escape. Such a body would neither radiate nor reflect light and would be truly black.

We can derive the same result by considering how the escape velocity from the surface of a star depends upon the size and mass of the star. The Newtonian expression for escape velocity is:

$$v_{esc} = \sqrt{\frac{2GM}{R}}$$

where R is the radius and M the mass of the star. If light is retarded by gravity then it will just fail to escape from the surface when the escape velocity equals the speed of light.

$$c = \sqrt{\frac{2GM}{R}}$$

This can be rearranged to give expressions for the radius R_s inside which a particular mass must be compressed to form a black hole, and for the critical density of spherically distributed matter that will collapse to form one. The symbol R_s is used to represent this radius since Newtonian theory gives the same result as the general relativistic analysis carried out by Schwarzschild in 1916. R_s is the Schwarzschild radius.

$$R_s = \frac{2GM}{c^2} \tag{4.14}$$

$$\varrho_c = \frac{3M}{4\pi R_s^3} = \frac{3c^6}{32G^3M^2\pi} \tag{4.15}$$

The significant point about the density equation is that critical density is inversely proportional to the square of the mass of the collapsing object. This means that, whilst objects of relatively low mass like the Earth or Sun would need to reach incredibly high densities to form black holes, much larger bodies like galaxies or clusters have a much lower critical density. If the object is large enough it may form a black hole at or below the density of ordinary matter. The Earth (approximate mass 6×10^{24} kg) must be compressed into a sphere of radius less than 1 cm to form a black hole. Its critical density is about 2×10^{30} kg m^{-3}. On the other hand, a galaxy of mass 10^{42} kg and radius 10^{21} m, will have a Schwarzschild radius of about 1.5×10^{15} m and a critical density of a mere 7×10^{-5} kgm^{-3} (less than one ten thousandth of the density of air at atmospheric pressure).

4.10.2 The Schwarzschild Solution.

General relativity was still hot off the press when Karl Schwarzschild published the first exact solution to the equations in 1916. The Schwarzschild solution gives the metric (i.e. the space-time geometry) outside a spherical non-rotating mass M. In particular it leads to an expression for the curvature r a distance x from the body which is just the expression we derived in section 4.6.3:

$$\frac{1}{r^2} = \frac{2GM}{c^2x^3}$$

To judge the significance of space-time curvature we rewrite this expression to compare x with r as is done below:

$$\frac{x}{r} = \sqrt{\frac{2GM}{c^2x}}$$

At the surface of the Earth this takes the value 3.7×10^{-5} and shows quite clearly that deviations from Euclidean results are going to be difficult to detect. This ratio is small everywhere in the solar system but is significant near the surface of collapsed stars. A typical value at the surface of a neutron star for example is 0.3, so space-time curvature is important here. At the Schwarzschild radius the ratio is one, space-time curvature is so extreme that it effectively wraps around the black hole preventing the escape of light or matter. The surface formed here is a 'trapped surface' no signals emitted from the internal region can be transmitted to the external space. The outermost trapped surface is the Schwarzschild surface or 'event horizon' of the black hole.

Light rays moving from lower to higher gravitational potential are red shifted. Rays leaving the Schwarzschild surface have an infinite red shift and so are never detected in the external universe. Another way to think about this is as if space-time in the vicinity of a black hole 'drains' into the hole. The velocity at which it flows inwards increases closer to the hole itself and reaches the speed of light at the Schwarzschild surface. Rays emitted radially outwards would make no progress relative to a distant observer. Rays emitted at any other angle would be bent back into the black hole. Rays emitted tangentially at 1.5 times the Schwarzschild radius move in closed orbits around the black hole and form what is called the 'photon sphere'.

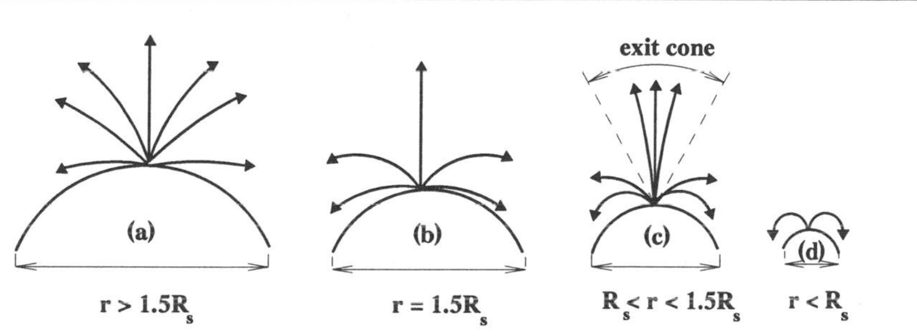

fig 4.44 Light will escape from any star with radius r greater than 1.5 times its Scwarzschild radius. This is the radius of the photon sphere, and light emitted tangentially at this radius will orbit the star. If r is between the photon sphere and the Schwarzschild radius there will be a limited exit cone for escaping light, rays outside this cone fall back to the surface of the star. No light can escape from a star that has collapsed inside its own Schwarzschild radius.

Although the picture of a 'space-time drain' is only an analogy, it gives a good idea of how light rays are affected in the space outside a black hole. This is often illustrated on a space-time diagram by drawing the light cone at various positions. The effect of space-time curvature is to tilt the future axis of the lightcone towards the black hole. At the Schwarzschild surface this tilting is so extreme that the outside edge of the cone becomes parallel to the worldine of a point on the surface. At a distance greater than R_s from the black hole the tilting has the effect of bending the light rays so that they are curved in the space-time of the distant observer. Close to R_s there will be a limited range of emission directions within which the rays escape. Rays emitted outside this 'exit cone' will be bent back into the black hole. This is illustrated in the diagram below (which is drawn from the point of view of a distant observer).

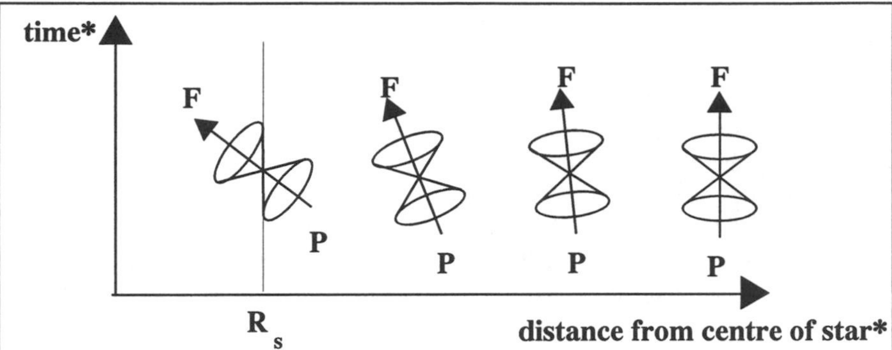

fig 4.45 Light cones in free falling reference frames at different distances from the centre of a massive star or black hole have their local future axes (F) tilted towards the centre of 'attraction'. At the Schwarzschild radius all rays emitted into the future light cone travel towards the black hole.

4.10.3 The Fates Of The Stars.

Black holes form if the mass of a body is so great that nothing is able to prevent gravitational collapse. An active star like our Sun is in a temporary equilibrium state where internal pressures from thermonuclear fusion reactions balance gravitational forces. Eventually nuclear fuel is exhausted and the star will collapse. What happens next depends on its mass. It may go through several stages of collapse followed by new equilibria as progressively heavier nuclei are fused in the core, but eventually its nuclear fuel must run out.

For stars up to about 1.25 solar masses the collapse will continue until the electron pressure produced as electrons resist being pressed closer and closer together is sufficient to oppose and balance gravitational forces. Such stars will end their days as 'white dwarfs', dead stars about the size of the Earth slowly cooling. It is estimated that about 10% of nearby stars are white dwarfs, but they are difficult to see because they have very low luminosities.

Stars between 1.25 and about 2 solar masses will become neutron stars. For these more massive stars electron pressure is not strong enough to prevent gravitational collapse. Electrons and protons combine to form neutrons and neutron pressure makes a last stand against gravity. The density of a neutron star is approximately 10^{18} kg m^{-3} and it has a radius of about 10 km. The magnetic field of a neutron star is about 10^{12} times greater than the earth's field and since the star spins rapidly on its axis (through conservation of angular momentum during collapse) it radiates strongly sending out regular pulses with the same period as its own rotation. These sources are pulsars and their characteristic frequency falls over millions of years as they radiate away their rotational kinetic energy.

We do not find neutron stars of three or more solar masses, even neutrons are unable to resist gravity in such a massive body. There is no other known mechanism to prevent collapse so massive stars are thought to collapse without limit and form black holes.

4.10.4 Into The Black Hole.

What would happen if we watched (from a safe distance) as a massive star collapsed to become a black hole? Assume that the star is not rotating so that the appropriate physics is described by the Schwarzschild solution. This gives expressions for the red shift of light and gravitational time dilation for signals transmitted from a point at distance r from the centre of the black hole.

$$\lambda' = \left(1 - \frac{R_s}{r}\right)^{-\frac{1}{2}} \lambda \tag{4.16}$$

$$f' = \left(1 - \frac{R_s}{r}\right)^{\frac{1}{2}} f \tag{4.17}$$

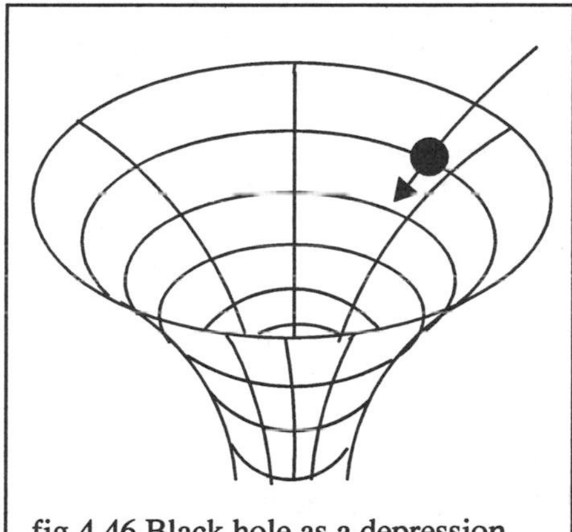

fig 4.46 Black hole as a depression in spacetime geometry.

In both the equations above the prime indicates values measured by the distant observer and the lack of a prime indicates the emitted values in the local free falling reference frame. If $r >> R_s$ we have no gravitational effects, emitted and observed frequencies and wavelengths are equal. At the other limit if $r = R_s$ then the wavelength becomes infinite and the frequency zero. This infinite red shift tells us no light escapes. The frequency of radiation relates to the rate of time (like frequency of a ticking clock) and this falls to zero at this distance. The outside observer will see events pass more and more slowly closer and closer to the event horizon and stop completely when it is reached.

Pulling all this together we see that the star's collapse will be seen clearly at first, the radiant cooling surface falling away from us toward its centre of gravity. However, as it approaches the event horizon at the Schwarzschild radius its radiation will be more and more red shifted and its intensity will become weaker and weaker. The rate of collapse will also appear to slow as time dilation becomes more severe. We shall observe the surface becoming redder and dimmer as it slows and slows, never quite reaching the event horizon itself. In *our* reference frame it would take an *infinite* time to complete the collapse, so the inside of the black hole is beyond the range of the external universe since objects only enter it after an infinite time has elapsed outside! We will never see an object enter a black hole unless we go with it.

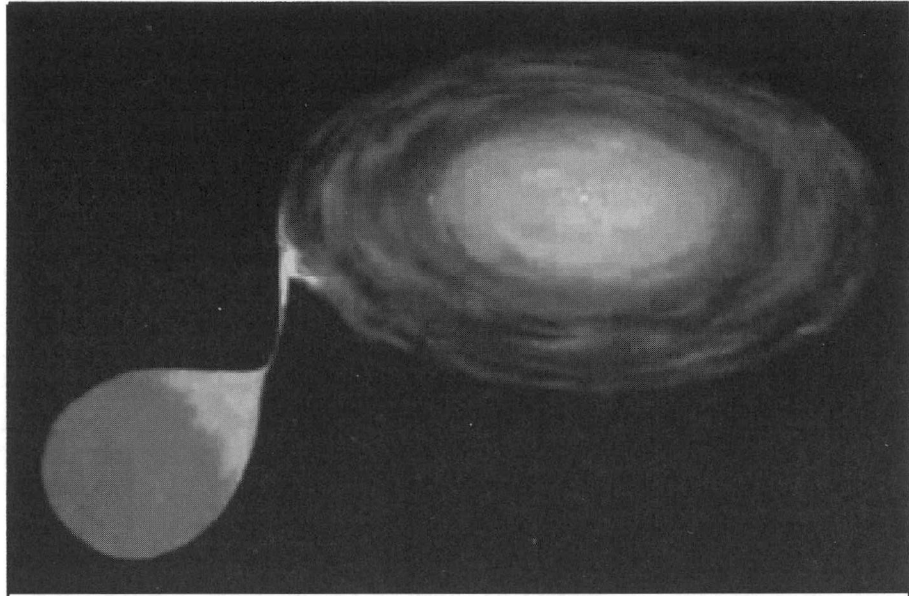

fig 4.47 An artist's impression of a massive black hole bound to a galaxy and gradually drawing matter from it. Although no radiation can escape from within the Schwarzschild radius, the acceleration of particles as they fall into the black hole results in powerful radiation before they reach the event horizon. Credit: NASA/ HST.

Space-time geometry outside the star will not be affected by its collapse - as the star shrinks it distorts local space-time more and more severely whilst leaving the curvature of distant space-time unchanged. This is illustrated using our 'rubber sheet analogy' in fig 4.48.

What would it feel like to fall into a black hole? At first it would feel like being in any other inertial reference frame. This is because of the equivalence principle, local free falling reference frames are inertial. However, this neglects the tidal effect of gravity, the effect which cannot be transformed away and which is responsible for the curvature of space-time. Far from the event horizon this is insignificant and goes unnoticed. Close to the event horizon it becomes severe and would be fatal for any human adventurer foolhardy enough to venture in. The difference in the strength of gravity between his feet and head would eventually produce internal forces strong enough to pull him apart. Before this happened he might notice other strange effects. Information transmitted to him from a base far from the black hole would be blue shifted and time in the outside world would run faster and faster. External hours would pass in minutes, centuries in hours as the outside universe races towards its end in the final seconds before the traveller falls through the event horizon itself. One other point, it would take a short finite time

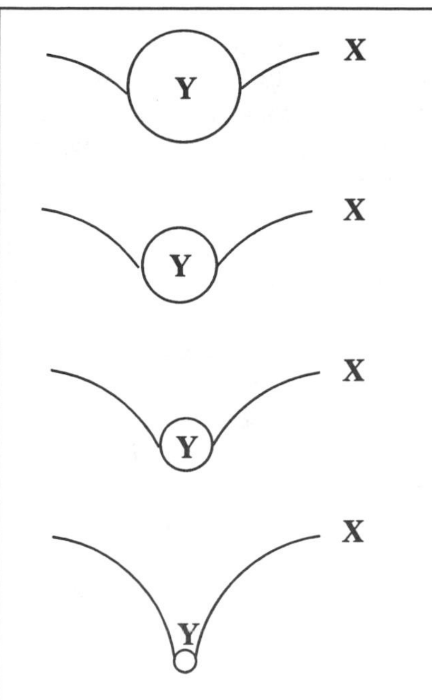

fig 4.48 Stages of collapse. The curvature at X is unaffected by the collapse of a star to form a black hole at Y.

for the traveller to fall from the event horizon to the singularity at the very centre of the black hole, not that he would be in a position to consult his wristwatch!

4.10.5 Heart Of Darkness.

The collapse of a massive star does not stop at the event horizon, even though it appears to do so as far as a distant external observer is concerned. With nothing to prevent ultimate collapse the density of matter rises without limit until an infinitely dense singularity is formed. Objects with infinite density are a bit of a problem to physicists so it is a relief to know that the event horizon effectively isolates the singularity from the rest of the universe for a non-rotating black hole. However, a non-rotating star would be a very rare phenomenon and even a slowly rotating star will form a rapidly rotating black hole as it conserves angular momentum during its collapse:

angular momentum $L = \frac{2}{5}MR^2\omega$ for a star of mass M and radius R

if this star collpapses to form a black hole of radius r without losing any mass

$\frac{2}{5}MR^2\omega = \frac{2}{5}Mr^2\omega'$ where ω' is the new angular velocity.

$$\omega' = \frac{R^2}{r^2}\omega$$

$r << R$ so $\omega' >> \omega$

A rapidly rotating black hole might 'balance' gravitational attraction against 'centrifugal forces' and allow the existence of a 'naked singularity'. This would be a real worry and Roger Penrose has suggested the cosmic censorship hypothesis that states that nature always conspires to cloak naked singularities with event horizons. Even so, rotating black holes have led to some remarkable predictions. Penrose himself has proposed a process whereby an object hurled towards a rotating black hole passes into a region called the ergosphere close to the event horizon, splits into two in such a way that one half falls into the hole against its direction of motion and the other is deflected back out of the ergosphere and off to infinity. The net effect is that the ejected matter increases its mass energy at the expense of the mass energy of the black hole which rotates more slowly after absorbing the jettisoned matter. The Penrose process could be used to extract energy from rotating black holes. It has also been suggested that a sufficiently massive rapidly rotating black hole could be used as a time machine to travel back to the past by following a particular trajectory close to its event horizon.

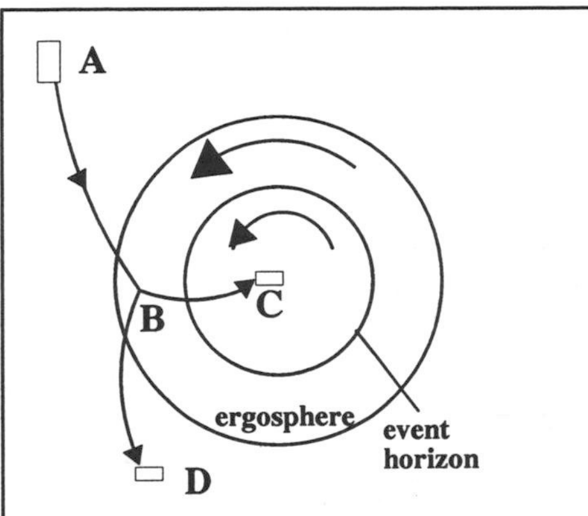

fig 4.49 The Penrose process. A mass at A falls into the ergosphere around a spinning black hole. At B it splits and one part falls into the hole at C. The remaining part is ejected, at D, with increased energy.

The ergosphere is a region in which space-time is 'dragged' around by the rotation of the black hole to such an extent that it would be impossible to avoid being dragged around with it. This is a severe example of an effect being tested with orbiting gyroscopes.

We have compared the distortion of space-time by matter to the distortion of a rubber sheet pressed down by some object. A black hole would cause the depression to grow steeper and steeper and narrower and narrower without limit. If we imagine such a model of a black hole in a rubber sheet which is not flat but curved and perhaps even folded back on itself then we can imagine one of the most exotic objects invented by physicists, the Einstein-Rosen bridge or 'wormhole': a space-time short cut between distant events. This is commonly used in science fiction as a device to overcome the problem of travelling the great distances that separate stars and galaxies without exceeding the speed of light.

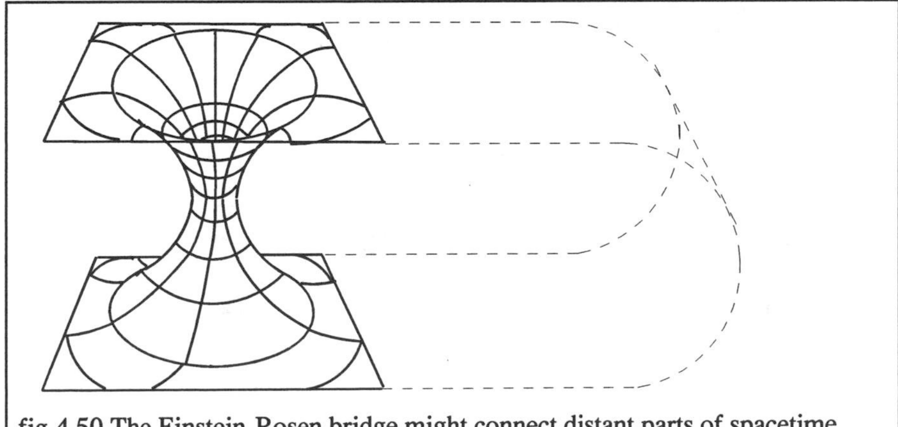

fig 4.50 The Einstein-Rosen bridge might connect distant parts of spacetime.

Wormholes are not just science fiction, they are certainly taken seriously by theoretical physicists, although they are divided as to whether it would be possible to survive a trip through them and also where and when you might emerge. There is no direct evidence for their existence as yet, but neither is there a good theoretical reason to rule them out. In the meantime there are plenty of researchers working out their properties in great detail!

fig 4.51 Stephen Hawking (b.1942)
'It is important to have a good name for a concept. It means that people's attention will be focused on it. I suppose that the name 'black hole' does have a rather dramatic overtone, but it is also very descriptive. It has a strong psychological impact. It could be a good image for human fears of the universe.'
Artwork by Nick Adams © 1996

There is no lower limit to the mass of a black hole, and it has been suggested that large numbers of mini-black holes might have formed in the extreme conditions following the Big Bang. Furthermore, in the 1970s, Stephen Hawking showed that black holes can radiate by absorbing one of a pair of virtual particles formed near the event horizon and ejecting the other. This radiation has a characteristic black body spectrum which allows us to associate a temperature with a black hole. Black holes are hot! If they have temperatures then they should be describable in terms of the laws of thermodynamics. This line of argument leads to the conclusion that the entropy (degree of disorder) of a black hole must never decrease and ultimately the entropy of the universe will be dominated by the entropy of black holes. The second law of thermodynamics states that the entropy of the universe never decreases. If black holes did not have entropy then it would be possible to violate the second law simply by throwing a disordered (high entropy) system into the black hole and reducing the entropy of the universe. Hawking and Bekenstein showed that the entropy of a black hole is proportional to the square of the area of its event horizon. When an object falls into a black hole its mass, charge and angular momentum are 'remembered', everything else is 'forgotten', and the increase in entropy of the black hole is greater than or equal to the loss of entropy of the original body.

The Hubble Space Telescope has now provided almost conclusive evidence for the existence of black holes. In 1994 it found one in the heart of galaxy M87, fifty million light years from our solar system. In December 1995 the HST produced dramatic images of a huge disc of gas whirling around a black hole in NGC4261. It is thought to contain 1.2 billion times more matter than the Sun and is growing as matter falls in a cataclysmic collision with another galaxy. Two giant plumes of gas generated by the in-falling matter are streaming away from the black hole and propelling it from the centre of the disc. The disc is 800 light years across and thought to be the remnant of a smaller galaxy that fell into the core of NGC4261. The rapid rotation of such a large disc requires an enormous centripetal force. This is provided by the gravitational field of the black hole at its centre.

4.11 GRAVITATIONAL WAVES.

4.11.1 What Are Gravity Waves?

Sound travels as longitudinal mechanical waves in a material medium. The speed of propagation is determined by the mechanical properties of the medium and the waves are possible because of the 'connectedness' of the medium, the fact that a disturbance at one point induces a delayed disturbance in an adjacent part of the medium. Without this property waves would never leave their origin. Electromagnetic waves do not require the presence of a mechanical medium and are described as disturbances of the electromagnetic field. The connectedness of the electromagnetic field is described by Maxwell's equations and the speed of light emerges naturally from these equations. In free space the speed c is given by:

$$c = \frac{1}{\sqrt{\varepsilon_o \mu_o}}$$

where ε_o and μ_o are the permittivity and permeability of free space respectively. The disturbances that invariably result in the emission of electromagnetic waves are when charges are accelerated. For example, the sudden deceleration of electrons striking a heavy metal target generates X-rays. Unlike sound, electromagnetic waves are transverse, and so can be polarized. Electromagnetic waves are more fundamental than sound waves because they depend directly on the ability of space to support and transmit the electromagnetic interaction whereas mechanical waves like sound depend in a much more complex way, through atomic and molecular interactions, on the same principles. The relatively slow velocity of sound compared to light is because we are not looking directly at the fundamental physics but at a large scale phenomenon sitting on top of it. It is analogous to diffusion, a painfully slow process despite the high molecular velocities that cause it.

General relativity is a field theory for gravity. The field itself is the spacetime metric and Einstein's equations describe how the metric depends on the distribution of mass, energy and momentum and how the curvature at one place joins smoothly onto the curvature adjacent to it. In this respect it is a similar theory to Maxwell's equations for the electromagnetic field. The gravitational field describes the global geometry of spacetime and displays the kind of connectedness necessary for waves to be propagated. Perhaps gravitational waves would result from some periodic disturbance of the gravitational field? Einstein was the first to investigate this possibility.

Einstein tried to construct a wave equation in general relativity like the one Maxwell derived for electromagnetism. However, there is an important physical difference between electromagnetism and gravity. Electric dipole radiation (produced by an oscillating charge) leads to observable effects on distant charges. If a charge oscillates at one place then other charges elsewhere are set into oscillation as the varying electric and magnetic fields reach them (this is the principle behind radio and TV). Gravitational dipole radiation however is undetectable (even in principle). The reason for this is that the gravitational field strength g is not an observable in local inertial reference frames. This seems strange, but a simple thought experiment illustrates that this is a consequence of the equivalence principle. Imagine what you would experience inside a freely falling lift on Earth and on the Moon. The physics in each of these reference frames would be indistinguishable and the value of g in either case remains unknown. So if an oscillating mass somewhere in space produces gravitational waves consisting simply of a fluctuating field strength these fluctuations will affect all bodies and measuring instruments in exactly the same way and result in no observable effects.

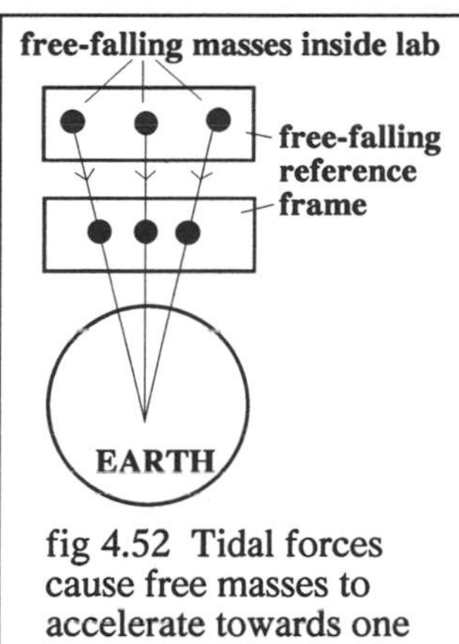

fig 4.52 Tidal forces cause free masses to accelerate towards one another inside a freely falling laboratory near the Earth.

There *is* one effect of the gravitational field that can be detected in a free falling reference frame, the tidal effect due to the *variation* of g in the frame. We have come across this before, it determines spacetime curvature and is really the only genuine effect of the gravitational field. Particles initially separated in the freely falling lift will move together or apart depending on their relative positions.

Einstein's paper on gravity waves was published in 1916. He showed that unlike dipole radiation, quadrupole radiation is detectable. Quadrupole radiation is a higher order disturbance than dipole radiation and results from systems that are changing their mass distribution (e.g. collapsing stars or binary systems) and producing observable tidal distortions of spacetime and matter. The distortions appear as a change in the separation of particles along lines perpendicular to the direction of wave travel.

Unfortunately the magnitude of the predicted effect is unimaginably small, even for the strongest expected sources of gravity waves. For example, the collapse of a rotating star into a black hole within our own galaxy would produce a change in separation of terrestrial atoms 1 m apart of only about 10^{-17} m. This is about 1% of the diameter of an atomic nucleus. Nonetheless, Einstein derived an expression for the rate of energy dissipation from a simple rotating dumbbell, a model which is very similar to that of a binary star system. He also showed that the waves would travel at the speed of light. Experiments designed to detect the effects of gravitational waves stretch our ingenuity to its limit.

4.11.2 Detecting Gravity Waves.

Since the 1960s two distinct types of gravity wave detector have been developed. The first is based on the Weber bar and attempts to measure the tiny simultaneous resonance vibrations of two separate metal cylinders as they are forced to oscillate by the passing waves. The second type is based on a large interferometer and measures the variation of interference fringes as the interferometer path lengths are changed by the geometric distortion as gravity waves pass through the apparatus. Although Weber himself claimed to have detected gravity waves as long ago as 1968 this was not verified by other groups and no subsequent experiments with either Weber bars or interferometers have succeeded in detection. This is not very surprising, the sensitivity of a detector is measured by a quantity h which is the smallest fractional change in separation per unit separation of particles that the detector can pick up. For the collapsing star described above this value would be 10^{-17}; for sources outside our galaxy it might be 10^{-22} or below. Weber's bars had a sensitivity of about 10^{-15} and modern low temperature versions (low temperatures reduce the thermal noise of vibrating atoms) might reach 10^{-18}. The latest interferometers might detect bursts at a level of 10^{-20} and continuous waves at 10^{-24} so it is quite possible that the positive detection of gravitational waves sources will occur in the near future.

4.11.3 The Weber Bar.

The original Weber bar was a cylinder of aluminium about 1.5 m long by 1 m in diameter suspended in a vacuum. Later bars were also cooled to low temperatures. The bars act as tuned receivers of gravity waves at a frequency of about 1660 Hz (a frequency expected to be contained within the spectrum of gravitational waves produced by collapsing stars of about the Sun's mass). This corresponds to the natural frequency for longitudinal oscillations of the bar and the apparatus was designed in such a way that it would have a very sharp resonance at this frequency. It was also isolated from external fields and sources of mechanical vibration. The induced mechanical strains are greatest at the centre of the bar, so Weber surrounded the central circumference with a ring of piezoelectric transducers that convert strains into electrical signals. Since the expected signal would be extremely weak and difficult to distinguish from sources of random noise in the bar and in the signal processing Weber set up two similar bars 1000 km apart and used a microwave telephone link to look for simultaneous signals in both. He then monitored the output of the detectors over a period of time. Any sources outside the solar system should produce simultaneous signals that fluctuate with a period equal to the sidereal rather than the solar day.

Figure 4.53 shows the experimental arrangement using a Weber bar. The lower diagrams show the expected distortions of the bar as the waves pass. These are converted into electrical signals using piezoelectric transducers.

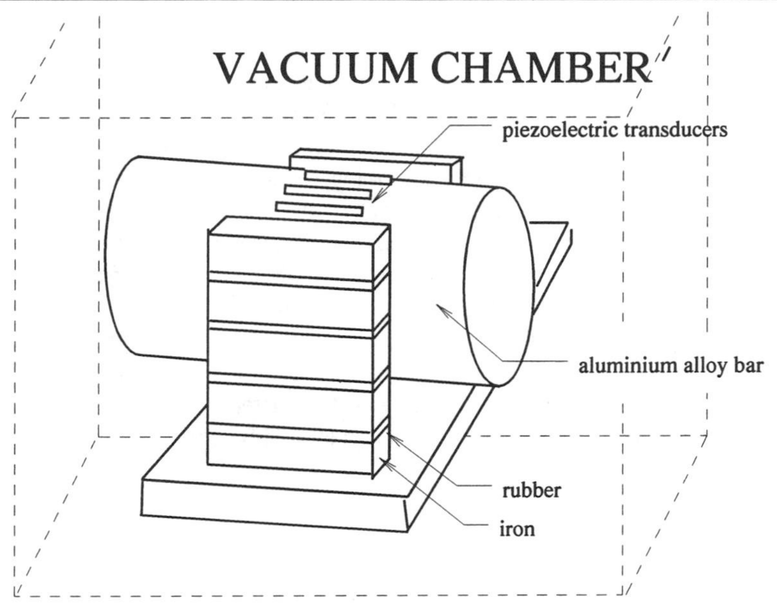

Pattern of oscillations for an expected resonance at 1.66 GHz

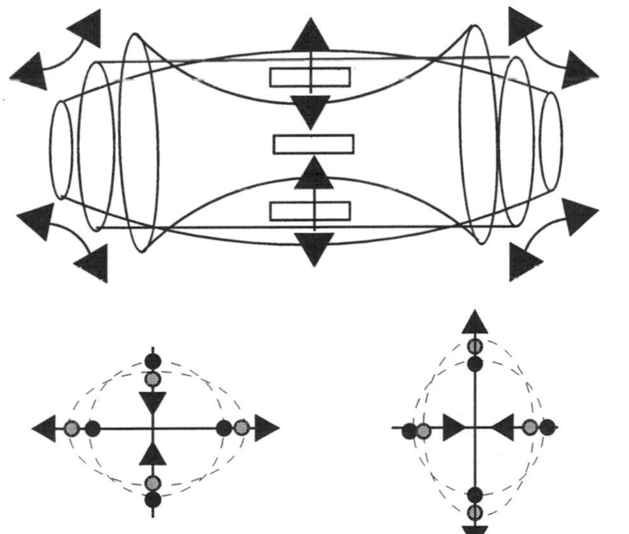

fig 4.53 The Weber bar - a gravitational wave detector. Fluctuating strains induced in the bar by gravitational waves are converted to electrical signals by piezoelectric transducers fixed to the sides of the bar. The bottom diagram shows how a pattern of particles lying on the circumference of a cylinder will oscillate as a gravitational wave moves along the axis of the cylinder. The circle flexes and alternates between a vertical and a horizontal ellipse.

4.11.4 Interferometers.

The periodic displacements of distant particles as a gravity wave passes can be used to alter the path difference in an optical interferometer. The overall effect is that the intensity of light detected where the two beams recombine fluctuates. Several research groups around the world are looking for gravity waves using this method with optical paths several kilometres long. In the absence of a gravity wave both arms are equal in length. As the wave passes it causes one arm to shorten and the other to lengthen. The phase difference between recombining waves at the photodetector changes and so does the signal strength.

The optical layout for one of these projects, GEO 600, is shown in the diagram. This experiment is a collaboration between the Universities of Hanover, Munich, Glasgow and Cardiff and will use an interferometer with arms 600 m long. It should have a sensitivity of 1 part in 10^{21} and should be completed in 1999.

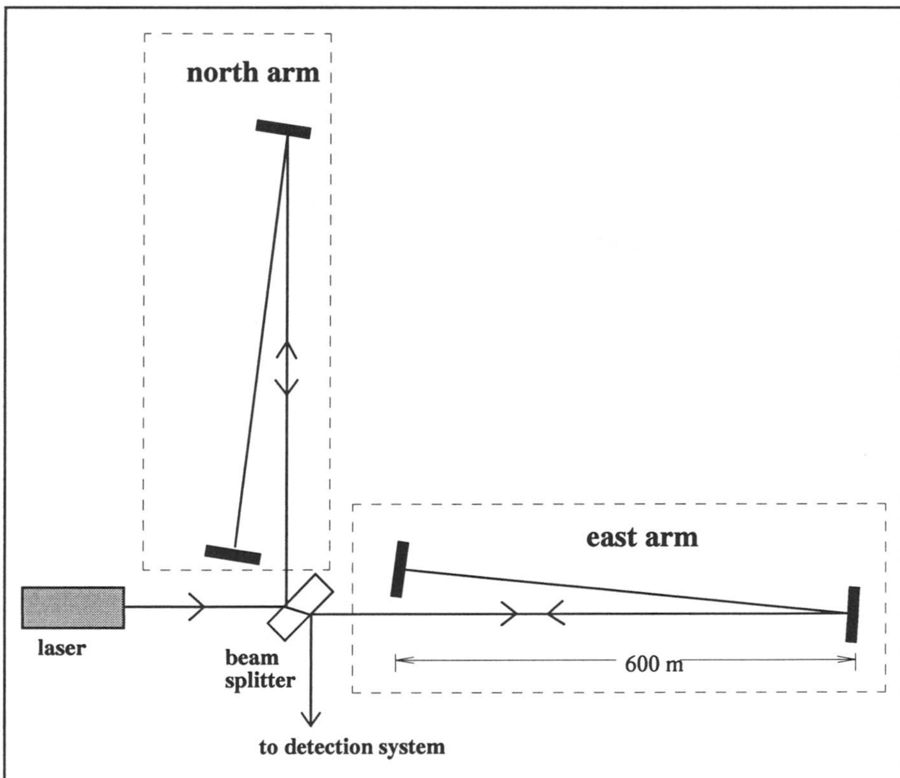

fig 4.54 GEO 600 is a gravitational wave interferometer being constructed at a site near Hanover. The diagram above shows the optical geometry. The main arms are 600m long. Reproduced with permission of Professor Hough, Gravitational Waves Group, University of Glasgow.

4.11.5 Pulsars.

In 1993 the Nobel Prize for physics was awarded to Russell Hulse and Joseph Taylor Jr. for their discovery and subsequent studies of a binary pulsar. They showed that the system lost energy at the rate predicted by Einstein for gravitational radiation from a binary system.

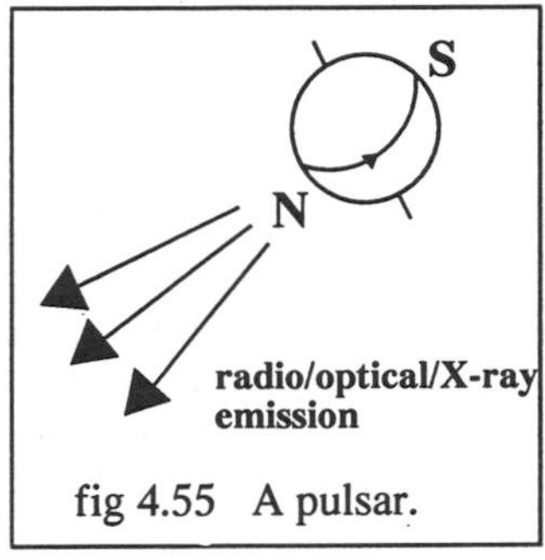

fig 4.55 A pulsar.

A pulsar is a rapidly rotating neutron star which signals its presence by beaming a regular series of pulses of electromagnetic radiation into space. As the star collapses its magnetic field intensifies and orients itself with poles lying on the equatorial circumference. Charged particles accelerated in this intense field emit powerful beams of electromagnetic radiation which are observed each time the pole sweeps past our line of sight. The precise regularity of the pulses from pulsars is comparable to that of terrestrial atomic clocks (the fastest known pulsar has a period of just 1.6 ms).

Pulsars were discovered by Hewish and Bell in 1967 but PSR B1913+16 discovered by Taylor and Hulse in 1974 was the first to be observed in a binary system. It is the interaction within this system that makes it so important for general relativity. The two stars are very close to one another, separated by about the radius of the Sun, and move rapidly around their common centre of mass with orbital speeds of about 0.001c. The proximity and speed increase relativistic effects and the regularity of the pulsar clock allows very precise measurements to be made. Surprisingly it has been possible to measure the mass of the pulsar and its companion more accurately than we have managed to measure the mass of any other object outside our own solar system. The researchers assumed general relativity and used the rate of advance of periastron (closest approach) to determine the total mass of the system. The rate of advance is 4.23 degrees per year, 36 000 times greater than the advance of perihelion of Mercury in our own solar system. The calculated total mass is 2.8275 solar masses.

Another relativistic effect allowed the masses of each star to be calculated. This is a periodic variation in the arrival times of pulses caused by a combination of special relativistic time dilation (from the moving source) and gravitational time dilation due to the potential of the pulsar's companion. The variation in period has an amplitude of 4.38 ms and leads to masses of 1.42 and 1.40 solar masses for the pulsar and its companion respectively. (Incidentally, the companion is also believed to be a neutron star. This is an important assumption because a larger star would produce additional tidal effects in the motion of the pulsar due to the companion's oblateness.)

According to Einstein a binary system should radiate some of its energy as gravitational waves. This reduces the orbital energy and the stars fall closer together, shortening the orbital period. In 1974 the period was 0.059029995271 s.

Since then it has reduced by about 1.5 ms, an observation in agreement with Einstein's prediction for the rate of slowing of about 75 μs per year due to gravitational radiation.

We may not have detected gravity waves with terrestrial detectors but the investigations of Taylor and Hulse leave little doubt that they are being radiated by PSR B1913+16 and its companion and, presumably, by an enormous number of other sources throughout the universe. Studies of this binary pulsar have also confirmed Einstein's calculations which were not universally accepted.

4.11.6 LISA

The tiny fractional change in length (h) produced by extra-galactic sources of gravitational waves make most such events almost impossible to detect on Earth in the foreseeable future. If $h = 10^{-20}$ the change of length in an interferometer arm of length 4 km is a mere 4×10^{-17} m, much smaller than the diameter of a single nucleus. One way this effect could be amplified would be to build an interferometer with incredibly long arms, and the LISA (laser interferometer space antenna) project has been proposed to do this.

The idea is to launch six spacecraft and place them in pairs at the corners of a giant equilateral triangle in space. All six spacecraft will be identical and pairs will be separated by a distance of 200 km and 'connected' by an optical reference beam. As gravitational waves pass through the apparatus one side of the triangle is compressed while another is stretched, this will alter the path of laser beams traveling along the sides of the triangle and so change their relative phase when they are recombined. Each corner of the triangle acts as the hub of an interferometer, so the whole apparatus consists of three interferometers. This gives the experiment some protection against unexpected equipment failure, measurements would still be possible even if two spacecraft failed to function (as long as they were not both at the same corner).

Despite the enormous size of the instrument, the change in length of each interferometer arm will still be tiny even for cataclysmic events like the collision of two black holes in a neighboring galaxy. With $h = 10^{-20}$ the maximum change in spacecraft separation over a distance of 5 million kilometers is still only 5×10^{-11}m, less than the diameter of an atom! As Peter Bender, a physicist at the Joint Institute for Laboratory Astrophysics in Colorado put it:

'We are talking about a scientific instrument a thousand times bigger than the Earth, located as far away as Venus, making measurements smaller than an atom.'

The LISA project has been proposed by the European Space Agency for launch sometime in the first part of the twenty-first century and will operate for two years. The biggest problem for the design team is to minimize all sources of noise so they will not swamp the tiny signal from gravitational waves.

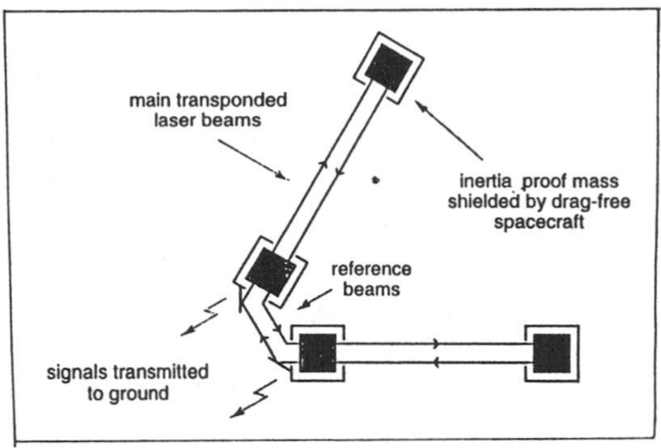

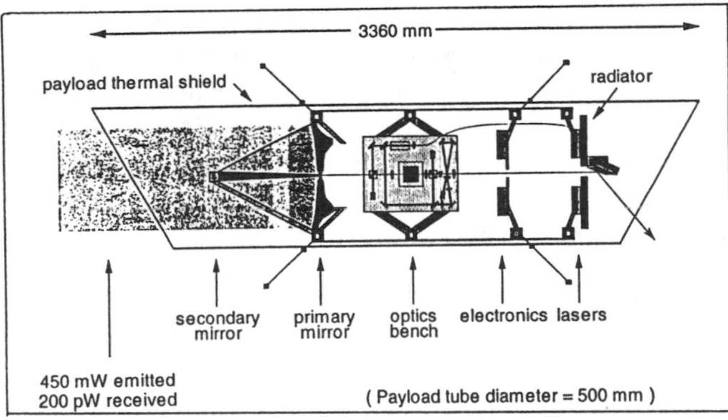

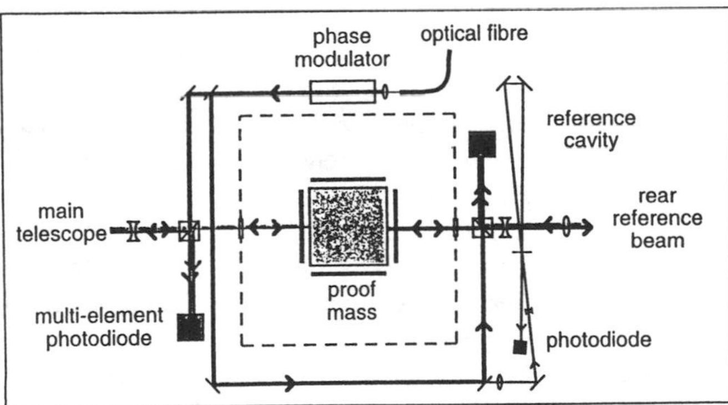

fig 4.56 LISA. The three diagrams above show (a) the layout for one of the interferometers (b) the science payload of one of the six spacecraft, and (c) details of the optical layout around the inertia-proof mass.
Reproduced with permission of Professor Hough, Gravitational Waves Group, University of Glasgow.

Each spacecraft will contain an infra-red laser and a telescope for focusing the received and transmitted light. Unlike terrestrial interferometers the 'reflectors' will not simply send back the signal they receive, the incoming light will stimulate the laser to emit a beam of identical frequency and phase to the incident beam (this is called phase-matched transponding). This is necessary since only 200 pW of the 450 mW emitted from the laser will be received at the far end of each interferometer arm. The optics involved is also interesting. The received light will be focused onto a small highly polished solid cube made from an alloy of platinum and gold (to minimize the effects of magnetic disturbances) which floats freely inside an evacuated titanium box. The position of the cubes relative to one another fixes the size and geometry of the three interferometers so it is essential that they be isolated from any external forces. One potential problem is the effect of radiation pressure from the Sun (about 10^{-15} Pa ± 1%) which would cause the spacecraft to drift and eventually make the cube collide with the side of its containing box. This is avoided by constantly measuring the position of the cube relative to the spacecraft and using weak smoothly adjustable thrusters to keep the spacecraft centred on the cube. The large scale of LISA makes it ideal to look for the low frequency gravitational waves emitted by large-scale violent events rather than the higher frequency (kHz) waves generated by smaller events like collapsing stars that may be picked up by terrestrial interferometers. LISA is one of the most ambitious experiments ever planned and this 'gravitational telescope' may well reveal unexpected objects that lead to new tests of general relativity and new challenges for theoretical physicists, astronomers and cosmologists.

4.12 SUMMARY OF IDEAS AND EQUATIONS IN CHAPTER 4

Inertial forces:	Introduced to explain the spontaneous acceleration of 'free' bodies in non-inertial reference frames.
Equivalence principle:	Physical processes obey the same laws in a uniformly accelerated reference frame as in a uniform gravitational field.
Free-fall frames:	Local gravitational fields can be transformed away by viewing physical processes from a reference frame in free-fall. Free-fall reference frames are local inertial reference frames.
Space-time curvature:	Macroscopic gravitational fields cannot be transformed away by extending any single freely-falling reference frame. Tidal forces remain and can be explained as a curvature of the local space-time geometry.
General relativity:	Einstein's equations linking space-time curvature to the distribution of matter in the universe.
Non-Euclidean geometry:	Self-consistent geometries developed by adopting

alternatives to Euclid's axiom of the parallels. In 2D these describe the intrinsic curvature of surfaces.

Gravitational red shift: Change of frequency and wavelength (to longer wavelengths) as photons lose energy escaping from a massive star or galaxy.

Expanding universe: Evidence from red shifts of receding galaxies suggests that the universe is in a state of expansion and originated in a hot Big Bang about 10 to 20 billion years ago.

Hubble Law: Red shift and recession velocity of the galaxies are proportional to their distance from us.

Cosmic background radiation: Present day remnant of the electromagnetic radiation created when most of the matter and antimatter created in the Big Bang annihilated.

Black hole: Singularity in space-time formed when a star or galaxy collapses under its own gravitation. The surface gravity becomes so great that nothing, not even light, can escape.

Gravity waves: Energy radiated from a system of accelerating masses in the form of periodic disturbances of space-time geometry. The wave speed is c.

Deflection of light:
$$\phi = \frac{GM}{Rc^2} \quad (4.1)$$

Focal length of a gravitational lens:
$$f = \frac{c^2}{4\pi G\sigma} \quad (4.2)$$

Gravitational frequency shift:
$$\frac{\delta f}{f} = -\frac{\delta\phi}{c^2} \quad (4.3)$$

Gravitational time dilation:
$$\frac{\delta T}{T} = \frac{\delta\phi}{c^2} \quad (4.4)$$

Gravitational red shift:
$$\frac{\delta\lambda}{\lambda} = \frac{\delta\phi}{c^2} \quad (4.5)$$

Space-time curvature:
$$\frac{1}{r^2} = \frac{2GM}{c^2 x^3} \quad (4.6)$$

Hubble Law:
$$v = H_o d \quad (4.8)$$

Hubble period:
$$T_h = \frac{1}{H_o} \quad (4.9)$$

Hubble constant and scale:
$$H_o = \frac{\dot{R}}{R} \quad (4.10)$$

Red shift and scale: $$\frac{\delta\lambda}{\lambda} = \frac{\dot{R}d}{Rc} = \frac{\delta R}{R} \qquad (4.12)$$

Critical density: $$\varrho_c = \frac{3H_o^{\,2}}{8\pi G} = \frac{3c^6}{32G^3M^2\pi} \qquad (4.13,15)$$

Schwarzschild radius: $$R_s = \frac{2GM}{c^2} \qquad (4.14)$$

Wavelength approximation: $$\lambda' = \left(1 - \frac{R_s}{r}\right)^{-\frac{1}{2}} \lambda \qquad (4.16)$$

Frequency approximation: $$f' = \left(1 - \frac{R_s}{r}\right)^{\frac{1}{2}} f \qquad (4.17)$$

4.13 PROBLEMS

1a. Use Newtonian equations to derive an expression for the gravitational field strength g_s at the Schwarzschild radius.
b. What would be the mass and Schwarzschild radius of a black hole if the 'surface gravity' at the event horizon is $g_s = 9.8$ N kg^{-1}?

2. The equivalence principle relates uniform gravitational fields to uniformly accelerated reference frames. Imagine you are standing on bathroom scales inside an elevator on Earth. Take your mass as m.
a. Write down an expression for your weight in the Earth's gravitational field.
b. Write down an expression for the reading (in N) on the scales in terms of the Earth's gravitational field, and the vertical acceleration of the elevator.
c. Explain why the reading on the scales is independent of the velocity of the lift.
d. Describe your experiences inside the elevator if it falls freely down the shaft. The word 'weightless' may enter your description - is it appropriate?

3. If aliens signal to us with radio waves they may well decide to use the hydrogen spin-flip frequency of 1.420 GHz. However, if their home system is moving relative to us these signals will suffer a Doppler shift. If our receiver detects a range of frequencies from 1.380 to 1.460 GHz calculate the range of velocities that the alien transmitter could have and still be detected from Earth.

4. This question is about tidal forces. A large room falls vertically towards the Earth. Inside the room there is a uniform distribution of dust suspended in the air.
a. Explain how the distribution of dust changes as the room falls.
b. Would the same thing happen if the room were in orbit around the Earth?
c. By treating the Earth as a freely falling body in the gravitational field of the Moon, explain the formation of the tides.

5a. Explain why light is deflected by gravity.

b. Why was Einstein's first calculation of the deflection of starlight by the Sun an under-estimate?

6. A useful measure of the 'strength of gravity' at a point or the significance of general relativistic effects there is the ratio $s = \left(\frac{v_{esc}}{c}\right)^2$ where the escape velocity is calculated from the point concerned.
a. Calculate s at the surface of the Sun. What does this value suggest about the significance of general relativistic effects in the solar system?
b. What is the value of s at the event horizon of a black hole?
c. Why is the advance of perihelion of Mercury greater than for any other planet in the solar system?

7. If a spacecraft accelerates at a constant rate in a gravitational field, clocks at the rear of the craft run slow with respect to clocks at the front.
a. Use the equivalence principle to explain why this is.
b. Derive an expression for the ratio of clock rates for a craft of length l accelerating at g.

8. In 1961 Hay, Schiffer, Cranshaw and Egelstaff measured clock rates at the centre and edge of a rapidly rotating disc.
a. Which clocks ran slow? Why?
b. The experiment involved transmitting signals (emissions from excited atoms) between the center and the edge and measuring the frequency shifts. Show that the change of frequency when a photon is transmitted between two points that differ in potential by $\Delta\phi$ is:

$$\Delta f = -\frac{\Delta\phi}{c^2} f$$

c. Hence show that when a photon moves through a gravitational field:

$$f + \frac{f\phi}{c^2} = \text{constant}$$

d. Use the equivalence principle to derive a potential function $\phi(r)$ in the rotating reference frame of this experiment.
e. Hence show that the frequency f received at radius r is related to the emission frequency at the centre by:

$$f(r) = \frac{f_o}{1 + r^2\omega^2/c^2}$$

where ω is the angular velocity of the disc.
f. Show that special relativity leads to the same result (for v<<c) if the time dilation of rotating clocks is interpreted as being caused by their motion.

9a. What is the ratio of clock rates at altitude $h<<R_E$ in the Earth's field and at the surface of the Earth (of radius R_E and mass M_E)? Which clock gains? Ignore the rotation of the Earth.
b. This gravitational time dilation could be counteracted by a special relativistic time dilation if one of the clocks were to orbit the Earth. How are the required orbital velocity and h related? Comment on your answer.

10. Two children, A and B, sit opposite one another on a roundabout. A third child, C stands beside the roundabout. The children are playing catch with a tennis ball. For each of the following situations describe the path of the ball as seen by each child.
a. A throws the ball to B who catches it.
b. B throws the ball to C who catches it.
c. C throws the ball so it passes over the centre of the roundabout but neither A nor B catches it.
d. How is it possible for an observer moving with the roundabout to account for the motion of the ball. (Assume this observer wishes to regard the rotating frame as an inertial frame.)

11a. Explain why a clock close to the surface of the Sun would run slow relative to a clock on the surface of the Earth.
b. How much time would pass on a terrestrial clock while 1 hour passes on a solar clock?
($M_E = 6.0 \times 10^{24}$ kg, $R_E = 6.4 \times 10^6$ m, $M_S = 2.0 \times 10^{30}$ kg, $R_S = 7.0 \times 10^8$ m)
c. How long would a spacecraft need to orbit the Sun at a distance of 10 R_S for the occupants to gain one hour relative to the Earth? (Consider only gravitational time dilation.)
d. Comment on the idea of achieving time travel by going to a black hole, orbiting at a safe distance and then returning to Earth.

12 Compare (roughly) the significance of time dilation effects due to velocity and time dilation effects due to gravitation for passengers in a passenger jet.

13 If time runs slow close to a massive body, does this mean that the speed of light will also slow down and so is not after all an invariant quantity?

14 The effect of gravitational time dilation can be likened to a series of concentric spherical shells surrounding a massive body. The smaller the radius of the shell, the more slowly time passes there.
a. Show how this will cause the deflection of light as it passes a massive body.
b. Compare the effect with the deviation of light by a glass prism.

15a. Calculate the fractional frequency shift for a photon travelling (i) down and (ii) up a 25 m lift shaft.

b. How fast and in what direction would the source have to be moving to produce the same frequency shift by the Doppler Effect?

16. Describe the metric on a cylindrical surface.

17a. Why do tidal forces arise inside a free-falling reference frame close to the Earth?
b. It is sometimes said that tidal forces are the only genuine effect of gravity, why is this?
c. If a dust-filled laboratory fell toward the Earth from a very great distance, where would the dust accumulate?
d. By treating the Earth as a freely falling body in the gravitational field of the Moon, explain the ocean tides.
e. Why is the effect of the Moon so much more important in forming ocean tides than the effect of the Sun?

18. This question shows how a 2D surface can have an intrinsic curvature that can be determined by measurements made within the surface itself.
a. Derive an expression for the circumference c of a circle of radius r drawn on the surface of a sphere of radius R.
b. Show that c approaches $2\pi r$ as r approaches zero.
c. Show that the expression derived in (a) reduces to $2\pi r(1 - \varepsilon r^2)$ when $r<<R$ and derive an expression for ε.
d. If the Earth were a perfect sphere of radius 6400 km and surveyors were able to measure the circumference of circles drawn on its surface to an accuracy of 1 part in 10^4, what is the smallest circle they could survey to convince themselves that the surface is not flat?

19a. Use the Newtonian equations for gravitation to calculate the escape velocity of a projectile fired from the Earth's surface.
b. What effect does the rotation of the Earth have on the actual escape velocity from the surface?
c. Derive an algebraic expression for the radius of the sphere into which the Earth must be compressed if it is to have an escape velocity equal to the speed of light and so become a black hole. Calculate this radius (the Schwarzschild radius).
d. The equation we derived for the radius of curvature of space at a distance x from a massive body was $r = \sqrt{\frac{c^2x^3}{2GM}}$. What is the significance of $x = r$ in this equation? Relate this to your answer to (c).
e. What will prevent either the Earth or the Sun from collapsing to become a black hole?

20 We showed in the text that the radius of curvature close to the Earth was around 1.7×10^{11} m, comparable to the radius of the Earth's orbit around the Sun. Does

this mean that the effects of space-time curvature should be obvious on the scale of the solar system?

21a. Distinguish between Doppler shifts due to motion, gravitation and expansion.
b. Are the galaxies really receding from us?
c. We have often used the equation $z = \frac{\delta\lambda}{\lambda} = \pm\frac{v}{c}$. Some quasars have extremely large red shifts, greater than $z = 4$. If this was a Doppler shift due only to motion would it imply that they are receding from us at greater than the speed of light?
d. Calculate the recession velocity of a quasar with $z = 4$.

22 The full special relativistic Doppler formula is $z = \left(\frac{c+v}{c-v}\right)^{1/2} - 1$.

a. Rearrange this formula to give $\frac{v}{c}$ in terms of z.
b. Calculate v when i) $z = 0.1$, (ii) $z = 1.0$ (iii) $z = 4.0$ (iv) $z = \infty$.

23 The gravitational red shift can be written as $z = \left(1 - \frac{R_s}{\mathrm{R}}\right) - 1$ where R is the radius from which light is emitted and R_s is the Schwarzschild radius.
a. What happens if $R = R_s$?
b. Calculate z for light traveling from the Sun to the Earth.
c. Show that the expression above reduces to $z = -\frac{GM}{Rc^2}$ if $R << R_s$.

24 The expansion red shift is given by $z = \frac{R_o}{R} - 1$. Explain the significance of the terms in this equation and the conclusions that can be drawn about an object with expansion red shift 2.0.

25 What would a red shift of z = 4 mean if it was due to:
a. expansion
b. gravitation
c. relative motion?

26 In an expanding universe the galaxies are all red shifted (ignoring local motions).
a. What would we see if the Universe had been contracting for a few billion years?
b. How would this view change with time as the universe approached a Big Crunch?

27 Our solar system is carried along with the local movement of the Milky Way galaxy at about 500 km s^{-1} in the direction of Leo. How must this be taken into

account when measuring the recession of distant galaxies?

28a. Is geometry an experimental or a theoretical discipline?
b. Where is the centre of the universe?

29a. Show that the Hubble Law implies a 'particle horizon', that is a maximum distance beyond which we cannot see any more stars or galaxies. Derive a simple expression for this 'Hubble length' in terms of the Hubble constant.
b. Apply the Hubble law to show that the relative recession velocity of galaxies beyond the Hubble length exceeds the speed of light. Does this violate the principle of relativity?
c. What happens to light emitted in our direction from a galaxy beyond the particle horizon?
d. How does the Hubble length change if the rate of expansion of the universe falls?
e. How does this affect the light from galaxies at present beyond the particle horizon?
f. Is it always true that galaxies beyond the particle horizon are 'unobservable' from the Earth?

30 The Hubble Law gives us the present distances to the galaxies (this is called the reception distance), but we see them at the time they emitted their light (this smaller distance is called the emission distance).
a. Use the equation for the expansion red shift to relate the reception distance to the emission distance.
b. Is it always true to say that galaxies with greater red shifts are further away?

31 This question is about the Olber Paradox. Assume the universe is infinite, static and uniform and filled with point-like stars with a number density n and equal luminosities L.

a. Explain why the apparent brightness of a star at distance r is given by $S = \dfrac{L}{4\pi r^2}$.
b. Show that the number of stars between r and δr is $4\pi r^2 n\delta r$.
c. Calculate the intensity δI at the Earth due to stars in this shell.
d. Hence show that, in this universe, the intensity of starlight at the earth's surface would be infinite.
e. How is this calculation affected if the stars have finite size?
f. What if the stars have a lifetime T?
g. What if the universe is expanding?

32. The time required to dissipate an energy comparable to the orbital energy of a binary system by gravitational radiation is of the order: $T \approx \dfrac{d^4}{cR_S}$ where d is the scale of the system (e.g. the diameter of the orbit) and R_s is the Schwarzschild

radius of the system. In this question we are going to make rough estimates of the significance of gravitational radiation losses on two systems, the Earth-Moon system and binary pulsar PSR1913+16 studied by Taylor and Hulse.

Data: Earth-Moon: $M_E = 6 \times 10^{24}$ kg; $M_M = 7 \times 10^{22}$ kg; $t_M = 27$ days

PSR1913+16: $M_1 \approx M_2 \approx 2.8 \times 10^{30}$ kg; $d \approx 7 \times 10^8$ m; $t = 29\,000$ s

a. Calculate the Schwarzschild radius and T for both systems.

b. Use the kinetic energy of the orbiting bodies as a rough measure of the orbital energy and calculate its value for each system.

c. Make a very crude approximation that each system dissipates all of its energy at a constant rate during time T and use this to estimate the power of gravitational radiation from each system. Comment on the values.

d. Now estimate the rate at which the period t of PSR1913+16 falls and compare its value with the measured value of about 2×10^{-12} ss^{-1}. (Accurate calculations are in very good agreement with this value).

(Hint: You could approach this by considering the energy lost in 1 s, finding its effect on kinetic energy and hence velocity and then calculating the new period assuming the radius of orbit is more or less unchanged. You could go into more detail than this, but we have neglected some important factors: for example, the eccentricity of the ellipse is 0.6).

5 APPENDIX

5.1 RELATIVITY AND ELECTROMAGNETISM.

5.1.1 Magnetism.

Coulomb's law gives the electrostatic force between charges. Faraday and Maxwell explained this by introducing the concept of an electric field. For a pair of charges q_1 and q_2 separated by a distance r the force F is given by:

$$F = \frac{q_1 q_2}{4\pi\varepsilon_o r^2}$$ where ε_o is the relative permittivity of free space.

If we consider q_1 to be the source of a field that acts on q_2 then the strength of this field at the position of q_2 is:

$$E = \frac{F}{q_2} = \frac{q_1}{4\pi\varepsilon_o r^2}$$

This is independent of q_2 and so should not depend on the motion of q_2. Experiment confirms this. The electric field of a static charge distribution is independent of the velocity of the charges that move through this field. We assume this in calculating the energy gained by charges in an accelerator. However, magnetic forces do depend on the velocity of a moving charge and are always perpendicular to the motion of the charge. How do these arise? A clue to the answer is glimpsed if we consider how static charges appear to a moving observer. Consider a pair of static charges in A's reference frame. B moves past A with a constant velocity v in a direction perpendicular to the line joining the two charges.

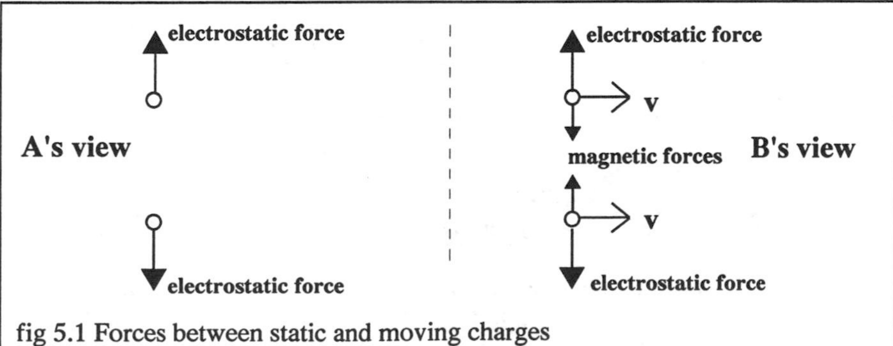

fig 5.1 Forces between static and moving charges

According to A the charges exert an electrostatic repulsive force on one another given by Coulomb's law. B, however, sees the charges moving through her reference frame and concludes that they represent parallel current elements. She would expect the moving charges to interact by a combination of electrostatic and magnetic forces. The currents are parallel so the magnetic force is one of attraction. Magnetism is a relativistic effect!

Another way to look at this is to use the Lorentz transformations to transform the electrostatic force in A's reference frame to B's. If this is carried through it reduces the electrostatic force between moving charges by exactly the calculated magnetic force. What we detect as a magnetic field is the difference between the electrostatic field of a static and a moving charge. What appears as a purely electric field in one reference frame is an electromagnetic field in another.

5.1.2 The Force Between Parallel Currents.

Consider two similar wires carrying equal parallel currents. They will attract one another with a magnetic force given by:

$$F = \frac{\mu_o I^2 l}{2\pi r}$$ where μ_o is the permeability of free space.

We can see how this force arises if we simplify the model of conduction in a neutral wire to a 'train' of equally spaced electrons drifting past another 'train' of equally spaced but oppositely charged static ions (see fig 5.2 below).

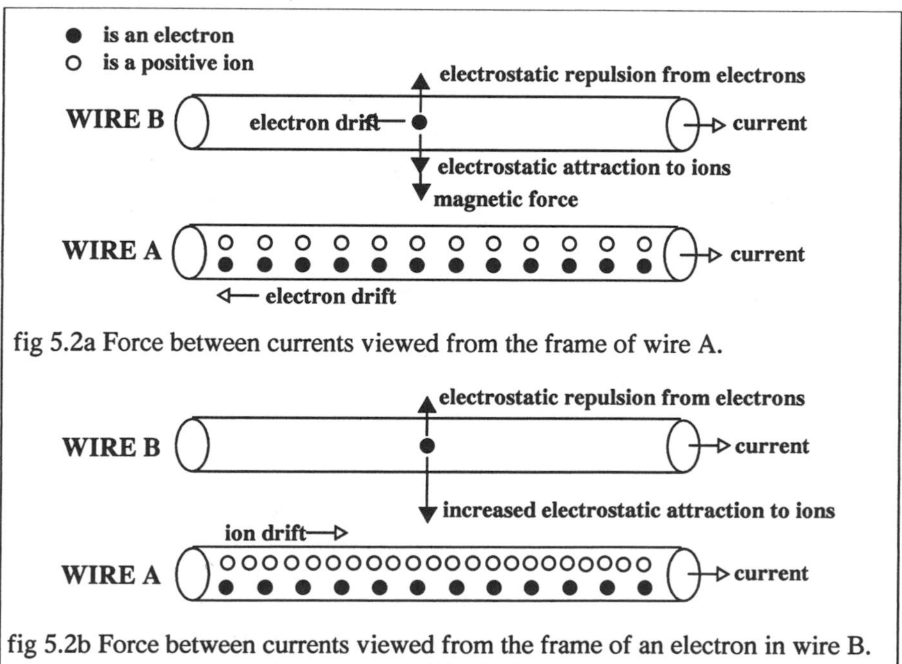

fig 5.2a Force between currents viewed from the frame of wire A.

fig 5.2b Force between currents viewed from the frame of an electron in wire B.

Consider forces on a single electron in wire B. First of all adopt the reference frame of the fixed ions in wire A. In this reference frame the ions in wire A will exert an attractive force on the electron in B but this is exactly balanced by the repulsion of the electrons in wire A (the spacings of ions and electrons in wire A are equal in this frame since the wire is neutral). This means there is no net electrostatic force. The only force on the moving electron is due to the magnetic field set up by the current in wire A. This will attract the electron (by Fleming's left hand rule).

Now switch to the rest frame of the electron in wire B. Since the electron at rest in this frame, it cannot experience a magnetic force, so how can we account for the fact that it is drawn towards wire A? Consider the two trains of charge in wire A. The electron train is at rest because wires A and B are identical and carry equal currents. The ions in wire A, however, are drifting at some velocity relative to the electron in B, so their separation is *reduced* by a factor γ (length contraction). Their charge per unit length is of greater magnitude and opposite sign to that of the electrons in wire A, so the electrons in wire B each experience a net attraction to wire A. In this frame the attraction is due to an imbalance of charge in wire A and is purely electrostatic!

5.2 RADIATION PRESSURE.

The relation between mass and energy follows from the expression for radiation pressure in classical electromagnetism. This expression can be derived quite simply by considering the force exerted on a conductor when electromagnetic waves are absorbed.

For a plane polarized electromagnetic wave traveling in the positive z-direction the electric and magnetic fields can be written:

$$E_x = E_o \sin\omega t$$

$$B_y = B_o \sin\omega t = \frac{E_o}{c}\sin\omega t$$

If these waves are incident on an aerial of length l parallel to the x-axis there will be an alternating p.d. along the aerial of:

$$V = E_o l \sin\omega t$$

and a current $I = \dfrac{V}{R} = \dfrac{E_o l \sin\omega}{R}$ in an external circuit attached to the aerial.

This current will be perpendicular to the incident magnetic field which will exert a 'motor effect' force on it. This force is given by:

$$F_z = B_y I_x l = \left(\frac{E_o}{c}\sin\omega t\right)\left(\frac{E_o l \sin\omega}{R}\right) l = \frac{E_o^2 l^2 \sin^2 \omega t}{Rc} \qquad (1)$$

At the same time the power dissipated in the aerial circuit is given by:

$$P = \frac{V^2}{R} = \frac{E_o{}^2 l^2 \sin^2 \omega}{R}$$

and its intensity is:

$$I = \frac{P}{A} = \frac{E_o{}^2 l^2 \sin^2 \omega}{RA} = \frac{F_z c}{A} \tag{2}$$

where A is the cross - sectional area of the collecting aerial in the xy plane. Comparing (1) and (2) we can see that:

$$p_z = \frac{I}{c} \quad \text{radiation pressure} = \frac{\text{intensity of radiation}}{\text{speed of light}} \tag{3}$$

5. 3 SOLUTIONS

5.3.1 Solutions to selected problems from Chapter 1.

1.1(a) 0.45 s, (b) 1.0 s, (c) 0.36 m up, (d) same time, 2.36 m down
1.2(a) $KE_1 = 3.2\times10^5$ J, $KE_2 = 0$, (b) $KE_1 = 6.4\times10^5$ J, $KE_2 = 3.2\times10^5$ J (c) Yes, but in (a) the collision is observed from the CM frame. In both cases 3.2×10^5 J is transferred to thermal energy. (d) The value of KE is relative.
1.4(a) Upthrust > weight so ball accelerates up toward surface. (b) Stays in the same place relative to the glass. Water, glass and ball are all in free-fall so accelerate downwards at the same rate. Another way to think about it is to say that there is effectively no gravitational field in a free-fall reference frame so nothing to distinguish one direction from another (there is no 'up' and 'down').
1.5 No. Consider the CM frame.
1.6(a) $\sqrt{2}$ (b) (–0.13, 2.23) and (0.23, 3.59) (c) $\sqrt{2}$
1.7(a)(b) Human jumping velocity is a few metres per second. Plane falls much faster than this, so jumping makes a negligible (but not zero) difference to the relative velocity of the passenger to the point of impact.
1.8 Depends on your reference frame and on how you interpret gravity. From an external inertial frame no - their weight provides the centripetal force that makes them orbit. However, in a free-fall reference frame there is no effective gravitational field. From a Newtonian point of view they are only 'apparently weightless', from an Einsteinian point of view they are actually weightless.
1.9(a) 56° (b) yes - stellar aberration.
1.10(a) 9.3×10^{-5} N (b) 4.7×10^{-5} N
1.11 Yes, tidal forces.
1.12(a) 234:4, (b) 234:4, (c) 4.5°, 88.4°, 92.9°

1.13 $\Delta T_{out} = T\left(1+\frac{v}{c}\right) \quad \Delta T_{ret} = T\left(1-\frac{v}{c}\right)$

1.15(a) 30 m, (b) 10 s, (c) 11.25 s, (d) 10.7 s.

1.17(a) About 7×10^8 ms^{-1}, (b) No. *Different* photons hit different points on the Moon's surface, all photons travel at the speed of light.

5.3.2 Solutions to selected problems from Chapter 2.

2.2(a) About 5 cm.

2.3(b) It is true that light on the parallel path is delayed in one direction and advanced in the other, but these do not cancel. The extra time spent moving directly against the 'ether wind' reduces the overall average speed.

(c) $t_1 = \frac{2l}{\sqrt{c^2 - v^2}} \quad t_2 = \frac{2cl_2}{(c^2 - v^2)}$

(d) $l_1 = \frac{l_2}{\sqrt{1-\frac{v^2}{c^2}}} = \gamma l_2$ The arm parallel to the motion through the ether must contract (Lorentz-Fitzgerald contraction).

2.4(b) $\frac{c}{\gamma^2}$ (c) 10^{-4}, $\delta t = \frac{2l}{c}\gamma^2\frac{v^2}{c^2} \approx 6.7\times10^{-16}\,\text{s}$ during which time the light moves about 2×10^{-7} m or about half the wavelength of visible light.

2.5 Use the velocity addition formula.

2.6(a) $\tan\alpha = \frac{v}{c} \approx 42''$ (b) 21″, (c) $\max = \pm\frac{v}{c} \quad \min = \pm\frac{v\sin\theta}{c}$

2.7(c) $\alpha = \pm\frac{\gamma v}{c}$ (d) Excellent if $v << c$.

2.8 c in every case.

2.9(a) Spherical, (b) $d = ct'$ (c) $s = \sqrt{x'^2 + y'^2 + z'^2}$ (e) $d = ct$

2.10(a) They are the same.

2.11(a) 0.06 m, (b) 0.30 m, (c) 0.98c, (d) 0.06 m

2.12(a) 1.6×10^6 m s^{-1} (b) $\gamma = 1.00003$, so relativistic effects are small.

2.13(a) 3.3 y, (b) 3.75 ly, (c) 2.48 ly, (d) Elliptical, semi-major axis R (perp. to motion), semi-minor axis $0.66R$ (parallel to motion). (e) 3 y, 2.7 y (f) 7 y

2.14(a) 2 s, (b) In the room's reference frame, yes. (c)(d) No. Wall ahead is approaching her and the *event* of first light emission (not to be confused with the light bulb's position, since this changes for our observer). This wall is illuminated first. Wall behind her is moving away from the event and so light reaches it later. Walls are illuminated about 1.4 s and 2.8 s after she passes the lamp. (e) Alarm is at rest in the room so it is effectively synchronized with the walls and room (think of signals sent from each wall to a central alarm). This alarm will go off.

2.15(a) $f' = f_o/\gamma$, (b)(c) $f' = \left(\frac{c-v}{c+v}\right)^{1/2} f_o$ $\lambda' = \left(\frac{c+v}{c-v}\right)^{1/2} \lambda_o$

2.16(a) About 1.4 parts in 10^{12}, (b) 5 ns
2.17(a) $0.099c$, (b) 1.005, (c) 512 keV, 2.5 keV, 514.5 keV (d) ±10 μm
2.18(a) 3.7×10^{-12} kg, (b) Source loses mass/energy, water gains an equal amount of mass/energy (if process is 100% efficient).
2.20(c) 0.00593 u, (d) 5.52 MeV, (e) $0.054c$. (5.52 MeV << rest energies of either alpha particle or nucleus, so motion is non-relativistic).
2.21(a) $1.15m_oc^2$, $1.15m_o$ (b) $0.15m_oc^2$ (c) $0.57m_oc$ (d) $3.15m_oc^2$, $3.15m_o$ (e) $0.30c$ (f) $3m_o$

2.23(a) $M_o = 2m_o(1+\frac{T}{m_oc^2})$ (b) $M_o = 2m_o\sqrt{1+\frac{T}{m_oc^2}}$

2.24. 290 MeV
2.25. 75 m s^{-1}
2.26 0.93c
2.27(a) $x = \gamma a, y = a$ (b) γa^2
2.28(a) $\gamma^{-3}a'_x$
2.29(a) 56 000 km, (b) 0.2 s (c) 1.3 s
2.31 Photons have zero rest mass.
2.32 Both are correct in their own reference frame. For an observer on the space station there is a moment when all of the rocket is inside the station. In this frame the rear end enters the station before the front end leaves. On the other hand, there is no such moment in the reference frame of the rocket. But there is no paradox. The two observers disagree on the relativity of simultaneity and so on the time order of events. They both observe the front of the rocket to leave the station and the rear to enter it, but they disagree on the time order of these two events. The observer on the space station will see a synchronization error between clocks at the front and rear of the moving rocket and the rocket observer will see a synchronization error between clocks at either end of the space station.

5.3.3 Solutions to selected problems from Chapter 3.

3.1(a) 4.8×10^7 J, (b) 5.3×10^{-10}J, (c) 1.9×10^9
3.2(a) 4.0×10^{26} W, (b) 4.4×10^9 kg s^{-1}, (c) 1.4×10^{13} y, (d) 10^{11} y
3.3(a) 5×10^{-12} kg,
(b) If projected by an external agent it would increase by this mass. If self-propelled and 100% efficient its mass would be unchanged (internal transfer). In practice its mass falls because of mass and energy loss to the surroundings (inefficiency generates and dissipates thermal energy and exhaust gases lost)

3.4(a) $EPE = \frac{e^2}{4\pi\varepsilon_o d}$ (b) $\frac{4\pi r^5 n^2 e^2}{15}$ (c) 6×10^{29} m^{-3} (d) 8×10^{16} J (e) 0.86 kg
(f) 4.2 kg (g) EPE is actually close to zero because of protons.

3.5(a) 1.7×10^{-8} N (b) 1.7×10^{-8} N; 3.3×10^{-8} N; 2.7×10^{-8} N
3.6(a) 9.8×10^4; 50 GeV; 50 GeV/c; 50 GeV/c^2
(b) 54; 50.9 GeV; 50.9 GeV/c; 50.9 GeV/c^2
(c) 7.1; 3.62 MeV; 3.6 MeV/c; 3.6 MeV/c^2
(d) 7.1; 6.66 GeV; 6.59 GeV/c; 6.66 GeV/c^2
3.8. 0.03 m s^{-1} (non-relativistic)
3.9(a) Use the first two terms in the binomial expansion.
(c) About 0.25c (consider the third term)
3.10 520 MeV
3.11(a) 3.45 h, 6.90 h, 13.35 h, (b) 1.15 h
3.12(a) 1.11×10^5 y; 4.8×10^4 y (b) 1.01×10^5 y; 1.4×10^4 y (c) 4.3×10^4 y; 1.4×10^4 y

3.14 $t_{1/2} = \dfrac{d \ln 2}{\gamma \ln f}$

3.15(a) 660 m (c) 14.8 km (d) 1.25 km
3.16(a) 2.19×10^{-15} s; 4.57×10^{14} Hz (b) 1.000002 (c) reduced by 2 parts in 10^3 (d) reduced by 2 parts in 10^6 (e) 4.56×10^{14} Hz.

3.17(a) $\dfrac{\gamma^2 m_o}{Al}$ (c) $\dfrac{m_o}{Al}$

3.18 An ellipse with semi-minor axis parallel to relative motion and equal to $r_0/3.2$.

3.19 Base (parallel to motion) l/γ , sides $\left(\dfrac{3}{4}+\gamma^2\right)^{\frac{1}{2}} l$, base angles $\arccos\dfrac{\gamma}{2}$, apex angle $\arcsin\dfrac{\gamma}{2}$.

3.20 Towards B at 0.30c.
3.23 6.3
3.24(a) 5 m (b) (1.866, 1.232) and (6.83, 1.83) (c) 5 m
3.26 $(x_2 - x_1) \le c(t_2 - t_1)$
3.27 9 192 631 770 Hz (exactly); 0.03261225 m (7 sig fig)
3.28(d) 1.8 s (e) 5.1 days
3.29(b) (±5 m, 1.7×10^{-8} s) (f) 8.6×10^7 m s^{-1}, (3.7 m, 1.3×10^{-8} s) and (– 6.5 m, 2.3×10^{-8} s).
3.30 0.30c, 9.85 light years.
3.33(b) 1.83 s

3.34(e) $k = \left(\dfrac{c-v}{c+v}\right)^{\frac{1}{2}}$

3.37(a) 934 MeV, 10.4 MeV, 0.511 MeV (d) 1.6×10^{-9} J, 1.78×10^{-26} kg
(e) 2.67×10^{-21} kg m s^{-1}
3.38 intervals are all 10 s (or 3×10^7 m); times are 10.01 s; 11.55 s; 70.90 s; separations are 1.2×10^8 m; 1.7×10^9 m; 2.1×10^{10} m.
3.39 (9.96, 10i) units are GeV/c

3.40 $(M_o - m_o)c^2$

3.41 $m_o = \dfrac{p^2c^2 - E_K}{2E_K c^2}$

3.42 $\left(\dfrac{E}{c}, \dfrac{i(E + m_o c^2)}{c}\right)$

3.43 $\left(\dfrac{E}{c}, \dfrac{iE}{c}\right)$

3.44(a) 0.314*c* (b) $(310.7, 989.6i)$ and $(0, 9350i)$ in GeV/*c* (c) 0.108*c*

3.45(a) 1.24×10^{-14} m, 1.30×10^{-14} m, 0.6×10^{-14} m (b) 63° (c) 1.0053 (d) 970 MeV/*c* (e) 54° (f) 117°

5.3.4 Solutions to selected problems from Chapter 4.

4.1(a) $g_s = \dfrac{c^4}{4GM}$; (b) 3.1×10^{42} kg; 4.6×10^{15} m
4.2(a) $W = mg$; (b) $R = mg + ma$
4.3 $v = \pm 0.028c$
4.6(a) 4.3×10^{-6}; (b) 1
4.7(b) $\left(1 - \dfrac{mgl}{c^2}\right):1$
4.8(a) In the rotating reference frame there is a centrifugal inertial force. This can be interpreted as a gravitational field acting outwards. The center is therefore at higher potential than the circumference and so clocks run faster at the center.
(d) $\phi(r) = -\dfrac{1}{2}r^2\omega^2$ (e) $f = \dfrac{f_o}{\left(1 - \dfrac{r^2\omega}{2c^2}\right)}$
(f) Expand both expressions to the first two terms using the binomial theorem.
4.9(a) $\left(1 + \dfrac{gh}{c^2}\right):1$ with the clock at higher altitude running fast. (b) $v \approx \sqrt{2gh}$
4.11(a) Sun's surface is at a lower gravitational potential. (b) 59 mins 59.9924 s. (c) 540 years.
4.13 As seen by the distant observer, *yes*. However, the speed of light is constant in a local inertial (i.e. free-falling frame).
4.15(a) $\pm 2.7 \times 10^{-15}$ (b) 8.2×10^{-7} ms^{-1}.
4.16 $\begin{pmatrix} 1 & 0 \\ 0 & 1 \end{pmatrix}$ The surface is flat - think how a cylinder can be made.

4.18(a) $x = R\sin\dfrac{r}{R}$ (angle in rads) (b) use small angle approx. (c) $\varepsilon = \dfrac{1}{3R^2}$

(d) 110 km.

4.19(a) $v = \sqrt{\dfrac{2GM}{r_E}}$ (b) 8.9 mm

4.20(c) No. This is local curvature, but radius increases further from masses and becomes much larger than the solar system further from the Sun and planets.
4.21(c) No, full relativistic Doppler formula must be used. (d) 0.92*c*

4.22(a) $v = \left\{\dfrac{1-(z+1)^2}{1+(z+1)^2}\right\}c$ (b) 0.095*c*; 0.6*c*; 0.92*c*; *c*

4.23(c) 2.1×10^{-6}
4.25(a) The light was emitted when the scale of the universe was 1/5 its present value. (b) Light was emitted from a collapsed body at 1.04 times its Schwarzschild radius (c) recession of 0.92*c*
4.26(a) Nearby galaxies blue shifted, far galaxies red shifted. (b) More and more galaxies appear blue shifted. All are blue shifted just before the Big Crunch.
4.27 $z \approx 6\times10^{-4}$ due to our own motion. This blue shifts light coming from directly ahead and red shifts light from behind. Light coming from other directions is affected by a component of our velocity. These affects must be allowed for when measurements of spectra are made.

4.29(a) $L = \dfrac{c}{H_o}$ (b) No, because local speeds remain sub-luminal. (c) It cannot reach us. (d) L increases if the rate of expansion falls. (e) Some galaxies that are at present invisible may become visible. If the universe recollapses all galaxies will become visible.

4.30(a) $\dfrac{d_{rec}}{d_{em}} = z+1$ (b) Greater *z* always implies a greater distance from us now (i.e. a greater reception distance) but it is possible that the emission distance of a higher z galaxy could actually have been less than that of a lower z galaxy. It all depends on how the expansion rate has changed over time.
4.32(a) For binary pulsar system $R_S = 8.3 \times 10^3$ m and $T = 1.4 \times 10^5$ s.
Comment - this very crude method does yield the correct order of magnitude for the rate of reduction of the binary period.

Further Reading

Background and Biography

I.Newton, 'The Principia' (Philosophiae Naturalis Principia Mathematica), London, 1686.
Newton's Philosophy of Nature: Selections from his writings, Ed. H.S.Thayer, Hafner Press, 1974.
R.Westfall, The Life of Isaac Newton, Canto, 1993.
G.Galileo, Dialogues Concerning Two New Sciences, University of California Press, Berkeley, 1967
E.Mach, The Science of Mechanics, Open Court,1960
Ed. A.Koslow, The Changeless Order: The Physics of Space, Time and Motion, George Braziller Publishing, New York, 1967.
E.A.Abbott, Flatland, Dover Publications, 1952.
A.S.Eddington, The Nature of the Physical World, Cambridge University Press, 1928.
J.Bernstein, Einstein, Viking press, 1973.
L.Pyenson, The Young Einstein, Institute of Physics Publishing, 1995.
A.P.French, Einstein: A Centenary Volume, Harvard University Press, 1979.
R.W.Clark, Einstein, The Life and Times, World Publishing Co., 1971.
A.Pais, Subtle is the Lord, The Science and the life of Albert Einstein, Oxford University Press, 1982.
E.Whittaker, From Euclid to Eddington, Dover Publications, 1958.
E.Whittaker, History of the Theories of Electricity and the Aether, Harper Torchbook, Harper and Row, 1960.
M.Jammer, Concepts of Space: The History of Theories of Space in Physics, Harvard University Press, 1954.
H.Reichenbach, The Philosophy of Space and Time, Dover publications, 1957.
G.E.Christianson, Edwin Hubble, Mariner of the Nebulae, Institute of Physics Publishing, 1997.

Relativity

A.Einstein, H.A.Lorentz, H.Weyl and H.Minkowski, The Principle of Relativity (original papers), Dover Publications, 1952.
A.Einstein, Relativity, The Special and the General Theory, Methuen, 1979.
A.Einstein, The Meaning of Relativity, Methuen, 1946.
H.Bondi, Relativity and Common Sense, Heinemann Educational Books Ltd., 1964.
A.P.French, Special Relativity, Nelson, 1981.

W.Rindler, Essential Relativity: Special, General and Cosmological, Van Nostrand Rheinhold, 1969.
E.F.Taylor and J.A.Wheeler, Spacetime Physics, W.H.Freeman and Co. Ltd, 1992.
Relativity Revisualised, L.C.Epstein, Insight Press, 1987.

Gravitation and Cosmology

A.S.Eddington, The Expanding Universe, Cambridge University Press, 1933.
H.Bondi, Cosmology, Cambridge University Press, 1961.
P.G.Bergmann, The Riddle of Gravitation, Scribners, 1968
E.Hubble, the Observational Approach to Cosmology, Oxford University Press, 1937.
S.Hawking, A Brief History of Time, Bantam, 1991.
E.Harrison, Cosmology: The Science of the Universe, Cambridge University Press, 1981.
S.Weinberg, The First Three Minutes: A Modern view of the origin of the Universe, Basic Books, 1977.
P.C.W.Davies, Space and Time in the Modern Universe, Cambridge University Press, 1977.
W.J.Kaufman, Relativity and Cosmology, Harper and Row, 1973.
R.Geroch, General Relativity from A to B, University of Chicago Press, 1978.
R.B.Rucker, Geometry, Relativity, and the Fourth Dimension, Dover, 1977.
J.Gribbin, In Search of the Big Bang, Corgi, 1987.
J.Gribbin, In Search of the Edge of Time, Bantam, 1993.
K.S.Thorne, Black Holes and Time Warps: Einstein's Outrageous Legacy, W.W.Norton and Co., 1992.
J-P. Luminet, Black Holes, Cambridge University Press, 1992.
M.Kaku, Hyperspace, Oxford University Press, 1994.
P.A.M.Dirac, General Theory of Relativity, John Wiley and Sons Ltd., 1975.
M.Berry, Principles of Cosmology and Gravitation, Institute of Physics Publishing, 1989.
C.M.Will, Was Einstein Right? Putting General Relativity to the Test, Oxford University Press, 1989.
J.A.Wheeler, A Journey into Gravity and Spacetime, Scientific American library, 1990.
M.R.Robinson, Ripples in the Cosmos, W.H.Freeman and Co. Ltd., 1993.
G.Smoot and K.Davidson, Wrinkles in Time, Little, Brown and Co., 1993.

Index